Collins

D0589422

Collins Revision

GCSE Foundation Science

Revision Guide

FOR AQA A + B

Contents

Coordination

The nervous system

- The **nervous system** enables the body to respond to changes in the surroundings and to coordinate behaviour.

- The brain and spinal cord make up the **central nervous system**. It is connected to the body's organs by **nerves**. Nerves carry information as electrical impulses which travel to and from the spinal cord and the brain.

- In the nervous system, electrical impulses travel along cells called **neurones**.

  - The neurones that carry information from receptors to the central nervous system are called **sensory neurones**.

  - The neurones carrying information from the central nervous system to effectors are called **motor neurones**.

brain — central nervous system
spinal cord
nerve

The human nervous system.

cell body
dendron — direction in which impulse travels
myelin sheath — axon

A sensory neurone.

cell body
direction of impulse
axon — myelin sheath
dendrite

A motor neurone.

G–E

D–C

Receptors

Receptors and effectors

- **Receptors** are special cells that detect stimuli. Stimuli include light, sound, changes in position, chemicals, touch, pressure, pain and temperature.

- Receptors are concentrated in the organs of our senses including the eyes, ears, nose and tongue. The skin all over our body has receptors for pressure, pain and temperature.

- Ear receptors sense a noise and send electrical impulses along nerves to the brain. The brain coordinates the response, sending impulses along nerves to neck muscles. The muscles contract and the head turns towards the noise.

- A muscle is an **effector**, an organ that does something in response to a **stimulus**.

- **Glands** are also effectors. Our salivary glands respond to a taste stimulus by secreting saliva.

G–E

Hormones

- For good health, the organs of the body must work together. Their processes are coordinated by chemicals called **hormones**. These are made by **glands** and **secreted** into the bloodstream to convey information to organs.

- Each hormone may coordinate processes in one or more **target organs**.

- Adrenaline from the adrenal gland affects the heart, breathing muscles, eyes and the digestive system. The hormones that control ovulation and menstruation target the ovaries.

pancreas secretes insulin — adrenal gland secretes adrenaline

ovary secretes female sex hormones (e.g. oestrogen) — testis secretes male sex hormones (especially testosterone)

Four glands that secrete hormones.

G–E

D–C

Questions

Grades G-E

1 Name the **two** organs that make up the central nervous system.

Grades D-C

2 Name the kind of neurone that carries impulses *away* from the central nervous system.

Grades G-E

3 Give **two** examples of stimuli.

Grades D-C

4 Give an example of a hormone and its target organs.

Reflex actions

Reflex actions and synapses

G–E

- A **reflex action** is a rapid, automatic response to a stimulus, often to avoid harm, such as the blink response to an approaching object.
    - A receptor detects the stimulus.
    - An electrical impulse travels along a **sensory neurone** to the spinal cord in the central nervous system.
    - The impulse passes from the sensory neurone to a **relay neurone**, inside the spinal cord.
    - The impulse passes from the relay neurone along a **motor neurone** to an **effector** (e.g. a muscle or gland), which responds to the stimulus.

D–C

- The route of the electrical impulse in a reflex action, the **reflex arc**, is along a sensory, a relay and a motor neurone.

- The time taken is less than if the impulse travelled to the brain and back.

- The gap between the ends of two neurones is a **synapse**.

- An electrical impulse stops at the end of one neurone, which then secretes a chemical into the gap.

- The chemical diffuses across to the next neurone, which produces an electrical impulse that continues along the neurone.

cell body of sensory neurone

spinal cord in cross section

a sensory neurone carries an impulse towards the spinal cord

motor neurone

relay neurone

when you stand on a pin, the stimulus is detected by a receptor

the leg muscle, the effector, contracts and pulls your foot away

A reflex arc.

impulse

neurone

neurone

What happens at a synapse.

transmitter chemical in vesicle

vesicle releases chemical

In control

Controlling water and ions

G–E

- For cell enzymes to work effectively, cells need a steady water content, ion content and temperature. Blood sugar concentration is kept constant, to give cells a constant supply of energy.

- The kidneys excrete excess water and ions. Water is also lost through the lungs.

- Sweat contains water, ions (including sodium and chloride) and urea.

- Water evaporates, takes heat from the skin and lowers body temperature.

D–C

- During digestion, starch and sugar in food are converted to **glucose** which travels throughout the body in the blood.

- When the blood glucose level is too high, the **pancreas** secretes **insulin** into the blood.

- Insulin reaches the **liver** and makes it remove glucose from the blood and store it. Between meals, the liver can release glucose back into the blood.

How Science Works

You should be able to: evaluate the claims of manufacturers about sports drinks containing water, ions and glucose.

water from lungs in breath

water and ions in sweat

water and ions from kidneys in urine

Ways the body loses water and ions.

3. The water in sweat evaporates, taking heat from the skin

2. The sweat lies on the surface of the hot skin

skin

1. The sweat gland secretes sweat

How sweating cools you.

Questions

Grades G-E

1 Name the **three** kinds of neurone involved in a reflex action.

Grades D-C

2 Explain what happens when a nerve impulse reaches a synapse.

Grades G-E

3 Explain why blood sugar content should be kept constant.

Grades D-C

4 Explain how sweating cools you down.

Reproductive hormones

The menstrual cycle and hormones

- An ovary releases an egg (one cell) every 28 days.

- The lining of the **uterus** (womb) thickens, ready to receive an embryo if the egg is fertilised.

- If the egg is unfertilised, the uterus lining breaks down and is lost through the vagina in **menstruation**. This begins the sequence of events called the **menstrual cycle**.

- **Hormones** control the menstrual cycle:
  - the pituitary gland just below the brain secretes **FSH** and **LH**
  - the ovaries secrete **oestrogen**.

G–E

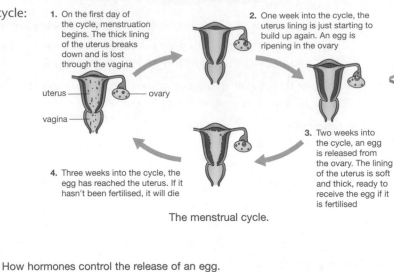

1. On the first day of the cycle, menstruation begins. The thick lining of the uterus breaks down and is lost through the vagina

2. One week into the cycle, the uterus lining is just starting to build up again. An egg is ripening in the ovary

uterus — ovary

vagina —

3. Two weeks into the cycle, an egg is released from the ovary. The lining of the uterus is soft and thick, ready to receive the egg if it is fertilised

4. Three weeks into the cycle, the egg has reached the uterus. If it hasn't been fertilised, it will die

The menstrual cycle.

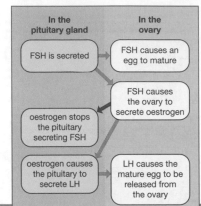

In the pituitary gland	In the ovary
FSH is secreted	FSH causes an egg to mature
	FSH causes the ovary to secrete oestrogen
oestrogen stops the pituitary secreting FSH	
oestrogen causes the pituitary to secrete LH	LH causes the mature egg to be released from the ovary

How hormones control the release of an egg.

D–C

Controlling fertility

You should be able to: evaluate the benefits of, and the problems that may arise from, the use of hormones to control fertility and IVF.

Fertility treatment

- A woman who cannot produce eggs can receive **fertility drugs** which contain hormones. FSH stimulates one or more eggs to mature in the ovaries.

G–E

- In **IVF**, a woman's eggs are fertilised outside her body using her partner's sperm.

- A woman undergoing **IVF treatment** receives fertility drugs to make her ovaries produce several eggs. These are removed under anaesthetic.

- *IVF* means *in vitro* (Latin for *in glass*) *fertilisation*, since eggs were first fertilised in a glass dish.

D–C

- A fertilised egg cell divides, making a tiny ball of cells, the **embryo**. One or two embryos are transferred to the woman's uterus. They become attached to the lining and each develops into a baby.

Oral contraceptives, benefits and problems

G–E

- To avoid pregnancy, women can take oral contraceptive pills containing hormones. These include oestrogen, which stops FSH production, so eggs do not mature.

- Oral contraceptives prevent unwanted pregnancies, reduce requests for abortions and, if widely used, may slow world population growth.

D–C

- But some cultures and religions outlaw their use. They may also encourage people to have more sexual partners and increase the danger of catching sexually transmitted infections.

Questions

(Grades G-E)
1 Name the organ that secretes oestrogen.

(Grades D-C)
2 Describe **two** effects of FSH.

(Grades G-E)
3 Explain how oral contraceptives stop egg production.

(Grades D-C)
4 State **one** advantage of oral contraceptives and **one** possible disadvantage.

Diet and energy

Energy and a balanced diet

G–E

- Food provides energy for the functions of cells and for muscle movement.

- Excess energy is stored as fat, which can be used later to provide energy.

- A healthy, **balanced diet** contains the right amounts of:
  - **carbohydrates** for energy
  - **fats** for energy and making cell membranes
  - **proteins** for growth, repair and energy
  - vitamins and minerals to keep healthy
  - roughage (fibre) for digestive system function
  - water.

- People with an unbalanced diet are **malnourished**. They are too fat or too thin, or have deficiency diseases.

D–C

- The rate of chemical reactions in cells – **metabolic rate** – is generally faster in men than in women, in young than in old people, and in people with a high muscle-to-fat ratio.

- Fit, active people, who exercise or do physical work, use more energy than inactive people. Metabolic rate stays high for some time after activity.

- Metabolic rate is higher in winter than in summer.

- It can depend on the genes we inherit.

- In babies, **brown fat** cells have a very high metabolic rate to maintain body temperature.

How Science Works

You should be able to: evaluate information about the effect of food on health.

Obesity

Obesity and illnesses

G–E

- People who are very overweight are described as **obese**.

- Obesity increases the chance of getting arthritis, diabetes, high blood pressure and heart disease.

- The **joints** of someone with **arthritis** are worn and painful. Many overweight people have arthritis in hip and knee joints, and have them replaced by surgery.

D–C

- People with **diabetes** cannot control their **blood glucose** level. If it rises too high, it harms cells by drawing water from them. In Type 2 diabetes, insulin fails to regulate blood glucose level.

- **Blood pressure** measures the effort the heart makes to pump blood. It is often too high in overweight people and those who eat too much salt. This can cause **heart disease** by straining the heart and damaging blood vessels.

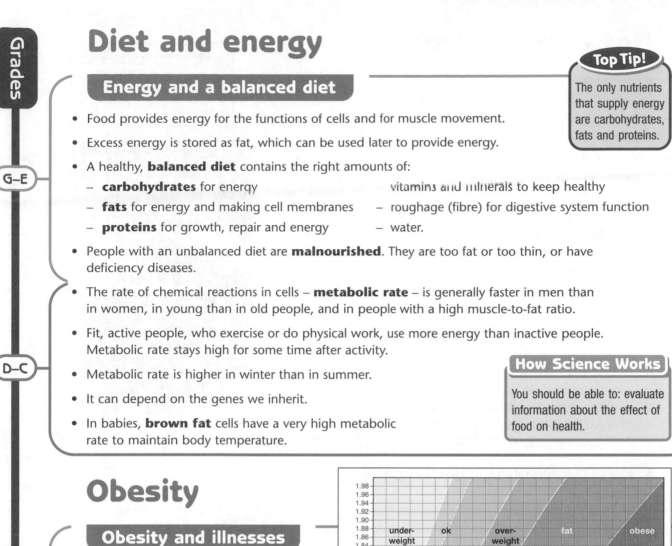

Chart of weight against height.

a normal knee joint has a thick cushion of smooth cartilage that allows the bones to slide easily over one another

an arthritic joint has thinner cartilage, and there are rough edges that grate against each other, causing pain and loss of mobility

A healthy knee joint and one with arthritis.

Questions

Grades G-E

1 List **seven** kinds of nutrient needed in a balanced diet.

Grades D-C

2 How does the proportion of muscle to fat in the body affect metabolic rate?

Grades G-E

3 Explain the meaning of the word 'obese'.

Grades D-C

4 Explain how arthritis makes movement difficult.

Not enough food

Eating too little or too much

How Science Works

You should be able to: evaluate claims made by slimming programmes.

- In some developing countries, drought, floods and war can prevent people from producing enough food.

- Resistance to diseases such as cholera or tuberculosis is lowered in people who do not have enough to eat. Women may have irregular or no periods.

- In rich countries, many people who over-eat try **slimming programmes** to lose weight. Eating less and exercising more would be more effective.

- Malnutrition affects small children most. Their bodies are growing fast and cannot combat infections well.

- Breastfeeding gives children a balanced diet with all the nutrients they need.

- Children between 1 and 3 years fed only on carbohydrates (e.g. maize or rice) may suffer from **kwashiorkor**. Without protein, their limbs are thin since muscle fails to develop, and their abdomen swells as fluid accumulates in the tissues. A balanced diet can bring full recovery.

G–E

D–C

A child suffering from malnutrition.

Cholesterol and salt

Cholesterol, salt and heart disease

- The bloodstream carries **cholesterol**. Too much cholesterol can block blood vessels and cause **heart disease**.

- **Saturated fats** in eggs, dairy products and meat can *raise* blood cholesterol level.

- **Monounsaturated** and **polyunsaturated fats** in plant oils can *lower* blood cholesterol.

- Cell membranes need cholesterol, and the liver makes cholesterol if the diet contains too little. The amount made seems to depend on genes.

- Blood carries cholesterol as **LDL** and **HDL** cholesterol. To be healthy, you should have more HDL than LDL in your blood.

- The body needs **salt**, but too much can cause high blood pressure and heart disease. Snacks and processed foods often contain a lot of salt.

- In the blood, cholesterol mixed with protein forms tiny balls of **lipoprotein**.

- **Low density lipoproteins** (LDLs) may form **plaques** in artery walls, reducing blood flow. The blood can then clot and stop blood flow in an artery altogether. In heart muscle, when the oxygen supply from blood stops, the heart stops beating properly, and the person has a heart attack.

a healthy artery has a stretchy wall and a space in the middle for blood to pass through

plaque

sometimes, a substance called plaque builds up in the wall. This is more likely to happen if you have a lot of LDL cholesterol in your blood

blood clot

the plaque slows the blood down, and a clot may form. Or a part of the plaque may break away

How a plaque develops in an artery.

G–E

D–C

Questions

1 List **three** factors that can stop people in a developing country getting enough to eat.

2 What causes kwashiorkor?

3 Which are best for your health – saturated fats or unsaturated fats?

4 Which kind of cholesterol should you have more of in your blood – LDLs or HDLs?

Drugs

Drugs, addiction and dependency

G–E

- For thousands of years, people have used **drugs** from plants and animals. Many people use medical drugs today, to combat disease and relieve pain.

- Fewer people use other, **recreational** drugs. They change chemical processes in the brain that affect the way you feel and behave.

- Cocaine, heroin and cannabis are **illegal** drugs. Alcohol and tobacco are **legal**, used by more people, and cause more deaths.

- People who use drugs can become **addicted** to or **dependent** on a drug. Eventually, the drug damages organs such as the brain and the **liver**, the organ that destroys harmful chemicals in the body.

D–C

- Problems in a person's life can cause them to start using a drug that makes them feel better. Then they cannot manage without it. They are addicted.

- If an addicted person stops taking the drug, they can experience bad **withdrawal symptoms** for days or weeks. They need help to cure their addiction and resume a normal life.

Trialling drugs

Testing and trialling a drug

G–E

- When developing a new drug, scientists **test** its safety on tissues or animals in the laboratory to check that it is not **toxic** (poisonous).

- Then they **trial** the drug on many human volunteers to check for any side effects.

- Before general use, scientists must show that the drug does what they claim – cure an illness or ease a symptom. The whole process can take five years or more.

- **Thalidomide**, a sleeping pill, was not tested on pregnant women, and such users had babies with deformed limbs. Thalidomide was banned. Now, it is used again, to treat leprosy.

A child affected by thalidomide.

Trialling the antiviral drug zanamivir

D–C

- Zanamivir is a new drug designed to lessen flu virus symptoms. It was tested on volunteer groups – children, adolescents, young soldiers, adults, people over 65, and family members.

- In a **double-blind trial**, half of each group received the drug, and the others received a **placebo**, which contained no drug. Who received either treatment is known only after results are collected.

	given zanamivir	given a placebo
no. of subjects	293	295
mean age in years	19	19
no. of days until their temperature went down	2.00	2.33
no. of days until they lost all their symptoms and felt better	3.00	3.83
no. of days until they felt just as well as before they had flu	4.5	6.3
average score the volunteers gave to their experience of the five major flu symptoms	23.4	25.3

Effect of zanamivir and a placebo on soldiers suffering from flu.

Questions

(Grades G-E)

1 Which of these drugs are illegal: heroin, cocaine, cannabis?

(Grades D-C)

2 Describe what happens when a person addicted to a drug stops taking it.

(Grades G-E)

3 What is done with a drug before it is tried out on human volunteers?

(Grades D-C)

4 Explain what is meant by a 'double-blind trial'.

Illegal drugs

Cannabis, heroin and cocaine

How Science Works

You should be able to: evaluate claims made about the effect of cannabis on health and the link between cannabis and addiction to hard drugs.

- **Cannabis** can cause bronchitis and lung cancer. It is believed to increase the risk of schizophrenia, a serious mental illness. People who use cannabis often move on to take hard drugs.

- **Heroin** and **cocaine** are **hard drugs**. They are illegal and dangerously addictive. Just one use of a hard drug can cause **dependency**.

G–E

More on hard drugs

- An addicted person may inject heroin or cocaine directly into a vein to speed up its effect on the brain. If several people use the same syringe, they can spread infections such as HIV/AIDS and hepatitis.

- Obtaining drugs and the money to buy them, often through crime, takes over an addicted person's life.

- An addicted person trying to break their habit can suffer **withdrawal symptoms**. These can include insomnia, sweating, vomiting and pain all over their body.

D–C

Alcohol

Effects of alcohol

- **Alcohol** affects the nervous system. It helps people to relax and slows down their reactions.

- People drinking to excess lose their self-control and may become **unconscious**, go into a **coma**, or even die.

- Alcohol is a poison. In the long term, alcohol causes brain cells to shrink and kills the liver cells that normally try to break it down to harmless substances.

alcohol is swallowed and then absorbed into the blood through the stomach

blood carries the alcohol to the brain

stomach

blood carries the alcohol to the liver

How alcohol reaches the brain and liver.

G–E

Is alcohol an acceptable drug?

- Because alcohol has been around for thousands of years, people do not think of it as a dangerous drug. If discovered today, it would be banned; some cultures and religions do ban it.

- Drinking alcohol in moderation causes little harm, but because of excess drinking:

  - 30 000 people a year are in hospital with liver damage

  - 150 000 people a year go to hospital for other drink-related conditions

  - many people are arrested for criminal damage

  - about 4000 women and 6500 men die each year.

- Some people are **dependent** on alcohol. Like hard drugs, alcohol ruins their lives and relationships.

If you drink and drive, you increase the chance of killing and being killed.

D–C

Questions

Grades G-E

1 List **two** hard drugs.

Grades D-C

2 Why do some people who use drugs have a high risk of developing AIDS?

Grades G-E

3 Describe how alcohol affects the nervous system.

Grades D-C

4 Which organ is likely to be damaged by drinking alcohol excessively?

Tobacco

Tobacco poisons and disease

- **Nicotine** in tobacco smoke affects the brain and is **addictive**.
- **Tar** contains **carcinogens**. They cause **cancer** of the lungs and other organs.
- Cancer cells divide uncontrollably, forming a lump or **tumour**. A quarter of all smokers die of cancer.
- **Carbon monoxide** gas stops red blood cells carrying oxygen that tissues and organs need. This results in heart disease and damaged blood vessels. Babies of mothers who smoke are often small at birth.
- Tobacco smoke irritates airways and lungs, which produce **mucus** that is coughed up. Smokers get **bronchitis** when bronchi become infected, inflamed and painful. In **emphysema**, air sacs in the lungs break down, causing an oxygen shortage and breathlessness.

The link between smoking and cancer

- Richard Doll first showed the **correlation** (link) between smoking tobacco and lung cancer. Today, no-one doubts that smoking shortens life.

How Science Works

You should be able to:
- explain how the link between smoking tobacco and lung cancer gradually became accepted
- evaluate the different ways of trying to stop smoking.

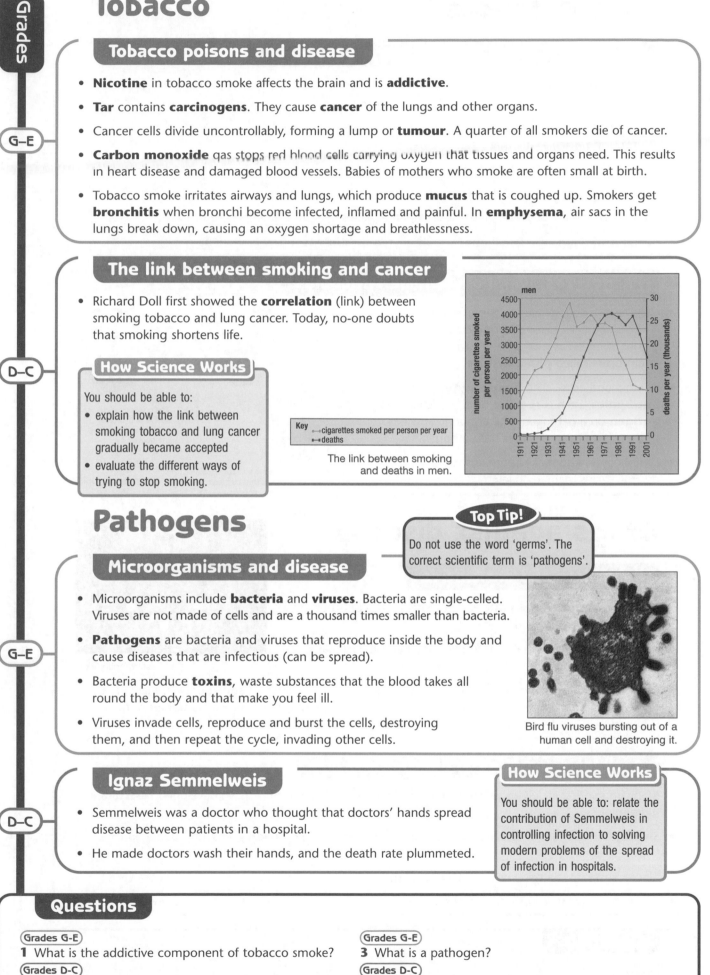

men

Key
— cigarettes smoked per person per year
— deaths

The link between smoking and deaths in men.

Pathogens

Top Tip!
Do not use the word 'germs'. The correct scientific term is 'pathogens'.

Microorganisms and disease

- Microorganisms include **bacteria** and **viruses**. Bacteria are single-celled. Viruses are not made of cells and are a thousand times smaller than bacteria.
- **Pathogens** are bacteria and viruses that reproduce inside the body and cause diseases that are infectious (can be spread).
- Bacteria produce **toxins**, waste substances that the blood takes all round the body and that make you feel ill.
- Viruses invade cells, reproduce and burst the cells, destroying them, and then repeat the cycle, invading other cells.

Bird flu viruses bursting out of a human cell and destroying it.

Ignaz Semmelweis

- Semmelweis was a doctor who thought that doctors' hands spread disease between patients in a hospital.
- He made doctors wash their hands, and the death rate plummeted.

How Science Works

You should be able to: relate the contribution of Semmelweis in controlling infection to solving modern problems of the spread of infection in hospitals.

Questions

Grades G-E
1 What is the addictive component of tobacco smoke?

Grades D-C
2 There is a correlation between smoking and getting lung cancer. What does 'correlation' mean?

Grades G-E
3 What is a pathogen?

Grades D-C
4 Explain why there was less disease in the hospital where Semmelweis worked, after he made doctors wash their hands.

Body defences

White blood cells, epidemics and pandemics

- White blood cells are part of the **immune system**. They protect the body against **pathogens** by:
  - **ingesting**: taking pathogens into their cytoplasm and destroying them
  - producing **antibodies** that destroy bacteria or viruses
  - producing **antitoxins** that neutralise bacterial toxins.
- Particular antibodies and antitoxins match particular pathogens.
- In an **epidemic**, many people are infected.
- A **pandemic** is a worldwide epidemic: Spanish flu killed over 50 million people in 1918.
- An epidemic takes hold when most people are not **immune** to a new pathogen.

Phagocytes, lymphocytes and antibodies

- Cells near a wound send a chemical signal to white blood cells called **phagocytes**, which gather to ingest any bacteria in the wound, in a process called **phagocytosis**.
- White blood cells called **lymphocytes** produce **antibodies**. Each is **specific** for a particular **antigen** on a pathogen. The antibody attaches to the antigen and destroys the pathogen.
- Lymphocytes also make antitoxins to destroy toxins made by bacteria.

1 A phagocyte moves towards a bacterium

2 The phagocyte pushes a sleeve of cytoplasm outwards to surround the bacterium

3 The bacterium is now enclosed in a vacuole inside the cell. It is then killed and digested by enzymes

Phagocytosis.

Drugs against disease

Testing antibiotics.

Painkillers and antibiotics

- Medical drugs called **painkillers** reduce pain symptoms but have no effect on the cause.
- **Antibiotics** are drugs taken into the body that kill bacterial cells but not body cells.
- In the experiment shown on the right, paper discs soaked in six different antibiotics were laid on a dish containing a layer of one type of bacterium. The bacterium failed to grow round four of the discs, showing that the discs contained antibiotics effective against the bacterium.

Antibiotics sources and antiviral drugs

- Many antibiotics, including **penicillin**, come from fungi. Some bacteria have developed **resistance** to antibiotics.
- Sources of newer antibiotics include frog skin. Others are made chemically, ensuring purity and known strength.
- Since viruses invade body cells, **antiviral** drugs are liable to kill the body cells as well as the virus.
- AZT is an antiviral drug that slows down the development of HIV/AIDS, but can cause severe side effects.

Questions

Grades G-E
1 What is an antibody?

Grades D-C
2 Which kind of blood cells produce antibodies?

Grades G-E
3 Explain why antibiotics are useful drugs.

Grades D-C
4 Explain why it is difficult to develop antiviral drugs.

Arms race

Resistance to antibiotics

G–E

- In the 1940s, **antibiotics** ended the death of millions from bacterial infections.

- By **natural selection**, bacteria develop **resistance** to antibiotics. Over-use of antibiotics increases the chance of this happening.

- **MRSA** is the strain of the bacterium *Staphylococcus **a**ureus* that is **r**esistant to the antibiotic **m**ethicillin. MRSA infections can kill weak and very young or old people.

1 A gene in one bacterium changes

2 This bacterium is not killed by an antibiotic that kills the others

3 It divides to produce an antibiotic-resistant strain

How a resistant strain of a bacterium develops

New viral diseases

D–C

- In 1993, a new, highly infectious **virus** disease, **SARS**, first appeared in China. With international travel and without drugs to combat it, a **pandemic** (worldwide epidemic) was feared.

- In an international effort, scientists developed a **vaccine** against it, and checked for people with SARS or those who had been in contact with it. By 2003, SARS was under control.

- **Bird flu** arose in 2004, killing people in Asia. It passes from poultry to humans, but human-to-human transfer was feared if the virus mutated. World health agencies monitor this disease closely.

Vaccination

How Science Works

You should be able to:

- explain how the treatment of disease has changed as a result of increased understanding of the action of antibiotics and immunity
- evaluate the consequences of mutations of bacteria and viruses in relation to epidemics and pandemics, e.g. bird flu
- evaluate the advantages and disadvantages of being vaccinated against a particular disease.

Immunisation

G–E

- Through **vaccination**, people are **immunised** (made immune) against dangerous diseases. A vaccine of harmless dead or inactivated viruses or bacteria (pathogens) is injected into the bloodstream.

- White blood cells form specific **antibodies**, which attach to **antigens** on the surface of the pathogen and destroy it. If the person is later infected with the live pathogen, the same antibodies are rapidly produced and combat the infection.

- **MMR** is a triple antiviral vaccine against measles, mumps and rubella.

The MMR controversy

D–C

- In 1998, a team of scientists claimed in an article that the MMR vaccine caused children to develop **autism**, a personality disorder.

- Some parents refused to let their children have the MMR jab. This meant that many children were not immune to the diseases, and so could pass them on and cause an epidemic.

- Extensive studies have now proved that there is no link between MMR vaccination and autism.

year	percentage of children having MMR jab	no. of mumps cases
1996	92	94
1997	92	180
1998	91	119
1999	88	372
2000	88	703
2001	87	777
2002	84	502
2003	82	1549
2004	(no figure available)	8104

When children failed to have the MMR jab, mumps cases increased.

Questions

Grades G-E

1 What kind of organism is MRSA?

Grades D-C

2 Explain why scientists feared that SARS might become a pandemic.

Grades G-E

3 How does an immunisation jab make you immune to a disease?

Grades D-C

4 Explain why the misinformation about the MMR jab had dangerous consequences.

B1a summary

Nerves and **hormones** coordinate body processes and help control the body's:

- water content
- ion content
- blood sugar concentration
- temperature.

Hormones are made by **glands** and **secreted** into the bloodstream to convey information to **target organs**.

The **menstrual cycle** is controlled by **FSH** and **LH** from the pituitary and **oestrogen** from the ovaries.
Glucose balance is maintained by **insulin** produced by the **pancreas**.

The **brain** and **spinal cord** make up the **central nervous system** (**CNS**).
Nerves made up of **neurones** connect the body's organs to the CNS.

Nerves and hormones

Receptors detect stimuli and are concentrated in:
– sense organs: eyes, ears, nose and tongue
– skin: for pressure, pain and temperature.
Electrical impulses from receptors pass along nerves:

- to the brain
- to effectors (**muscles** and **glands**), which respond.

Reflex actions are rapid, automatic responses to a **stimulus**.
An electrical impulse travels along a **reflex arc**: sensory neurone → spinal cord → **relay neurone** → motor neurone → **effector**

Sensory neurones carry information from **receptors** to the CNS.
Motor neurones carry information from the CNS to **effectors**.
The gap between the ends of two neurones is a **synapse**.

A **balanced diet** contains the right amounts of:

- **carbohydrates**
- **fats**
- **proteins**
- **vitamins, minerals, fibre** and **water**.

Diet

Too much **cholesterol** or **salt** increases the chance of getting heart disease. Cholesterol levels are:

- *raised by* **saturated fats**
- *lowered by* **monounsaturated** and **polyunsaturated fats**.

Obesity increases the chance of getting:
– arthritis
– diabetes
– high blood pressure
– heart disease.

Medicinal drugs combat disease and relieve pain.
Recreational drugs can be:

- **legal** (e.g. alcohol, tobacco)
- **illegal** (e.g. cannabis, cocaine).

Drugs

Hard drugs such as heroin are very **addictive** and cause severe health problems.

Tobacco smoke contains:

- **nicotine** (addictive)
- **carcinogens** (cause cancer)
- **carbon monoxide** (prevents blood from carrying oxygen)
- **irritants** (cause bronchitis, emphysema and diseases of the heart and blood vessels).

Pathogens are microorganisms that cause infectious diseases.
Bacteria and **viruses** produce **toxins** and damage cells, which makes people feel ill.

Body defences:
white blood cells

- ingest bacteria
- produce **antibodies**
- produce **antitoxins**.

Disease

Drugs against disease:

- **painkillers** relieve symptoms but don't cure diseases
- **antibiotics** can kill bacteria, but bacteria can develop resistance to them
- **immunisations** and **vaccinations** offer protection from infectious diseases.

Hot and cold

Animals and plants are adapted to different temperatures

G–E

- Where it is hot, some animals lose heat from blood circulating through bare skin and large thin ears.
- Where it is cold, thick fur **insulates** mammals from heat loss. Small ears and compact bodies reduce **surface area**.
- Where it is hot, cacti reduce water loss with swollen stems, a thick surface, spines not leaves and a spreading, shallow root system to catch limited rainfall.
- Where it is cold, plants grow low to absorb heat from the ground warmed by sunshine and to avoid cold winds.

Examples of adaptations for extreme cold and heat

D–C

- Caribou can stand in snow at –50 °C but their legs do not freeze. Arteries and veins are close together, so arterial blood warms the blood in veins.
- Camels are adapted to a very dry desert **habitat**.

How a camel is adapted to desert life.

fat in hump avoids heat-insulating fat layer on body

no insulating fat under the skin

stomach can hold over 20 litres of water

can drink 20 litres of water in just a few minutes

very little urine

long legs keep it above the hot ground

38°C   37°C
20°C   19°C
9°C   8°C

Heat exchange in a caribou's leg.

Adapt or die

Competition

G–E

- To **survive**, organisms take materials from their surroundings and from other living organisms.
- Animals **compete** for food, territory and a mate. Only those that are well **adapted** to their **environment** are likely to survive.
- Moles have a good sense of smell for finding food underground. The long mouth tube of butterflies reaches the nectar in flowers. A male peacock's tail attracts the female.
- Plants compete with each other for light and for water and nutrients from the soil.
- Plant leaves are arranged so that the greatest number receive sunlight. Their root systems are large to maximise water and nutrient intake.

How Science Works

You should be able to:

- suggest how organisms are adapted to the conditions in which they live
- suggest the factors for which organisms are competing in a given habitat
- suggest reasons for the distribution of animals or plants in a particular habitat.

Special adaptations against predators

D–C

- Sharp **thorns** protect plants from grazing animals.
- Some plants and animals produce **poisons**, and have a bright warning colour, to ensure that **predators** do not eat them. Tiny, brightly-coloured poison dart frogs in Central and South America produce poisons used by forest dwellers on their darts.
- Harmless hoverflies **mimic** (are adapted to look like) wasps that attack predators.

Questions

(Grades G-E)

1 An animal has thick fur and small ears. Is it adapted to live in a hot climate, or a cold climate?

(Grades D-C)

2 Name the habitat of a caribou, and state **one** way in which it is adapted to live there.

(Grades G-E)

3 List **three** things that animals compete for.

(Grades D-C)

4 How does the bright colour of a poison dart frog help it to survive?

Two ways to reproduce

Top Tip!

There can be sexual reproduction with only one parent – for example in some plants where a single flower produces both male and female gametes.

On to the next generation

- **Genes** in the nucleus of cells carry information about characteristics from parents to offspring. Different genes control different characteristics.

- In **sexual reproduction**, an individual **inherits** half its genes from each parent. Both parents and offspring share characteristics, but different mixtures of genes make each individual differ from all others.

- **Asexual reproduction** involves only one parent. The genes of parent and offspring are identical: they are all **clones**.

G–E

Differences between sexual and asexual reproduction

- In sexual reproduction, each parent produces **sex cells** or **gametes**: females produce **eggs** and males produce **sperms**. However, in asexual reproduction, there are no gametes.

- In sexual reproduction, a female gamete fuses with a male gamete. This is **fertilisation**. However, in asexual reproduction, there is no fertilisation.

- The offspring of sexual reproduction have different mixes of genes. The parents and offspring show **variation**. However, in asexual reproduction, all the offspring are genetically identical to each other and to their parent. They are **clones**.

the sperm cell has a set of genes from the father

the egg cell has a set of genes from the mother

these two gametes fuse – this is called fertilisation

the fertilised egg contains genes from both parents

How an egg is fertilised in sexual reproduction.

the parent cell copies all of its genes

a set of genes goes to the two new cells

each cell is a clone – they all have the same genes

Asexual reproduction – the first steps.

D–C

Genes and what they do

Genes and chromosomes

- A **gene** is a stretch of **DNA** on a **chromosome** which contains many genes. The **nucleus** of each cell contains a set of chromosomes.

- At **fertilisation**, when sex cells (gametes) fuse, the chromosomes come together. The fertilised egg has a set of genes from both parents.

- The mix of genes helps to determine an organism's characteristics.

- Other characteristics are not caused by genes. They result from events in an organism's life. Examples are tattoos in humans or a tree losing a branch.

- Some characteristics, such as an unattached earlobe, result from just one gene.

- The monk Gregor Mendel was the pioneer of inheritance studies 150 years ago. He bred generations of pea plants, and by good chance chose to observe characteristics that were each determined by one gene on a different chromosome.

- From his observations, he correctly proposed how genes were passed from parents to offspring.

the nucleus contains chromosomes

chromosome - a long chain of genes

DNA contains genes

gene: these control the way we grow and develop

Looking at the detail – from genes to nucleus.

G–E

D–C

Questions

(Grades G-E)

1 The offspring of asexual reproduction are 'clones'. What does this mean?

(Grades D-C)

2 Describe **one** difference between sexual reproduction and asexual reproduction.

(Grades G-E)

3 Where are genes found?

(Grades D-C)

4 Explain how the shape of your earlobes is determined.

Cuttings

Growing plants from cuttings

- Plants can be produced asexually by taking **cuttings**. **Hormone rooting powder** speeds up new root growth.

- The new plants and the parents are **genetically identical**, so they are clones.

- Cuttings are best taken of the **stem**, **leaf** or **root** of a young, actively growing plant:
  - all new plants are genetically identical to the original plant and to each other
  - all are likely to grow at the same rate
  - taking cuttings is easy, cheap and guarantees that all plants show the same characteristics.

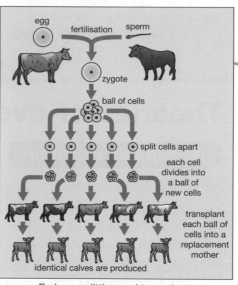

parent plant

this stem should have leaves on it

take a healthy plant and cut off a small length of stem

dip the end of the cut stem into hormone rooting powder

put the stem into a flowerpot full of damp compost

cover the pot with a plastic bag to keep it moist

this will grow into a new plant

How to take cuttings.

Clones

Tissue culture and embryo transplants

- In plant **tissue culture**, small groups of cells from a plant grow on a jelly containing all the nutrients a plant requires. The cells divide and grow into new plants, all with the same genes: they are **clones**.

- In animal **embryo transplantation**:
  - a fertilised egg (a **zygote**) is taken
  - it divides to form a ball of cells, all genetically identical
  - the cells are separated before they become **specialised**
  - each cell divides to form a ball of new cells (an **embryo**)
  - each identical embryo is **transplanted** into a host mother where it develops into a new individual.

egg
fertilisation
sperm

zygote

ball of cells

split cells apart

each cell divides into a ball of new cells

transplant each ball of cells into a replacement mother

identical calves are produced

Embryo splitting and transplants – many offspring from one zygote.

Adult cell cloning and replacing damaged skin

- In **adult cell cloning**, the nucleus of an egg cell is removed and replaced with the nucleus of an adult **body cell**. The egg cell develops into an individual which is a clone of the body cell donor.

- New skin can be grown to form **skin grafts**. A victim of burns often needs skin grafts, but may not have enough undamaged skin for them. Instead, a few healthy skin cells from the victim are **cultured** in a **growth medium**, to produce large sheets of skin which can then be transferred to burnt areas by plastic surgery.

Skin ready for plastic surgery.

Questions

Grades G-E

1 What is a cutting?

Grades D-C

2 List **two** advantages of growing new plants from cuttings.

Grades G-E

3 Name the process by which a whole plant is grown from a little group of cells.

Grades D-C

4 Describe how adult cell cloning is done.

Genetic engineering

GM soya beans and insulin

- **Genetic engineering** is the technique of transferring a gene from the chromosome of one organism to another, which is then **genetically modified** (**GM**).

- A gene for a desired characteristic from one organism is inserted into embryos of a plant. When they have grown into adult plants, they may show the desired characteristic.

- Soya beans are a rich protein source. Soya plants compete with weeds. The **herbicide** that kills weeds also kills soya. A gene for **resistance** to the herbicide was inserted into a chromosome in soya, making it a **GM crop**.

- People with **diabetes** lack the hormone **insulin** to regulate their blood sugar levels.

- By genetic engineering, the gene for producing insulin was cut from human DNA by a special **enzyme**, and inserted into bacteria which provide insulin on an industrial scale.

The genetic engineering of bacteria to produce insulin.

G–E

D–C

Theories of evolution

Darwin and evolution

- Evolution is the gradual change in a species' characteristics over time.

- **Charles Darwin**'s book *On the Origin of Species* was based on his knowledge of fossils and observations of living organisms. He suggested that **species** gradually changed from one form to another as a result of **natural selection**. Darwin's theory replaced others, including the belief that all organisms were created by God in the same form as they are today, and Lamarck's idea that characteristics **acquired** in an organism's lifetime can be passed on to the next generation.

G–E

- Darwin collected finches from several different islands in the Galapagos. He noticed that the different kinds of finches all looked quite similar to each other, but had different kinds of beaks.

- He suggested that they had all evolved from one kind of finch that had arrived on the islands from the mainland. Different kinds of beaks had gradually evolved as adaptations to eat different kinds of food.

The characteristics of finches adapted to eat different kinds of food.

type of finch		food sources
ground finch		eat seeds which they crush with their strong beaks
cactus finch		long, slender beaks suck up nectar from cactus flowers
vegetarian tree finch		curved beaks, like a parrot, eat buds and fruit
insectivorous finch		stubby beaks to eat beetles and other insects
woodpecker finch		use a cactus spine to prise small insects out from cracks in tree bark

D–C

Questions

Grades G-E

1 Explain what is meant by a 'GM organism'.

Grades D-C

2 What kind of GM organism is used to produce insulin?

Grades G-E

3 Name **two** people who put forward different theories of evolution.

Grades D-C

4 Why do the Galapagos finches have different kinds of beaks?

Natural selection

Genes and survival

- Organisms compete for survival. Those with characteristics best suited to their surroundings are more likely to survive, while others are more likely to die. This is **natural selection**.

- Different characteristics are controlled by different **genes**. Occasionally a gene undergoes a change or **mutation** and so the characteristic changes. If the changed characteristic increases an organism's survival and breeding chances, it is more likely to pass to later generations.

- This can cause a **species** to change, sometimes rapidly.

Natural selection in action

- Peppered moths are usually pale with dark speckles. They are **camouflaged** from predator birds.

- In cities, the industrial revolution blackened buildings and trees with soot. Dark moths – a **variation** in the species – were better camouflaged. The dark moths survived, while birds ate the more visible, pale moths. Dark moths were being favoured by **natural selection**.

- Now, with cleaner air, most peppered moths are pale again.

Top Tip!

Remember that organisms like the moths do not change on purpose – it happens through natural selection.

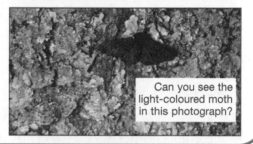

Can you see the light-coloured moth in this photograph?

Fossils and evolution

How Science Works

You should be able to: suggest reasons why scientists cannot be certain about how life began on Earth.

Fossils and the origins of life

- A **fossil** forms when a dead organism is buried in **sediment**, more sediment piles on top, the pressure (and temperature) increases and the organism turns to stone.

- Comparing **fossil** remains with present-day plants and animals shows us that, since life began, some organisms have changed (**evolved**) a lot, and some only a little.

- Fossils and the **theory of evolution** support the idea that all present-day species evolved from the first **life-forms** that existed about 3.5 billion years ago.

- We can learn about the evolutionary relationships and ecological relationships between organisms by looking at similarities and differences between them.

What fossils tell us

- All present-day species have evolved from simpler life-forms.

- There are gaps in the fossil record for most species.

  The fossil record for the horse shows much change in 60 million years.

years ago	60 million	40 million	10 million	modern horse
shape, size and front leg bones	height: 0.4 m	height: 0.6 m	height: 1.0 m	height: 1.6 m
how it lived	on soft ground near water; its feet could support its weight without sinking into the mud	in drier conditions; needed to be able to run away from predators	in very dry conditions; it was a fast runner	on grassland; it is a fast runner

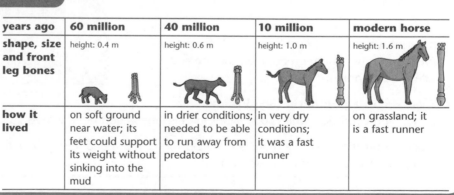

Questions

1 What is the name for a change in a gene?

2 Explain why the black variety of the peppered moth became more common in the industrial revolution.

3 How does a fossil form?

4 What do fossils tell us about the habitats in which the very first horses lived?

Extinction

Causes of extinction

- A species that cannot adapt to change may become **extinct**; if:
  - a change occurs to its **physical** (non-living) **environment**
  - a new **predator** arrives   – a new **disease** appears
  - a more successful **competitor** moves into its **habitat**.

- Dodos became extinct on Mauritius when rats arriving with people ate dodo eggs.

- The 30 m high giant club-moss became extinct 300 million years ago when the climate cooled and its swampy habitat dried out.

The dodo became extinct 300 years ago.

- When there is an **environmental** change, such as temperature change, organisms that adapt or move away will survive; others will die out.

- Building on green sites destroys habitats so species become locally extinct.

- A **predator** newly arrived in a habitat may kill and eat all individuals of a species not adapted to escape it.

- A species may not be adapted to withstand a new **disease**.

- Successful **competitors** for food may survive while less successful competitors become extinct.

- Hunting and habitat destruction by humans has caused many extinctions.

This is the last St Helena olive tree, a species made extinct by a fungal disease.

More people, more problems

Human populations and population growth

- By 2035, the world's **population** may reach 10 billion, straining the supply of **raw materials** and affecting the **environment**.

- As living standards and the number of people rise, **non-renewable energy resources** such as **crude oil** and **coal**, and minerals such as **metal ores** and **rocks**, are rapidly used up.

- **Waste** and **pollution** will increase if they are not regulated by laws.

- Until the 1800s, the human population was small. Disease, bad housing and insufficient food limited the length of people's lives.

- Then, industrial **development** in western countries brought wealth and improved conditions. People lived longer with better food, medical care and safe water supplies.

2000 6.1 billion

less-developed countries

more-developed countries

Human population growth – will it rise as predicted?

- There is now a **population explosion** in many developing countries, while populations in developed countries are roughly stable.

- Bigger populations require more land for buildings, roads and food production, so the habitats of plants and other animals are destroyed.

Questions

Grades G-E

1 Explain why the dodo became extinct.

Grades D-C

2 Explain how a new competitor could cause a species to become extinct.

Grades G-E

3 Why may pollution by waste increase if we do not regulate it?

Grades D-C

4 Look at the graph above. What is happening now to the *rate* of population growth?

Land use

Loss of the natural environment

- Expanding populations lead to natural landscapes being lost:
  - people need more **buildings** for homes, workplaces and shops
  - **building materials** are cut from quarries and dredged from sand and shingle beds
  - more **food** is needed, so natural habitats become farmland
  - **waste** increases and is disposed of in landfill sites (or in incinerators).

Intensive farming

- To maximise food output and minimise prices, **intensive farming** methods are used:
  - herbicides and pesticides (toxic chemicals) and fertilisers increase the production of crops and livestock
  - hedgerows are removed to enlarge fields and make it possible to use combine harvesters
  - hens and other livestock are reared in vast battery sheds.
- Populations of wild plants and animals are wiped out by these processes.

Intensive farming replaces small fields of different crops with a huge field of one crop.

Pollution

Top Tip!

Properly treated sewage is absolutely safe – it is untreated sewage that causes problems.

How water is polluted

- **Pollution** occurs when damaging materials are put into the environment.
  - **Toxic chemicals**, for example in an oil spill, can poison living organisms.

name of disease	its symptoms
gastroenteritis	stomach cramps, diarrhoea and vomiting
Weil's disease	liver and kidney damage, it can be fatal
allergic alveolitis	fever, breathlessness, cough

Diseases you can get from untreated sewage.

  - **Sewage** contains human waste and rubbish including toilet paper, industrial waste and microorganisms. If not properly treated at sewage works and suitably disposed of, it can cause human diseases and pollute rivers, killing aquatic organisms.
  - Excess **fertiliser** applied to farmland can reach streams and rivers. Surface water plants grow rapidly and blanket the waterways. Light for photosynthesis fails to reach the plants below and they die. Bacteria feed on the dead plants, respire and use up oxygen, killing organisms such as fish and snails.

Water pollution: effects on humans and other animals

- Because **sewage** is a danger to health, all who work in sewers, toilets and water treatment plants and who transport sewage sludge, should wear protective clothing.
- In one or two very rare cases, very high concentrations of nitrates from **artificial fertilisers** reached the drinking water supply and caused blue baby syndrome, when the blood of babies cannot transport sufficient oxygen.
- The artificial pesticide **DDT** was widely used 50 years ago, especially against mosquitoes and other parasites. It did not break down and passed up the food chain to reach lethal concentrations in higher animals. Now its use is banned.

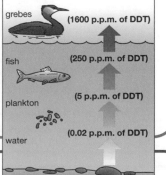

How DDT is taken up the food chain.

grebes (1600 p.p.m. of DDT)
fish (250 p.p.m. of DDT)
plankton (5 p.p.m. of DDT)
water (0.02 p.p.m. of DDT)

Questions

1 List **four** ways in which humans reduce the amount of land available for other organisms.

2 What is meant by 'intensive farming'?

3 List **three** causes of land and water pollution.

4 Explain why DDT killed many predatory birds.

Air pollution

Pollution by smoke, smog and sulfur dioxide

- Power stations and traffic produce **smoke**, tiny particles suspended in the air.

- Smoke particles plus water droplets in fog form **smog**.

- Burning fossil fuels, especially coal in power stations, forms **sulfur dioxide**, an acidic gas. Sulfur dioxide helps form **acid rain**.

- Smoke particles and sulfur dioxide damage the lungs.

London, 1952, when over 3500 people died because of the smog.

coal is burned	→	sulfur is released	→	sulfur combines with oxygen to form sulfur dioxide	→	sulfur dioxide dissolves in rainwater to form acid rain

How acid rain is formed.

G–E

Effects of air pollution, and remedies

Smoke can turn buildings black.

- **Sulfur dioxide**:
  - causes bronchitis, particularly in the elderly and young, and triggers asthma attacks
  - turns plant leaves yellow, so they cannot photosynthesise
  - **reacts** with limestone on buildings.

- **Smoke**:
  - causes coughing, lung damage and asthma attacks
  - settles as black **soot** on buildings.

- To reduce pollution by particles and harmful chemicals:
  - many factories fit filters to clean the smoke they emit
  - power stations use gas scrubbers
  - cars are fitted with catalytic converters.

D–C

What causes acid rain?

Top Tip!

Do not confuse acid rain with the greenhouse effect! One is caused by sulfur dioxide, and the other by carbon dioxide.

More about acid rain

- Several acidic gases are present in polluted air.
  - Power stations that generate electricity by burning **fossil fuels** produce sulfur dioxide. This gas reacts with oxygen and water in the air to produce **sulfuric acid**.
  - Oxides of nitrogen in traffic exhaust form **nitric acid**.

- Rainwater is naturally weakly acidic: **carbon dioxide + water → carbonic acid**

- The acids dissolved in rainwater form **acid rain** that damages plants and **erodes** stone.

G–E

Why trees and lakes die and stone erodes

- **Trees** die because acid rain:
  - strips the protective waxy surface from leaves, the cells beneath shrivel and so cannot photosynthesise
  - washes useful nutrients and minerals out of the soil, and releases aluminium which is toxic to tree roots.

- Acid rain causes **lakes** to become so acidic that fish and other creatures are killed. **Lime**, a base, is added to acidic lakes to neutralise the acid.

- **Limestone** and **sandstone** statues and buildings are eroded because acid rain reacts with the calcium carbonate they contain and washes away their surfaces.

D–C

Questions

Grades G-E
1 Name a gas that helps to cause acid rain.

Grades D-C
2 How can sulfur dioxide affect people's health?

Grades G-E
3 List **three** acids that may be present in acid rain.

Grades D-C
4 How does acid rain harm limestone buildings?

Pollution indicators

How Science Works

You should be able to: analyse and interpret scientific data concerning environmental issues.

Organisms that indicate pollution levels

- **Lichens** are often used as indicators of sulfur dioxide **air pollution**.

- **Invertebrate animals** are often used as indicators of **water pollution**.

Investigating the effects of sewage

- This table shows the results of a study of river pollution by **sewage**, using **invertebrates** as pollution level indicators.

distance downstream from sewage entry point (m)	what the water is like	invertebrates found	oxygen levels
sewage enters here 0–10	dark and cloudy; very smelly	*Chironomus* larva; rat-tailed maggot	falling quickly
10–100	cloudy; bad smell	tubifex worm (sludge worm); mosquito larva	very low
100–200	beginning to clear; slight smell	flatworm; caddis fly larva	gradually rising
200+	clear	freshwater shrimp; mayfly larva; stonefly larva	back to normal

The effects of untreated sewage in a river.

Deforestation

Deforestation and its effects

- **Deforestation** is the permanent clearing of forests, particularly of **tropical rainforests**.

- Trees are cut for **timber** or cleared (often burned) to make way for **agriculture**.

- **Photosynthesis** uses **carbon dioxide** from the air to produce carbon compounds in plant tissue.

- Much more carbon is 'locked up' in a tree trunk than in crops occupying the same area. When wood burns, carbon dioxide returns to the air.

- Deforestation reduces the rate of carbon dioxide removal from the air.

- Deforestation destroys **habitats** and reduces **biodiversity**: species become **extinct** and survivors show reduced **variation**. Humans may have had uses for some lost organisms.

The carbon cycle and the demand for wood

- All living things **respire** and give out carbon dioxide. In photosynthesis, plants take in carbon dioxide, but deforestation 'unlocks' it.

- In 20 years, half the tropical rainforests of Borneo have been destroyed – an area larger than Great Britain – to grow palms for oil to make soap, cosmetics and other products, and to export timber.

The carbon cycle.

- This raises CO_2 levels in the air and threatens over 40 species of mammals, including the orang-utan.

Questions

1 Name a group of organisms that can be used as indicators of sulfur dioxide pollution.

2 If a river contains rat-tailed maggots, is it polluted or unpolluted?

3 State how deforestation affects carbon dioxide levels in the atmosphere.

4 How can rotting trees affect carbon dioxide levels in the atmosphere?

The greenhouse effect – good or bad?

The greenhouse effect and global warming

- The **atmosphere** ensures that not all the Sun's heat is reflected back into space. It traps (absorbs) some heat in the **greenhouse effect**.

- **Global warming** occurs when the greenhouse effect *increases* and less heat escapes into space. Flooding, storms and droughts will increase.

- Global warming increases with an increase in the level of the heat-absorbing gases: **carbon dioxide** from burning fossil fuels, and **methane** from cattle and rice fields.

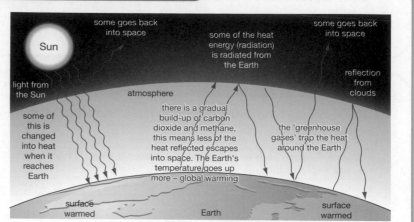

How an increase in the greenhouse effect leads to global warming.

- Air from the past trapped in Antarctic ice tells us that 250 years ago the atmosphere contained 270 parts per million of carbon dioxide. This is expected to double by 2025.

- The Earth's average temperature has risen in step with this increase. It is expected that:

  - climates throughout the Earth will change

  - Earth's great ice caps, including those at the poles, will melt, raising sea levels and flooding coasts.

Top Tip!

The greenhouse effect is good because without it the Earth would be too cold for life. But if it increases, that is *not* good.

G–E

D–C

Sustainability – the way forward?

Means of sustainable development

- As populations around the world increase and expect a better life, the Earth's **natural resources**, including fossil fuels, are being used up faster than they can be replaced or substituted.

- To ensure the long-term survival of humans, ways of **sustainable development** could include:

  - **recycling** rubbish and making it a **resource**, not waste

  - **insulating** buildings to reduce the heat energy lost

  - **walking** and **cycling** to reduce the use of cars that burn fuel

  - using **public transport** where possible.

- Fossil fuels used to generate electricity could be replaced by energy sources that do not produce carbon dioxide, such as nuclear power, wind and wave power and hydroelectric power.

How Science Works

You should be able to:
- evaluate methods used to collect environmental data and consider their validity and reliability as evidence for environmental change
- weigh evidence and form balanced judgements about some of the major environmental issues facing society, including the importance of sustainable development.

G–E

D–C

Questions

Grades G-E
1 Name **two** gases that cause the greenhouse effect.

Grades D-C
2 What has happened to CO_2 levels and temperature in the past 250 years?

Grades G-E
3 Explain what is meant by 'sustainable development'.

Grades D-C
4 State **two** alternative energy sources for generating electricity that do no produce CO_2.

B1b summary

Animals and plants are adapted:
- to **survive** in different **habitats**
- to **compete** for resources
- for **protection** against predators.

Animals compete for food, territory and a mate.

Animal and plant adaptations

Plants compete for light, water and nutrients from the soil.

Caribou and camels have adapted to extreme environments.

Reproduction can be:
- **sexual** (using **gametes**, variation in offspring)
- or **asexual** (one parent, identical offspring).

DNA contains **genes** on **chromosomes** that control our **inherited characteristics**.

Reproduction and genes

Genetically identical **clones** can be produced:
- from **cuttings** and **tissue culture** in plants
- by **embryo transplantation** in animals.

In **genetic engineering**, a gene from one organism is inserted into another organism.

Genetically modified organisms include soya plants resistant to herbicide and bacteria that make human insulin.

Charles Darwin proposed the theory of **evolution** of **species** by **natural selection**.

Mutation of a gene to cause an advantageous change to a characteristic can increase a species' chance of **survival**.

Evolution

Comparing **fossils** with present-day plants and animals can show us how species have evolved.

Factors that contribute to **extinction** include:
- environmental change
- new predators or diseases
- successful competitors
- adverse human impact on habitats.

Increase in the human **population** has affected the environment by:
- using up resources more rapidly
- increasing waste and pollution.

Land is used up for:
- buildings and quarrying materials to make them
- producing food
- waste disposal.

Intensive farming methods and **deforestation** destroy habitats, reducing **biodiversity**.

Water pollution is caused by toxic chemicals, raw sewage and excess fertiliser.

Human impact on the environment

Air pollution caused by burning fossil fuels results in smoke, smog and acid rain.

Sulfur dioxide and **oxides of nitrogen** dissolve in rainwater to give **acid rain** which damages plants, pollutes lakes and erodes limestone and sandstone structures.

Lichens and invertebrates can be used as **pollution indicators**.

Increases in the heat-absorbing gases responsible for the **greenhouse effect** can contribute to **global warming**.

Sustainable development includes using renewable energy sources, recycling and conserving energy and fuels.

Elements and the periodic table

The elements and patterns in properties

- All substances are made of tiny **particles** called **atoms**. An **element** is made of only one sort of atom.

- Atoms of each element are given a **chemical symbol**. O represents an oxygen atom and Na represents a sodium atom.

- Elements are arranged in the **periodic table** below. Each vertical column is a **group**. Elements in a group have similar properties.

- In 1869, Mendeleev first made a similar table of the elements known then. He left gaps for unknown elements, which scientists later discovered.

G–E

D–C

The periodic table of the elements.

Atomic structure 1

Particles in an atom

- Each atom has a small central **nucleus**. It contains **protons** and **neutrons** (except hydrogen; its nucleus is just one proton).

- **Electrons** orbit around the nucleus.

- Protons, neutrons and electrons are unimaginably small and impossible to weigh. But we use numbers, **atomic mass units**, to express their masses relative to one another (**relative masses**).

The structure of an atom.

G–E

particle	relative mass	charge
proton	1	+
neutron	1	neutral
electron	almost 0	−

- In an atom, the number of protons = number of electrons = **atomic number** (unique to each element). An atom has no net charge.

- The **mass number** = number of protons + number of neutrons

- The electrons are in **shells**. Two electrons fill shell 1 and eight electrons fill shell 2.

D–C

Questions

Grades G-E

1 How many different kinds of atoms are there in an element, such as oxygen?

Grades D-C

2 Use the periodic table to find the chemical symbol for zinc.

Grades G-E

3 Name the **two** kinds of particles found in the nucleus of an atom.

Grades D-C

4 Use the periodic table to find the atomic number of magnesium. What does this tell you about the structure of a magnesium atom?

Bonding

Bonds in elements and compounds

G–E

- Few elements exist naturally as single atoms. Two atoms of the *same* element can join to form a **molecule**:
  - nitrogen and oxygen molecules, N_2 and O_2, each have two atoms
  - the atoms *share* a pair of **electrons**, one from each atom, in a **covalent bond**.

- When atoms of two (or more) *different* elements join, they form a **compound**.
  - The atoms in water, H_2O, are arranged as HOH. Each hydrogen atom *shares* a pair of electrons with the oxygen atom. Water is a covalent compound.
  - A magnesium atom and an oxygen atom join to form magnesium oxide, MgO. Magnesium *gives two* electrons, and oxygen *takes* them. Both magnesium and oxygen then have a charge – they are **ions**. Their opposite charges attract, forming an **ionic bond**.

D–C

- Atoms with a full outer shell of electrons (see page 26) are **stable**. The atoms of most elements have an incomplete outer shell.

- An atom will react chemically with another atom that either gives, takes or shares electrons to give it a complete outer shell.

All atoms in a water molecule have full outer electron shells: the hydrogens have two electrons in shell 1 and oxygen has eight electrons in shell 2.

Extraction of limestone

Limestone: its source and uses

G–E

- **Limestone** is the chemical compound **calcium carbonate**. It is the main compound in chalk and marble, which, like limestone, are mined from quarries.

- Limestone is a building material. It is also used in the making of cement, concrete and glass and in extracting iron from its ore.

How Science Works

You should be able to: consider and evaluate the environmental, social and economic effects of exploiting limestone and producing building materials from it.

Limestone extraction: for and against

D–C

- Advantages of a local **limestone quarry** include:
  - jobs for local people
  - a better road system
  - local income to pay for health care, recreational and other facilities.

- Disadvantages include:
  - scarred landscape and loss of natural habitats
  - heavy traffic; lorries shake and damage buildings and roads
  - noise from blasting and dust pollution.

- Remedies to minimise the disadvantages include: use of rail transport, daytime working on weekdays only, and landscaping spent quarries.

Quarries can cause environmental damage.

Questions

Grades G-E

1 Nitrogen gas in the air exists as pairs of nitrogen atoms joined together. Is nitrogen gas an element or a compound?

Grades D-C

2 An atom of oxygen has six electrons in its outer shell. How many does it need to make this shell full?

Grades G-E

3 Name the chemical compound in limestone.

Grades D-C

4 List **two** advantages and **two** disadvantages of having a limestone quarry near a small town.

Thermal decomposition of limestone

Reaction of limestone on heating

- The **formula** of a compound shows the number and type of atoms joined together.

- Limestone is mainly **calcium carbonate**, formula **CaCO₃**. It contains 1 calcium atom, 1 carbon atom and 3 oxygen atoms.

- When limestone is heated strongly, a **thermal decomposition** reaction occurs – a breakdown into simpler compounds:

  calcium carbonate → calcium oxide (quicklime) + carbon dioxide

- On heating, other metal carbonates break down similarly:
  magnesium carbonate → magnesium oxide + carbon dioxide

- Calcium oxide (quicklime) reacts with water to produce calcium hydroxide (slaked lime):

  calcium oxide + water → calcium hydroxide

G–E

The equation for the thermal decomposition of limestone

- The **symbol equation** for the chemical reaction of limestone on heating is: $CaCO_3 \rightarrow CaO + CO_2$

- The equation is **balanced**, having the same number of each type of atom on either side.

- In this and all reactions: no atoms are lost or made; the mass of the products equals the mass of the reactants (in this case, one reactant).

D–C

Uses of limestone

Limestone in building materials

- **Cement** is formed when crushed **limestone** is strongly heated with **clay** in a **kiln**. Cement reacts with water and sets gradually, and eventually becomes very hard.

- To make **concrete**, sand and rock chippings are added to cement. The mixture reacts with water and sets hard.

- Bricks are joined together with **mortar**, a mixture of sand, cement and water that sets hard.

- **Slaked lime** (**calcium hydroxide**), once widely used for building, has been mostly replaced by cement which lasts longer.

- To make **glass**, **sand** and **sodium carbonate** are heated with limestone.

How Science Works

You should be able to:
- evaluate the developments in using limestone, cement, concrete and glass as building materials
- evaluate their advantages and disadvantages over other materials.

G–E

Mortar and concrete

- When mixed with water, the cement in **mortar** 'hydrates', hardening and binding the sand particles together.

- **Concrete** is a mixture of cement, sand and rock chippings, materials of different sizes. When cement sets, these materials together have a far greater strength than mortar.

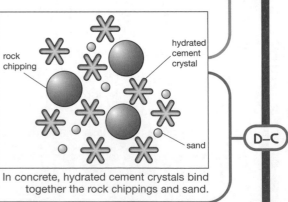

In concrete, hydrated cement crystals bind together the rock chippings and sand.

D–C

Questions

Grades G-E

1 Write down the word equation for the thermal decomposition of limestone.

Grades D-C

2 Write the balanced equation for the thermal decomposition of limestone.

Grades G-E

3 How is glass made?

Grades D-C

4 What is the difference between concrete and cement?

The blast furnace

Raw materials for making iron

- **Iron** is a strong, very widely used metal, often in the form of **steel**. It is used in cars, trains and ships, and in countless other products (see page 30).

- The raw materials for making iron are abundant and economical to extract and use:

  - **haematite**, iron ore, mainly **iron(III) oxide**, Fe_2O_3

  - **coke**, purified carbon-rich coal

  - **limestone**, mainly calcium carbonate, $CaCO_3$

  - **oxygen** in air.

Haematite contains the compound iron(III) oxide (Fe_2O_3).

Extracting iron from its ore

- Iron is extracted from its ore in a blast furnace.

- Iron is extracted by heating it with carbon, because carbon is more reactive than iron.

- The carbon takes oxygen from the iron ore, so **reducing** it.

- Iron ore, coke (carbon) and limestone go into the top of the blast furnace, and hot air into the bottom.

- The limestone helps to remove impurities from the iron ore.

- **Molten iron** is produced in the blast furnace.

Using iron

Different types of iron

- Iron from the blast furnace is 96% iron. It is remelted and moulded to form **cast iron** objects which require **hardness** and **strength** (e.g. drain covers).

- **Impurities** that make cast iron **brittle** and limit its use are removed to give **wrought iron**. This is pure iron which is softer and easily bent and shaped.

- Most iron produced is purified and then made into types of **steel**. **Carbon** and other **metals** are added in different quantities, according to the properties the steel needs to have.

The structure of wrought iron

- **Wrought iron** is pure iron metal. The atoms are bonded together in a three-dimensional, **regular**, closely packed arrangement called a **lattice**.

- Wrought iron is relatively soft and easily shaped because **layers** of atoms can slide over each other. This makes it suitable for gates and railings but too soft for many other uses.

The regular arrangement of atoms in wrought iron.

Top Tip!

Pure metals are soft and easily shaped because the atoms form a regular arrangement. The layers of atoms can then pass easily over each other.

Questions

1 What is the name of the iron compound found in iron ore (haematite)?

2 Iron ore is *reduced* in the blast furnace. What does this mean, and why does it happen?

3 What is done to the iron from a blast furnace, to make it into wrought iron?

4 Explain why pure metals, such as iron, can easily be shaped.

Using steel

Types of steel

- Steel is an **alloy** of iron. Different types of steel contain other metals and carbon (a non-metal) in different amounts.

- Compared with pure iron, steel is harder, much stronger, resists corrosion and is not brittle.

The hammer must be strong (not brittle). The scissors must be hard to retain its cutting edge, and be rust-proof.

Low, medium and high carbon steels

- The different elements in alloys have different sized atoms that distort the layers of atoms in the pure metal. They cannot slide easily over each other, so alloys are **hard**.

- *Low* carbon steels (0.4% carbon or less) are soft and easy to shape, properties suitable for car body making.

- *Medium* carbon steels (about 0.8% carbon) are harder and stronger, suitable for hand tools.

- *High* carbon steels (1.0–1.5% carbon) are very hard, suitable for knives and razor blades, but may snap when bent.

- *Stainless* steel is 70% iron, 20% chromium and 10% nickel. It is very resistant to corrosion (rust).

Transition metals

The properties of transition metals

Top Tip!

Remember that metals are on the left-hand side of the periodic table, non-metals are on the right.

- In the periodic table (see page 26), the **transition metals** are all the elements in the nine columns headed scandium (Sc) to copper (Cu).

- Transition metals are strong and make excellent **structural** materials.

- They are good **conductors** of **heat** and **electricity**, and can be **drawn** into wires and **hammered** into sheets.

How properties of metals relate to their structure

- The outer shells of transition elements contain **delocalised** electrons *free* to move through the metal lattice (see page 29).

- They leave the metal atoms with a positive charge, so they are actually ions.

- Delocalised electrons and ions attract each other, holding the structure together.

- The ordered layers of atoms in pure metals enable them to be **drawn** into wires and hammered into shapes.

- Metals conduct **heat** well. The mobile delocalised electrons carry thermal energy rapidly through the metal, and transfer it to the ions.

- Delocalised electrons make metals good conductors of **electricity** because they carry charge and are mobile.

Copper can be drawn into wires.

Questions

Grades G-E

1 What is mixed with iron, to make steel?

Grades D-C

2 Explain why steel containing a lot of carbon is used for making knives.

Grades G-E

3 Use the periodic table to find out whether these are transition metals or not: calcium; cobalt; rubidium.

Grades D-C

4 Explain why copper is a very good conductor of electricity.

Aluminium

Aluminium extraction and uses

- Aluminium ore, **bauxite** (aluminium oxide), is very abundant.
- Extraction involves **electrolysis**, a process with high energy costs.
- Pure aluminium is relatively soft. Adding small amounts of similar metals makes **alloys** that are hard, strong, light (low density), easy to shape, good thermal and electrical conductors and resist corrosion.
- Aluminium alloys are used in drink cans, bicycles, cars, pylons and cooking utensils.

Electrolysis, properties and atomic structure

- Aluminium (unlike iron) is more reactive than carbon, so carbon will not reduce aluminium oxide.
- For electrolysis, reactants are molten to allow ions to move. Adding cryolite reduces bauxite's melting point.
- **Cathode: aluminium ions + electrons → aluminium atoms**. Molten aluminium sinks and flows out.
- **Anode: oxide ions – electrons → oxygen molecules**. Oxygen reacts with the graphite (carbon) anodes which are replaced regularly.
- Aluminium atoms pack less closely than iron atoms, so aluminium is less **dense**.
- Alloys are **hard** because different-sized atoms stop layers of atoms easily sliding over each other.
- A thin film of **aluminium oxide** quickly forms on the surface. It is a barrier to oxygen reaching deeper.

> **Top Tip!**
>
> Remember 'oil rig':
> **O**xidation **I**s **L**oss
> **R**eduction **I**s **G**ain
> (of electrons).

graphite anodes

cathode   bauxite dissolved   molten
            in cryolite   aluminium

Aluminium is extracted by electrolysis. Aluminium forms at the cathode. Oxygen forms at the anode.

Aluminium recycling

> **How Science Works**
>
> You should be able to: consider and evaluate the social, economic and environmental impacts of exploiting metal ores, of using metals and of recycling metals.

Advantages of recycling

- Recycling aluminium:
  - saves the high **energy** cost of electrolysis (recycling costs are one-twentieth of this)
  - reduces the demand for **bauxite** from mining that damages the **environment**
  - reduces **waste** buried in **landfill** sites
  - can be repeated more than once.

The impact of bauxite mining

- Bauxite is mined from huge open-cast **pits**. Large areas of **trees** are felled and burned, producing carbon dioxide that contributes to **global warming**.
- In some countries, indigenous people are **moved** from areas that are their livelihood. From mine workers they contract **diseases** to which they have no immunity. The area becomes a dump for **pollutants** such as oil. There are also major dust problems in some bauxite plants in the Caribbean – people say they get asthma attacks.

Questions

(Grades G-E)
1 Name the ore from which aluminium is obtained.

(Grades D-C)
2 Why cannot aluminium be extracted from its ore by heating with carbon, as iron can?

(Grades G-E)
3 List **two** advantages of recycling aluminium.

(Grades D-C)
4 List **two** ways in which bauxite mining can harm the environment.

Titanium

Properties and extraction of titanium

- Titanium is light, strong, withstands high temperatures, resists corrosion and is easily shaped.

- Because of high energy costs for extraction of titanium from its ore, titanium dioxide, its use is limited to high-value items including replacement joints, aircraft and missiles.

Why is the F22 made from a titanium alloy?

- Titanium is more **reactive** than carbon, so carbon cannot reduce titanium dioxide (TiO_2). Because titanium dioxide is covalently bonded, so does not conduct electricity, titanium cannot be extracted by electrolysis.

- Instead, titanium dioxide is converted to titanium chloride. Then, the more reactive metal magnesium **displaces** titanium from titanium chloride.

ore heated with
carbon and chlorine   titanium chloride reacts
      with molten magnesium

$$TiO_2 \longrightarrow TiCl_4 \longrightarrow Ti$$

titanium dioxide   titanium chloride   titanium

Titanium alloy is used in a replacement hip to make the ball and shaft that fits into the top of the femur (leg bone).

- Titanium is very reactive but does not corrode because a thin layer of titanium dioxide forms on its surface, protecting the metal beneath.

G–E

D–C

Copper

Properties and extraction of copper

- Copper is a dense, **unreactive** metal with a high melting point. It is a good conductor of heat and a very good **conductor of electricity**.

- Copper wire is used for electrical wiring and plumbing.

- Ores include **chalcocite** (Cu_2S) and **chalcopyrite** ($CuFeS_2$).

- There are only limited amounts of copper-rich ores.

- The ores, extracted from open-cast pits, are low in copper, so large quantities of ore produce little copper but cause extensive damage to the landscape.

- In a traditional extraction process, the ore is crushed, washed and heated in air.

- Drawbacks are that it produces copper that must be purified, and sulfur dioxide which pollutes the atmosphere.

- In a method to increase the yield of copper from low-grade ore, the copper is first **leached** out of the ore – chemically treated to form a solution containing copper. The solution then undergoes electrolysis, and copper collects at the cathode.

G–E

D–C

Questions

Grades G-E

1 How do the properties of titanium make it especially useful for making aircraft?

Grades D-C

2 Explain why titanium cannot be extracted from its ore (titanium dioxide) by electrolysis.

Grades G-E

3 Why is copper used for making electrical wires?

Grades D-C

4 Explain why people are researching new ways of extracting copper from low-grade ores.

Smart alloys

Summary: metals and alloys

- Most **metals** are reactive, existing naturally as **compounds** in ores from which the pure metals are extracted. Gold is unreactive, occurring naturally as an **element**.

- Pure metals are usually too soft for use. Adding other metals gives **alloys** with improved properties. Steels are iron alloys (page 30) with carbon and often other metals. For aluminium alloys see page 31.

How Science Works

You should be able to:
- evaluate the benefits, drawbacks and risks of using metals as structural materials and as smart materials
- explain how the properties of alloys (other than smart alloys) are related to models of their structures.

- **Smart alloys** have a **shape memory**. They can be **deformed**, but heating returns them to their original shape.

- Two types are: the cheaper copper-zinc-aluminium alloys, and the more expensive nickel-titanium alloys used in spectacle frames.

- General properties of **transitional metals**: they are strong, can be drawn into wires and hammered into shape, and are good conductors of heat and of electricity.

metal (cost)	ore	extraction	properties and uses
iron (cheap to extract)	haematite, iron(III) oxide, Fe_2O_3	Reduction of ore with carbon monoxide: carbon monoxide + iron oxide → carbon dioxide + iron	Wrought iron: easily shaped. Steels are rigid, very strong and hard, so used for construction, bodywork, tools.
aluminium (costly to extract)	bauxite, aluminium oxide, Al_2O_3	By electrolysis of molten ore. Aluminium is more reactive than carbon, so cannot be extracted by heating with carbon (coal).	Light, does not corrode, so used for aircraft, spacecraft, drinks cans, bike frames.
titanium (very costly)	rutile, titanium dioxide, TiO_2	Many-stage process. Titanium cannot be extracted by heating ore with carbon (coal) or by electrolysis (since covalently bonded).	Very strong and light, does not corrode, so used for replacement joints, rockets, missiles and aircraft.
copper (moderately costly)	chalcocite, copper sulfide, Cu_2S	Electrolysis of solutions containing copper compounds.	Ductile (can be drawn into wires), very good conductor of electricity. Used for electrical wiring and water piping.

Fuels of the future

How Science Works

You should be able to: evaluate developments in the production and uses of better fuels, for example ethanol and hydrogen.

Fuel use: impact and remedies

- Oil is a **non-renewable** fuel source. Its supply is **finite** – it will run out eventually.

- **Petrol** and **diesel** burn to give **carbon dioxide** and **water vapour**.

- Incomplete combustion releases **unburnt hydrocarbons**, **carbon monoxide** and **nitrogen oxides**. Each causes problems for health and the environment (see also page 35).

- **Air pollution** by vehicle exhaust gases and particles is worst in cities with traffic congestion.

- New cars have a **catalytic converter** to reduce harmful emissions by converting: carbon monoxide to carbon dioxide, unburnt hydrocarbons to carbon dioxide and water, and nitrogen oxides to nitrogen. A catalytic converter works only when hot and only with unleaded petrol.

- **Ethanol** (**alcohol**), a **renewable** resource, produces less carbon monoxide than petrol. Brazil adds ethanol made from sugar cane to petrol.

- Burning **hydrogen** as a vehicle fuel causes no pollution, producing water vapour only. But hydrogen is produced by electrolysis (see page 31), a costly process.

Questions

Grades G-E

1 Explain why gold can be found naturally as an element, not a compound.

Grades D-C

2 Explain what is meant by a 'smart alloy' and give an example.

Grades G-E

3 Why is oil said to be a *non-renewable* fuel source?

Grades D-C

4 Why would using hydrogen as a fuel cause less air pollution?

Crude oil

Crude oil and fractional distillation

- **Crude oil** is a **mixture** of hundreds of compounds. Most are **hydrocarbons** whose molecules contain **hydrogen** and **carbon** atoms only.

- The size of the molecules influences the **properties** of hydrocarbons and how they are used.

- Crude oil **fractions** contain molecules with a *similar* number of carbon atoms (see table).

fraction	no. of carbon atoms	uses
petroleum gas	1–4	heating and cooking
naphtha	5–9	making other chemicals
petrol	5–10	motor fuel
kerosene	10–16	jet fuel
diesel	14–20	diesel fuel and heating oil
oil	20–50	motor oil
bitumen	over 50	bitumen (tar)

- The hydrocarbons in a fraction have similar boiling points, and each fraction has its own range of boiling points. **Fractional distillation** uses the physical process of boiling crude oil to separate the fractions.

- Thermal energy (heat) weakens intermolecular forces of attraction.

- Forces of attraction between smaller hydrocarbon molecules are less than forces between larger hydrocarbon molecules.

- So smaller molecules boil at lower temperatures than larger molecules.

- Crude oil is heated to **evaporate** all its compounds. The vapour rises up the **fractional distillation column**. As it cools, the fractions **condense** at different temperatures and leave the column.

A fractional distillation column. Temperatures are highest at the foot of the column. The smallest molecules reach the top.

G–E

D–C

Alkanes

The alkane family

- The **alkane** family of hydrocarbons has a **general formula** of C_nH_{2n+2}. n is the number of carbon atoms.

- The first two alkanes are: **methane** CH_4; **ethane** C_2H_6.

- Like all hydrocarbons, the atoms in alkane compounds are held together by strong **covalent** bonds (see page 27).

- Unlike some other hydrocarbon families, all the bonds in alkanes are **single** bonds. Therefore, alkanes are **saturated** compounds.

- All alkanes share a similar structure and have similar properties.

The atomic model of methane.

Top Tip!

Remember that all members of the alkane family have the formula C_nH_{2n+2}.

name	formula	structure
methane	CH_4	H–C–H with H above and below
ethane	C_2H_6	
propane	C_3H_8	
butane	C_4H_{10}	
pentane	C_5H_{12}	

G–E

D–C

Questions

1 Name the **two** elements that are present in hydrocarbons.

2 Which hydrocarbon fractions come out at the top of the fractionating column – those with big molecules, or those with small molecules?

3 Write down the chemical formula for ethane.

4 Write down the chemical formula for an alkane with **three** carbon atoms.

Pollution problems

The consequences of burning fossil fuels

G–E

- **Coal** (carbon), **oil** and **natural gas** (hydrocarbons) are fossil fuels. When burnt in air:

  carbon + oxygen → carbon dioxide

  hydrocarbon + oxygen → carbon dioxide + water vapour

  – Carbon dioxide is a **greenhouse gas** (see page 24). In excess, it contributes to **global warming**.

- All fossil fuels contain small amounts of sulfur. When burnt:

  sulfur + oxygen → sulfur dioxide

  – Sulfur dioxide causes acid rain (see page 22).

- Tiny particles from burnt fuels may cause **global dimming**.

Acid rain, global warming and global dimming

D–C

- Normal rainwater is slightly acidic (pH 5.5):

  carbon dioxide + water → carbonic acid

- Sulfur dioxide makes rainwater more acidic (to as low as pH 2.5):

  sulfur dioxide + water → sulfuric(IV) acid (= sulfurous acid)

  – As **acid rain**, it damages vegetation and pollutes lakes and rivers, killing fish and other organisms.

- In excess, carbon dioxide unbalances the **greenhouse effect** which sustains life on Earth (see page 24), and contributes to **global warming**.

- Atmospheric smoke particles reduce the amount of sunlight reaching the Earth's surface (**global dimming**), possibly lowering its temperature.

Top Tip!

Remember that pH *decreases* as rainwater becomes acidic.

Reducing sulfur problems

How Science Works

You should be able to:
- evaluate the impact on the environment of burning hydrocarbon fuels
- consider and evaluate the social, economic and environmental impacts of the use of fuels.

Capturing the sulfur in fossil fuels

G–E

- **Sulfur** is a valuable chemical **resource**. When fossil fuels are burned, sulfur is lost to the atmosphere as sulfur dioxide.

- The sulfur dioxide formed when **coal** burns can be collected from the waste gases.

- Sulfur compounds in **natural gas** and **oil** can be removed before the fuels are burned.

D–C

- Sulfur occurs as **hydrogen sulfide** gas. A **solvent** is used to dissolve it out of natural gas and oil.

Coal-burning power stations harvest sulfur

D–C

- Coal burned in power stations produces sulfur dioxide in its waste gases:

  sulfur + oxygen → sulfur dioxide

- In **scrubbers**, the waste gases are sprayed with a mixture of powdered limestone and water. About 95% of the sulfur dioxide reacts to form solid calcium sulfate. This is collected and can be sold as a soil improver.

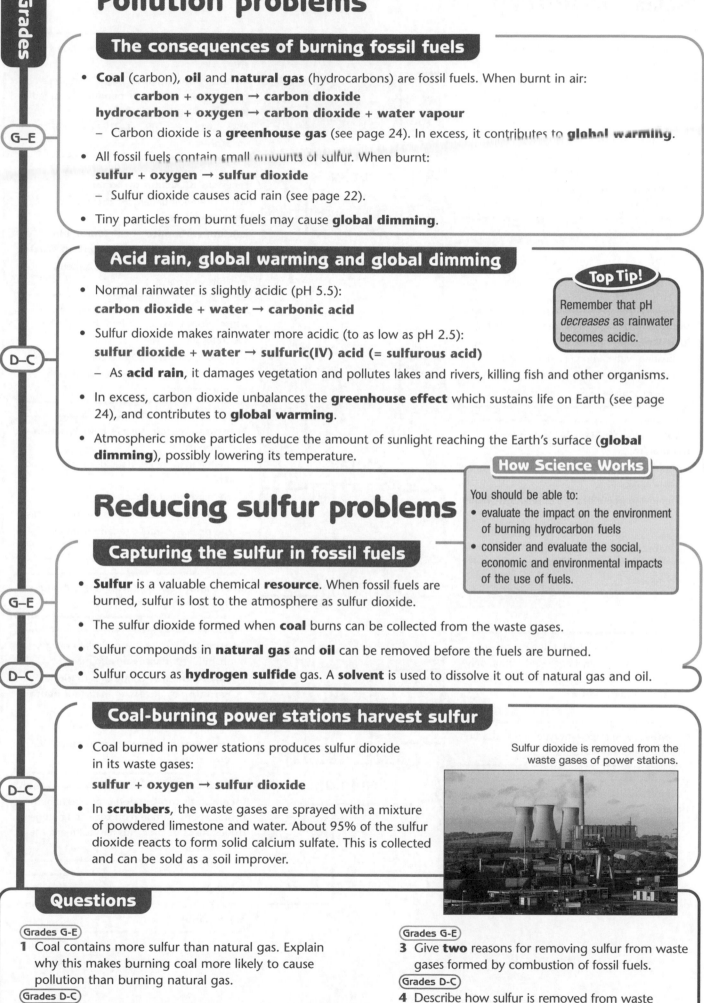

Sulfur dioxide is removed from the waste gases of power stations.

Questions

(Grades G–E)

1 Coal contains more sulfur than natural gas. Explain why this makes burning coal more likely to cause pollution than burning natural gas.

(Grades D–C)

2 Why is normal rain, formed in unpolluted air, slightly acidic?

(Grades G–E)

3 Give **two** reasons for removing sulfur from waste gases formed by combustion of fossil fuels.

(Grades D–C)

4 Describe how sulfur is removed from waste gases at coal-burning power stations.

C1a summary

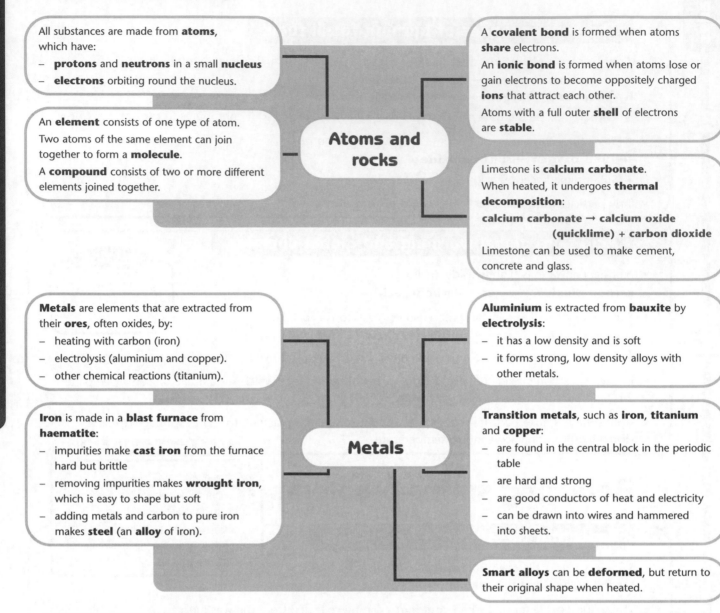

All substances are made from **atoms**, which have:

- **protons** and **neutrons** in a small **nucleus**
- **electrons** orbiting round the nucleus.

An **element** consists of one type of atom. Two atoms of the same element can join together to form a **molecule**.
A **compound** consists of two or more different elements joined together.

Atoms and rocks

A **covalent bond** is formed when atoms **share** electrons.
An **ionic bond** is formed when atoms lose or gain electrons to become oppositely charged **ions** that attract each other.
Atoms with a full outer **shell** of electrons are **stable**.

Limestone is **calcium carbonate**.
When heated, it undergoes **thermal decomposition**:
calcium carbonate → calcium oxide
(quicklime) + carbon dioxide
Limestone can be used to make cement, concrete and glass.

Metals are elements that are extracted from their **ores**, often oxides, by:

- heating with carbon (iron)
- electrolysis (aluminium and copper).
- other chemical reactions (titanium).

Iron is made in a **blast furnace** from **haematite**:

- impurities make **cast iron** from the furnace hard but brittle
- removing impurities makes **wrought iron**, which is easy to shape but soft
- adding metals and carbon to pure iron makes **steel** (an **alloy** of iron).

Metals

Aluminium is extracted from **bauxite** by **electrolysis**:

- it has a low density and is soft
- it forms strong, low density alloys with other metals.

Transition metals, such as **iron**, **titanium** and **copper**:

- are found in the central block in the periodic table
- are hard and strong
- are good conductors of heat and electricity
- can be drawn into wires and hammered into sheets.

Smart alloys can be **deformed**, but return to their original shape when heated.

Crude oil is a **fossil fuel**. It is a mixture of **hydrocarbon** (i.e. hydrogen and carbon only) compounds that can be separated by **fractional distillation**.

Alkanes are an important type of hydrocarbon with the **general formula C_nH_{2n+2}**.

All bonds in alkanes are **single, covalent** bonds.

Crude oil

Burning fossil fuels releases useful **energy**, but harmful substances are also made:

- **carbon dioxide** contributes to **global warming** by increasing the **greenhouse effect**
- **sulfur dioxide** causes **acid rain**
- **smoke** particles contribute to **global dimming**.

Incomplete combustion produces unburnt hydrocarbons, carbon monoxide and nitrogen oxides. All cause problems for health and the environment.

Scientists are working to reduce pollution and develop alternative, cleaner fuels for the future, e.g. **ethanol**.

Cracking

New molecules from crude oil

- Many of the hydrocarbons in crude oil are long-chain **alkanes**. They are separated into **fractions** of similar-sized molecules.

- We describe alkanes as **saturated** because all their carbon atoms are joined by **carbon-carbon single bonds**, C–C (see page 34). This makes alkanes relatively unreactive.

- Short-chain alkanes are more useful as **fuels** – petrol and diesel (motor vehicles), fuel oil (ships) and kerosene (aircraft).

- The **cracking** process makes shorter-chain molecules from long-chain ones:
  - crude oil is heated to form a vapour that passes over a hot catalyst
  - the molecules 'crack', splitting into shorter molecules that include **alkenes**

    long-chain alkanes → shorter-chain alkanes + alkenes
  - this reaction is an example of **thermal decomposition**, in which large molecules are broken into smaller ones by heating.

Top Tip!

Monkeys Eat Peanut Butter helps you count the number of carbons in molecules.

1 = meth-
2 = eth-
3 = prop-
4 = but-

Why hydrocarbons are cracked

- The hydrocarbons required for petrol are 5 to 12 carbon atoms long. Fractions containing longer-chain molecules are cracked to make more petrol.

- Cracking a hydrocarbon molecule produces an alkane and an alkene. Alkenes are a family of more reactive hydrocarbons that are the starting materials for making many useful compounds, including **plastics**.

- Plastics are **polymers**, formed when alkene molecules join in very long chains.

Alkenes

Alkenes

- **Alkenes** are a family of hydrocarbons.

- The smallest alkene is **ethene**, chemical formula C_2H_4. The next is **propene**, C_3H_6.

- The **general formula** for the alkenes is C_nH_{2n}.

ethene

propene

Why alkenes are more reactive than alkanes

- All alkenes are **unsaturated** because they contain a **carbon-carbon double bond**.

- Both double-bonded carbons in alkenes are joined to *three* other atoms. They can react and join to *four*. So alkenes are much more **reactive** than alkanes.

- The general formula for alkenes – C_nH_{2n} – shows that for every carbon atom, an alkene has two hydrogen atoms.

Top Tip!

In all molecules, a carbon atom always forms four bonds. This is a good way to check if your structures are correct.

Questions

Grades G-E

1 Alkanes are saturated molecules. What does this mean?

Grades G-E

2 Describe how long-chain alkanes in crude oil are cracked.

Grades G-E

3 Alkenes are unsaturated molecules. What does this mean?

Grades D-C

4 Explain why alkenes are more reactive than alkanes.

Making ethanol

Making ethanol from crude oil

- Ethene is a gas produced by cracking crude oil (see page 37).

- Ethene is an **alkene**, and so it is reactive. When a mixture of ethene and steam is passed over a **catalyst**, the colourless liquid **ethanol**, C_2H_5OH, is formed.

- Crude oil is a **non-renewable** source of ethene, and therefore of ethanol.

- Ethanol is in the **alcohol** family of compounds, containing an –OH group.

- The –OH group makes ethanol reactive. Ethanol is highly **flammable**. It burns to give carbon dioxide and water.

ethanol

G–E

Making ethanol from plants

- Throughout history, sugar in grapes has been **fermented** to make ethanol ('alcohol'). The sugar is dissolved in water with **yeast**. The yeast changes the sugar to ethanol.

- Crops, including sugar cane and sugar beet, also provide sugar that is fermented to ethanol.

- In some countries including Brazil (see page 33), ethanol is added to petrol, or replaces it as a motor fuel. Ethanol is produced from **renewable** crops. It is a **biofuel**.

D–C

Top Tip!
Ethanol is the chemical name for alcohol. It has the formula C_2H_5OH. Ethanol is not a hydrocarbon because it contains oxygen as well as carbon and hydrogen.

How Science Works
You should be able to: evaluate the advantages and disadvantages of making ethanol from renewable and non-renewable sources.

Plastics from alkenes

How polymers are formed

- **Polymers** are made when small molecules called **monomers** are joined together to make very long chains.

- The polymer **poly(ethene)** (**polythene**) is made from molecules of the gas ethene formed in **cracking** (page 37). Ethene is put under very high pressure and heated with a catalyst. The reaction 'uses up' the double bond in each ethene monomer:

- The **carbon-carbon double bond** makes alkenes very reactive (see page 37). In a reaction, one of the bonds between the two carbon atoms breaks. This allows both carbon atoms to join to an additional atom.

- This is called **polymerisation**, when many thousands of monomers of one compound join up like a string of beads.

- The alkene, **propene** can polymerise to form poly(propene):

G–E

three ethane molecules   poly(ethene)

D–C

three propene monomers

a section of a strand of poly(propene)

Questions

(Grades G-E)
1 Describe how ethanol is made from ethene.

(Grades D-C)
2 Explain why ethanol is reactive.

(Grades G-E)
3 Poly(ethene) is a polymer. What is a polymer?

(Grades D-C)
4 Explain why alkenes can form polymers.

Polymers are useful

Special properties of polymers

- An artificial hip replaces the ball and socket of a worn hip joint (see page 32). The socket is made of the polymer **Teflon®**:
  - its surface has almost no friction, so the ball of titanium alloy swivels freely
  - it is non-toxic and inert (non-reactive), essential inside the body
  - it can withstand high temperatures (it is also used in non-stick pans).
- Teflon® is like poly(ethene) (polythene) but a fluorine atom replaces every hydrogen atom. The fluorine gives Teflon® its special properties.

- **Slime** is made from aqueous solutions of the polymer **poly(ethenol)** (PVA) and of the salt borax. When they mix, borax forms **cross-links** between the polymer chains.
- Slime forms a ball which seems to melt when tipped, and breaks when stretched quickly.
- The **viscosity** of slime depends on the concentrations of the solutions and the length of the polymer chains.

> **Top Tip!**
>
> Viscosity is a measure of the thickness or gooeyness of a liquid. The thicker it is, the higher its viscosity.

Uses of polymers

- Chemists choose monomers that give polymers (plastics) with the **properties** required for a particular purpose:
  - packaging materials must be strong, light and flexible
  - waterproofing for tent fabrics must repel water
  - dental polymers for crowns and dentures are designed to harden when exposed to ultraviolet light
  - wound dressings must breathe and hydrogels must retain moisture for rapid healing.

Disposing of polymers

Reasons for recycling

- Most polymers are produced from **non-renewable** crude oil. They are not **biodegradable** (microorganisms cannot break them down). They take up space in waste dumps and landfill sites.
- Polymers can be **recycled**. They are shredded, melted and formed into pellets used to make new objects such as fleeces.
- It is becoming cheaper to recycle polymers than to throw them away.

Alternatives to recycling

- Polymers can be burned to generate heat energy, but they can produce particle **pollution** and toxic fumes that damage the **environment**.
- To make them **biodegradable**, polymers are mixed with cellulose or starch that microorganisms will break down.

> **How Science Works**
>
> You should be able to:
> - evaluate the social and economic advantages and disadvantages of using products from crude oil as fuels or as raw materials for plastics and other chemicals
> - evaluate the social, economic and environmental impacts of the uses, disposal and recycling of polymers.

Questions

Grades G-E

1 State **two** useful properties of the polymer Teflon®.

Grades D-C

2 Slime is a polymer. State **one** property of it that varies according to the conditions under which it is made.

Grades G-E

3 Why are there problems with the disposal of waste polymers?

Grades D-C

4 Describe how polymers can be made biodegradable.

Oil from plants

Sources of vegetable oils

- The fruits, seeds and nuts of some plants are rich in **vegetable oils**.

- Trees with oil-rich **fruits** include palm, olive and avocado.

- The **seeds** of cotton, rape, mustard, sunflower and sesame supply oils.

- Oil-rich **nuts** include coconuts and groundnuts.

- Fruits, seeds and nuts are crushed in a press to break any hard coatings and to extract the oils. Then water and impurities are removed.

G–E

Purification and importance of vegetable oils

- To **purify** an oil containing water and other impurities, the mixture can be shaken with a hydrocarbon solvent (which does not mix with water) and distilled. Water molecules are small, while the solvent and oil molecules are larger and have higher boiling points. So, on heating, the water boils first and is discarded. Then the oil and solvent boil off together and are collected. Dust and other rubbish remain in the flask.

- Natural oils are high in chemical **energy**. They are used widely in foods and for cooking.

- Some nuts, particularly almonds, brazil nuts and pistachios, are excellent sources of **vitamins** and **trace elements**.

Top Tip!

Distillation works because liquids have different boiling points.

thermometer   water out

condenser

oil, solvent, water and dust   ↑ HEAT   cold water in   water, then oil/solvent

Distillation apparatus. The water boils off first, then the oil/solvent mixture.

D–C

Green energy

Biofuels and fossil fuels

- Some vegetable oils and other substances obtained from plants can be used as **fuels**, known as **biofuels**.

- Biofuels, either alone or mixed with petrol or diesel, are sometimes called **green fuels**. They:

  - include rapeseed oil, sunflower oil and wood

  - are **renewable fuels** because new plants will supply more of them

  - are 'greenhouse neutral' ('carbon neutral') because the amount of carbon dioxide *given out* when they burn is *taken in* by plants grown to replace them.

- As **non-renewable** fossil fuels run out, the need for renewable fuels will increase.

- As the supply of fossil fuels diminishes, their price rises. World economies needing alternative energy sources see biofuels as a major solution.

- To increase the use of green fuels for motor vehicles:

  - biofuels too thick for some engines are mixed with traditional fuels

  - new engines are being designed that prevent soot and the poisonous gas carbon monoxide from forming.

G–E

How Science Works

You should be able to:

- evaluate the effects of using vegetable oils in foods and the impacts on diet and health
- evaluate the benefits, drawbacks and risks of using vegetable oils to produce fuels.

D–C

Questions

Grades G-E

1 Give **two** sources of vegetable oil.

Grades D-C

2 Describe how a vegetable oil can be purified.

Grades G-E

3 Give **two** examples of biofuels.

Grades D-C

4 Explain why we will need to use more biofuels in the future.

Emulsions

Making an emulsion

G–E

- Oil and water do not mix. The oil forms a separate layer on top.

- When shaken hard, oil forms tiny droplets in water. This mixture is an **emulsion**. An emulsion is formed when one liquid is dispersed in another.

- Adding an **emulsifier** stops the oil from separating out from the water.

- Emulsions are thicker than oil or water alone, and have useful properties:
  - sauces and dressings are easy to pour slowly
  - emulsion paints cover surfaces well without dripping.

- Milk is an emulsion of fat droplets spread out in water.

- Butter is an emulsion of water droplets spread out in fat.

oil (dispersed phase)

water (continuous phase)

This emulsion consists of tiny oil droplets suspended in water.

Top Tip!

Oils are liquids at room temperature, and fats are solids.

More about emulsions

D–C

- The liquids in an emulsion do not normally mix – they are **immiscible**.

- The liquid forming droplets is the **dispersed phase**. The liquid between the droplets is the **continuous phase**.

- Droplets of the dispersed phase are **suspended** in the continuous phase.

- Emulsions are opaque or milky because the tiny droplets scatter light.

Top Tip!

Liquids that do not mix together are called immiscible liquids. Those that do mix together are called miscible liquids.

continuous phase

incident light rays

dispersed phase

Light rays are scattered by the droplets in an emulsion, making it look opaque.

Polyunsaturates

Saturated and unsaturated oils and fats

G–E

- Unsaturated fats and oils have one or more double bonds between carbon atoms.

- Olive oil has one double bond, so it is a **monounsaturated** oil.

- Fats with more than one double bond are **polyunsaturates**.

- Fats and oils with single bonds only between carbon atoms are **saturated**.

D–C

- To find if a fat or oil is saturated or not, the **bromine water** (or iodine water) test is used.

- If there are no double bonds, the orange colour of the bromine is unchanged.

- In the presence of one or more double bonds the bromine becomes colourless. This is because bromine reacts with the carbons at the double bond.

Top Tip!

Do not say that the bromine becomes clear. It is clear anyway, even when it is orange! Say that it changes its colour from orange to colourless.

Questions

Grades G-E

1 Give **one** example of an emulsion.

Grades D-C

2 Explain why emulsions are not transparent.

Grades G-E

3 Explain the difference between a 'saturated' fat and an 'unsaturated' fat.

Grades D-C

4 Describe how you test a liquid to find out if it contains unsaturated compounds.

Making margarine

Making margarine

- **Margarine** is made from vegetable oils such as corn, sunflower and olive oils.

- The *liquid* vegetable oil is **unsaturated**, containing **carbon-carbon double bonds**. It is reactive and is chemically hardened to form a *soft solid* fat:
  - hydrogen gas passes through the oil warmed to about 60 °C
  - a nickel catalyst speeds up the reaction
  - hydrogen reacts with the carbon-carbon double bonds in the oil.

  The reaction is **hydrogenation**.

G–E

How hydrogenation changes vegetable oils

- The molecules of fats and oils contain long-chain molecules called **fatty acids** with carbon-carbon double bonds.

- At a double bond, fatty acid molecules bend in a V-shape. They cannot pack tightly, and the oil remains liquid at room temperature.

- **Hydrogenation** replaces some of the double bonds in the fatty acids with single bonds.

- These fatty acid molecules straighten out and can then pack closely in parallel.

- This raises the density and the melting point of the vegetable oil, and it becomes a solid fat suitable for spreading and cake-making.

Oleic acid has one double bond.

D–C

Food additives

What are additives?

- An **additive** is a chemical put in food to keep it fresh or make it appetising.

- Some additives are natural and some are synthetic. Small amounts are safe, but large amounts of some additives could be **toxic**.

- Each permitted additive has been tested for safety and is allocated an **E-number** which identifies its chemistry and what it does.

G–E

- The use of additives is regulated by the European Union and the UK government. Additives must be listed on food labels.

- Permitted types of additives include:
  - antioxidants
  - colourings
  - emulsifiers, stabilisers, gelling agents and thickeners
  - flavourings
  - preservatives
  - sweeteners.

How are sweets made to look as colourful as this?

D–C

- Permitted additives are not harmful, but none of the synthetic additives is essential for a healthy diet.

How Science Works

You should be able to: evaluate the use, benefits, drawbacks and risks of ingredients and additives in food.

Questions

Grades G-E
1 What is margarine made from?

Grades D-C
2 Explain why hydrogenation makes liquid oils become solid margarine.

Grades G-E
3 What is a food additive?

Grades G-E
4 Explain why additives listed on food labels have an E-number.

Analysing chemicals

Analysing the chemicals in food

- **Additives** in foods, such as artificial colourings, can be identified by techniques of **chemical analysis**.

- **Paper chromatography** separates substances in a mixture. Samples removed from the paper can be chemically analysed and identified. Further studies can check whether they are safe.

Separating dyes in a food dye using paper chromatography

- The preparation of a chromatogram of artificial food dye is as follows:

  - the unknown food dye is spotted on the pencil line alongside samples of known dyes

  - the bottom edge is placed in a solvent (e.g. water or ethanol) that dissolves all the dyes

  - the solvent rising up the paper carries the known dyes and separates out the food dyes

  - when the spots are well up the paper, the paper is removed from the solvent and dried

  - the food dye spots are matched to the known dyes, and unidentified spots are analysed further.

Chromatogram of known dyes and an unknown food dye.

The Earth

land forms about 30% of the surface

diameter 13 000 km

water covers about 70% of the surface

The size and features of the Earth.

The Earth and its atmosphere

- The Earth is one of the rocky planets of the Solar System (see page 67) and is about 4.6 billion years old. The Moon is Earth's satellite.

- The Earth is a slightly squashed sphere with a diameter of 13 000 km surrounded by an atmosphere about 30 km deep.

- Seas and oceans cover roughly 70% of the surface. Life evolved in water some 3.4 billion years ago. Humans appeared 30 000 years ago.

The Earth's interior

- The Earth has three main layers

  - the **crust**, the **mantle** and the **core**.

- The outermost **crust** is a thin layer of solid rock.

- The **mantle**, largely of solid rock, occupies half the diameter. Heat from radioactive decay drives convection currents that circulate the 'plastic' rock beneath the crust.

- The **core** occupies the centre and about half the Earth's diameter:

  - the *outer core* is a fluid mixture of iron and nickel; the Earth's magnetic field is thought to be generated here

  - the *inner core* is solid iron and nickel.

crust: cold, thin, solid rock layer

mantle: hot 'plastic' rock layer moved slowly by convection currents

core: hot iron and nickel, liquid outer part, solid central part

The layers of the Earth.

Questions

Grades G-E

1 Name a technique that can be used to separate a mixture of artificial colours in a food.

Grades D-C

2 How does this technique separate the colours?

Grades G-E

3 How old is the Earth?

Grades D-C

4 Name the **three** main layers that make up the Earth.

Earth's surface

Old theory of Earth's formation

You should be able to:
- explain why the theory of crustal movement (continental drift) was not generally accepted for many years after it was proposed
- explain why scientists cannot accurately predict when earthquakes and volcanic eruptions will occur.

- About 4.6 billion years ago, the Earth's surface solidified.
- Scientists once thought that:
  - the landmass then shrank, breaking into the pattern of **continents** and oceans that exists today
  - heat in the Earth's interior was just the heat left over from the Earth's formation.

G–E

How Earth's surface features were formed

- In 1915, the German Alfred Wegener suggested that:
  - the first landmass to form was one **supercontinent**, which he named **Pangaea**
  - Pangaea broke up 200 million years ago, and the pieces (plates) 'drifted' to their present positions.

- Knowing more about the Earth's interior, today's scientists agree with Wegener except for how he thought continental drift happens.

- The Earth's crust and the semi-solid upper mantle form the **lithosphere**. It is cracked into **tectonic plates**.

- Below the lithosphere is the 'plastic' rock of the **mantle**. Heat from radioactive processes inside the Earth drives **convection currents** in the rock. The tectonic plates ride on this circulating material.

- Movement is only centimetres a year, but, for example, the landmass of India has been carried across an ocean to collide with Asia.

Pangaea, over 200 million years ago.

D–C

Earthquakes and volcanoes

Earthquakes, volcanoes and subduction zones

- As the mantle moves underneath the plates, it exerts huge pressures on them. Eventually, plates are forced to move relative to one another where weakest – along **plate boundaries** – and an **earthquake** occurs. This usually happens very suddenly.

- A plate can push:
  - into another, driving them both upwards
  - along the boundary in a sliding action
  - under an adjacent plate which then overlaps it.

- **Volcanoes** form at points of weakness in the Earth's crust. Hot molten **magma** from the mantle is forced up through a **vent** in the crust, and blasts out as **lava** and gases. Solidifying lava forms the volcano's cone shape round a crater.

- A region where one plate moves under another is a **subduction zone**. Subduction zones are at high risk of earthquakes and volcanic eruptions.

The main plate boundaries and active volcanoes.

D–C

Questions

Grades G-E

1 What was the first explanation that scientists suggested for why the Earth has continents and oceans?

Grades D-C

2 What makes the inside of the Earth hot?

Grades D-C

3 Explain why earthquakes happen very suddenly.

Grades D-C

4 Explain why earthquakes often happen at subduction zones.

The air

Composition of the air

- The air is a **mixture** of colourless gases. It is mostly **nitrogen**, **oxygen** and a variable amount of **water vapour**.

- Other gases, in very small amounts, include **carbon dioxide** and the **noble gases**.

- The noble gases, in Group 0 of the periodic table, are unreactive because they have a full outer shell of electrons.

- They are used in filament lamps and in electrical discharge tubes.

- Helium is used in air balloons because it is much less dense than air.

- The gases in Earth's early atmosphere were very different from those of today.

- Developing life forms changed the **composition** of the atmosphere.

Gases in the air.

Evolution of the air

Gases of the present atmosphere

- The table shows the percentage by volume of the ten most abundant gases in air today.

gas	nitrogen	oxygen	argon	carbon dioxide	neon	helium	methane	krypton	hydrogen	xenon
% of air	78.08	20.95	0.93	0.035	0.0018	0.00052	0.0002	0.00011	0.00005	0.00001

Evolution of the atmosphere

- The graph shows the change during Earth's history in levels of nitrogen, oxygen and carbon dioxide.

- Intense volcanic activity of the early Earth produced **water vapour** which **condensed** to become the first seas.

- **Carbon dioxide** and **methane** also came from volcanoes.

- The very high level of carbon dioxide dropped as it dissolved in the seas.

- Early life forms evolved in warm shallow water which protected them from the Sun's harmful ultraviolet rays. Some of these photosynthesised. Their waste product, **oxygen**, built up in the atmosphere.

Changes in the levels of nitrogen (N_2), oxygen (O_2) and carbon dioxide (CO_2) since the Earth's formation.

- Some oxygen was converted to **ozone**, O_3, forming a shield against ultraviolet rays. This enabled plants to evolve on land.

- Plants photosynthesised, removing carbon dioxide and adding more oxygen. Animals evolved, taking in oxygen and giving out carbon dioxide.

- **Ammonia**, NH_3, from volcanoes, reacted with oxygen gas and produced **nitrogen**, which gradually built up in the atmosphere. Bacteria also produced nitrogen from ammonia.

- The gases in the atmosphere eventually became balanced about 200 million years ago.

Questions

(Grades G-E)
1 List the **three** most abundant gases in our atmosphere.

(Grades G-E)
2 What is the approximate concentration of carbon dioxide in the atmosphere?

(Grades G-E)
3 Which of the gases shown in the table above are noble gases?

(Grades D-C)
4 Explain why there is oxygen in our atmosphere.

Atmospheric change

The atmosphere is heating up

- The atmosphere is getting warmer. This is called **global warming**, and its effect on the weather is **climate change**.

- In the past 100 years, the **average temperature** of the Earth's atmosphere has increased by 1 °C.

- As atmospheric temperature increases:
  - ice sheets melt and sea level rises, leaving low coasts and islands under water
  - the energy in the atmosphere increases and weather becomes stormier and more extreme. Generally, summers are hotter and, in some places, winters are colder.

Global warming

- Atmospheric temperature remains stable while the heat gained from the Sun is balanced by the heat lost to space (see page 24).

- Temperature rises when the level of **greenhouse gases** increases. **Carbon dioxide** is a greenhouse gas.

- Carbon dioxide circulates between the atmosphere and living and non-living things (see page 78).

- The burning of **fossil fuels** has added more carbon dioxide to the atmosphere than can leave it.

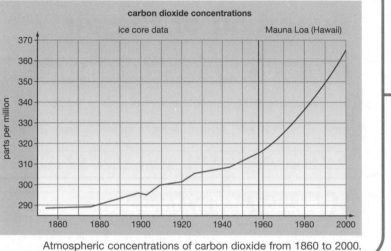

Atmospheric concentrations of carbon dioxide from 1860 to 2000.

Carbon sinks

- Carbon in carbon dioxide leaves the atmosphere when it is deposited in **carbon sinks**. These include **sedimentary rocks**, **plants** and **fossil fuels**.

- Most of the carbon in the Earth's original atmosphere is locked up in carbon sinks.

- Carbon dioxide is *water soluble*.
  - It dissolves in rainwater and seawater. It forms **carbonic acid** that reacts with calcium-containing materials in seawater to form **calcium carbonate**. Being insoluble, this builds up as sediment on the sea floor. Over millions of years, sediments become **sedimentary rocks** including **limestone** and **chalk** which can contain the shells of microscopic organisms.

- Carbon dioxide is taken up by plants in **photosynthesis** (see page 73):
  - in a given area, trees 'lock up' far more carbon than smaller plants.

- Over millions of years, plant and animal material can be fossilised:
  - plants are buried and form **coal**
  - microscopic marine animals and plants form **oil**.

How Science Works

You should be able to:
- explain and evaluate theories of the changes that have occurred and are occurring in the Earth's atmosphere
- explain and evaluate the effects of human activities on the atmosphere.

Questions

Grades G-E

1 What has happened to the average temperature of the atmosphere over the past 100 years?

Grades D-C

2 How does carbon dioxide contribute to global warming?

Grades D-C

3 What is a 'carbon sink'? Name **three** carbon sinks.

Grades D-C

4 Most of the carbon from carbon dioxide in the air is locked up in sedimentary rocks. How does this happen?

C1b summary

Crude oil contains **hydrocarbons** including **alkanes** with the general formula C_nH_{2n+2}. Alkanes can be cracked to form a new family called the **alkenes**.

Alkenes are **unsaturated**. They contain carbon-carbon **double bonds**.
The **general formula** for alkenes is C_nH_{2n}.

Polymers are made when small molecules called **monomers** are joined together to make very long **chains** in a process called **polymerisation**. Alkenes can be made into polymers.

Polymers can be used to make many **useful substances**, e.g. Teflon® and plastics for packaging materials, dental compounds, waterproof materials, etc.

Many polymers are **not biodegradable** and are produced from **non-renewable** crude oil. To dispose of polymers, they can be:
- **recycled** to make other products
- burnt to generate heat (but this can result in pollution)
- mixed with starch or cellulose to make them biodegradable.

New substances from crude oil

Many **plants** contain useful **natural oils** that can be **extracted**, e.g. from fruits, seeds and nuts.

Biofuels can be made from plant oils and other substances. These '**green fuels**':
- include rapeseed oil, sunflower oil and wood
- are **renewable**
- are '**greenhouse neutral**'.

An **emulsion** is formed when one liquid is **dispersed** in another.

Fats and **oils** can be **saturated** (containing only single bonds) or **unsaturated** (containing C=C double bonds). **Hydrogenation** is used to **harden** unsaturated liquid vegetable oils in the manufacture of **margarine**.

Oil from plants

The **Earth** has three main layers: the **crust**, the **mantle** and the inner and outer **core**.

The Earth's **lithosphere** is cracked into **tectonic plates** which ride on the plastic part of the mantle.
Earthquakes can occur at the **plate boundaries** where they collide with each other.
Volcanoes form at weak points in the Earth's crust.

The Earth's **atmosphere** has changed over millions of years. Many of the gases that make up the atmosphere came from volcanoes, e.g. water vapour, carbon dioxide and methane.
Ozone (O_3) forms a shield against ultraviolet rays.

The **air** is a mixture of mainly **nitrogen**, **oxygen**, variable amounts of **water vapour** and very small amounts of other gases such as **carbon dioxide** and the **noble gases**.

The levels of the **greenhouse gases** in the atmosphere are rising, resulting in **global warming** and **climate change**.

Sedimentary rocks, **plants** and **fossil fuels** are **carbon sinks**, locking up the carbon from carbon dioxide that was once in the atmosphere.

The Earth

Heat energy

Heat and temperature

- **Heat** is **energy** that can flow from one material to another.

- **Temperature** indicates how hot something is.
  A thermometer measures temperature.

Heat energy flows from the warm surrounding air to the cold lolly.

Thermal radiation

- **Thermal (infrared) radiation** is the **transfer** of **heat energy** by rays.

- The rays are **electromagnetic waves**. They travel:
  - in straight lines
  - at the speed of light
  - through a vacuum: the Sun's radiation passes through space.

- Thermal radiation makes particles in an object vibrate.

- All materials **absorb** (take in) and **emit** (give out) thermal radiation. Hot objects emit more than cold ones.

- An **infrared camera** detects thermal radiation.

The image produced by an infrared (thermal imaging) camera shows that objects emit different amounts of thermal radiation.

Thermal radiation

Top Tip!
Remember that 'cold' is not transferred! It is always heat energy that is transferred, from a hotter object to a colder one.

Measuring emitted thermal radiation

- Hot objects always emit more thermal radiation than cold objects.

- The greater the difference between the temperatures of an object and its surroundings, the faster the rate of heat transfer.

- The amount of thermal radiation an object emits depends on its shape, dimensions and type of surface.

- A **thermopile** (e.g. an 'ear thermometer') measures thermal radiation.

- More thermal radiation is emitted by: black surfaces than white surfaces; dull surfaces than shiny surfaces.

Emission and absorption of thermal radiation

- Hot water in the dull black beaker in the diagram loses heat – **emits** thermal radiation – faster than the silver beaker.

- Filling the beakers with ice-cold water in warm surroundings shows that the dull black surface also **absorbs** heat faster than the silver surface.

- The black grid behind a refrigerator emits thermal radiation fast, while silver 'space blankets' slow down emission.

- White buildings in hot countries minimise the thermal radiation absorbed.

- As polar ice melts, the dark-coloured sea will absorb more thermal radiation and may promote **global warming**.

dull black beaker · shiny silver beaker
thermometer
metal beaker of hot water

Equipment to compare the emission of thermal radiation by different surfaces.

Questions

(Grades G-E)

1 When you hold a cup of hot coffee in your cold hands, in which direction is heat energy transferred?

(Grades D-C)

2 Explain how an infrared camera can detect a person hiding among trees.

(Grades G-E)

3 Do hotter objects emit more thermal radiation or less thermal radiation than cold objects?

(Grades D-C)

4 Explain why silver objects heat up *and* cool down more slowly than black objects.

Conduction and convection

Conductors, insulators and convection

G–E

- **Heat energy** flows easily through **thermal conductors** (e.g. metals).

- It does not flow easily through **thermal insulators** (e.g. glass, plastic, air).

D–C

- A metal saucepan conducts heat rapidly from cooker to food, but a plastic handle conducts heat very slowly.

- Double-glazed windows reduce heat loss because glass layers and the air between are poor conductors.

G–E

- Warmed liquids and gases rise by **convection** because they are less dense than cooler liquids (see diagram) and gases.

streaks of purple dye moving through clear water

The dye shows convection in the heated water.

The particle model and conduction and convection

D–C

- Particles in **solids** vibrate in fixed positions. When one end of a solid rod is heated, the particles vibrate more. They **transmit** vibrations to their neighbours, **transferring** heat energy along the rod.

- When heated, particles in **liquids** and **gases** move more vigorously. They take up more space. The heated material **expands** (becomes less dense), causing **convection currents**.

Heat transfer

Heat transfer in liquids and gases

G–E

- Liquids are poor thermal conductors – they transfer heat slowly, as the diagram shows.

- Gases are very poor thermal conductors. Air trapped in a cold-weather jacket greatly slows heat transfer from a person's body.

boiling tube

metal gauze

ice cubes

Bunsen burner

Top Tip!

Do not say that particles in a solid 'start to' vibrate when heated. They are already vibrating! Heat energy makes them vibrate *more vigorously*.

It takes a long time before the ice cube starts to melt. (The less dense hot water remains at the top.)

Energy transfer and the particle model

D–C

- Fixed **particles** in a solid **transfer energy** to each other by **vibrating**.

- Only particles on the surface of a solid can transfer energy to the **surroundings**. Therefore, the rate of heat transfer:
  - is fast for a small volume with a large surface area
  - is slow for a large volume with a small surface area.

- Metals are good conductors because they have **free electrons** that can move in all directions between the fixed particles. They help to transfer energy to the fixed particles.

- Particles in liquids are not firmly held together, so energy is less easily transferred from one particle to another. Heat transfer by conduction between particles is slow.

- Particles in gases are far apart. Heat energy is transferred only when particles collide, so heat transfer by conduction is very slow.

How Science Works

You should be able to evaluate ways in which:
- heat is transferred in and out of bodies
- the rates of these transfers can be reduced.

fixed atoms vibrate making neighbours vibrate   energy from vibrations

free electrons this end have lots of energy and move fast

free electrons this end have less energy and move slower

free electrons move easily, transferring energy to atoms far away

Free electrons in metals help to transfer heat energy.

Questions

(Grades G–E)

1 Why does hot air move upwards?

(Grades D–C)

2 Explain how heat energy is conducted along a metal rod.

(Grades G–E)

3 Why does wearing several layers of thin clothes keep you warmer than wearing one thick layer?

(Grades D–C)

4 Explain why gases are very poor conductors of heat.

Types of energy

Energy, forms and measurement

- **Energy** enables things to happen. Without energy, nothing happens.

- We experience energy in different **forms** (types), as **heat**, **light** or **sound**, **chemical** energy, **microwave** and **radio wave** energy, and **kinetic** and **potential** energy.

How many types of energy have you used today?

- Energy is measured in **joules (J)**. Lifting a small apple from floor to table takes about 1 J. Walking 1 km on level ground uses about 280 000 J.

- A teenager uses about 10 million J of energy a day, mostly for life processes (e.g. maintaining body heat).

- Energy can be stored for use when required. **Energy stores** include fuels, batteries, stretched elastic bands and tightly wound springs.

Energy changes

Energy transfers and transformations

- **Energy transfers** occur when the *same* form of energy moves from *one place to another*.
  For example:

  - a cyclist's **kinetic energy** is transferred to become the bicycle's **kinetic energy**.

- **Energy transformations** occur when *one* form of energy *changes to another*.
  For example:

  - a light bulb transforms **electrical** energy to **light** energy

  - a music system transforms **electrical** energy to **sound** energy

  - a battery transforms **chemical** energy to **electrical** energy.

Examples of energy transformations.

Wasted energy and the Law of Conservation of Energy

- Whenever a device transfers or transforms energy for a useful purpose, some energy is always **wasted** – it is in a form that is not useful:

  - a light bulb also produces unwanted heat

  - a rocket motor that provides kinetic energy also produces unwanted heat, light and sound.

- Energy never disappears. The total amount of energy before and after an event is always the same. The **Law of Conservation of Energy** says:

  Energy cannot be created or destroyed.

  It can only be transferred or transformed from one form to another.

  The total energy always remains constant.

Questions

Grades G-E

1 What form of energy do the people have because of their position on the high wall?

Grades D-C

2 If you use 10 million J of energy in a day, how much chemical energy should there be in the food you eat each day?

Grades G-E

3 A girl hits a gong with a mallet. Kinetic energy in her arm becomes kinetic energy in the gong and then sound energy in the air. Which is an energy transfer? Which is an energy transformation?

Grades D-C

4 When you ride a bike, not all the kinetic energy in your legs is transferred to kinetic energy in the bike. Where does the wasted energy go?

Grades

G–E

D–C

G–E

D–C

Energy diagrams

Diagrams showing energy changes

Energy changes for a light bulb.

G–E

- A **Sankey diagram** shows the **energy input** and **energy output** of a device.

- The energy output includes **useful energy** and **wasted energy**. Together, they add up to the value of the energy input. The total amount of energy stays the same.

- The second diagram shows this for an engine. Input and output are both 1000 J.

- You can use Sankey diagrams to calculate missing values for energy input and output.

1000 J chemical energy input
? useful kinetic energy
600 J wasted heat energy

A Sankey diagram for an engine.

Measuring wasted energy

D–C

- The top diagram shows that the candle wastes the most energy and the fluorescent light wastes the least.

- A device may have several **energy transfers** or **transformations**. At each one, energy is wasted. The bottom diagram shows this for a petrol engine.

Energy transformations in different light sources.

incandescent bulb	fluorescent tube	candle
60 J — 6 J light / 54 J heat	12 J — 6 J light / 6 J heat	0.2 J light / 60 J — 59.8 J heat

A Sankey diagram for a petrol engine.

1000 J chemical energy input
785 J kinetic energy in engine
275 J useful kinetic energy in car
? wasted heat in exhaust gases
? wasted heat energy in moving engine parts

Energy and heat

Wasted heat energy and where it goes

G–E

- In any energy transfer or transformation, wasted energy eventually appears as **heat energy**.

- Wasted heat energy is not 'lost' but is transferred to the **surroundings**, making them warmer. For example, a runner who transforms **chemical energy** in glucose to **kinetic energy** for muscle contraction, will also transfer heat to the surrounding air.

- Wasted heat is transferred to the surroundings by conduction, convection and thermal radiation.

The particle model: warming and energy spread

D–C

- Inside all objects, the particles are **vibrating**. When an object absorbs any sort of energy, its particles vibrate more vigorously – it heats up. The more energy it absorbs, the more vigorously its particles vibrate and the hotter it gets.

How could you decide which is the more concentrated store of energy – the coal or the wood?

- In all energy transfers and transformations, energy spreads out and becomes more difficult to use for further energy transformations.

- Energy is more useful when **concentrated**, as in fuels and batteries, which are useful **stores of energy**.

How Science Works

For a range of devices, you should be able to:
- describe the intended energy transfers/transformations
- describe the main energy wastages that occur.

Questions

Grades G-E

1 How much useful kinetic energy is provided by the engine in the second diagram?

Grades D-C

2 Calculate how much heat energy is wasted in the exhaust gases and as heat in the moving parts in the engine in the fourth diagram.

Grades G-E

3 Name the **three** ways in which wasted heat energy can be transferred to the environment.

Grades D-C

4 Explain why a bath full of warm water is said to be a 'less concentrated' store of energy than a kettle full of hot water.

Energy, work and power

Work and power

- The **work** that a machine does is measured in **joules**.

- The **power** of a machine is measured as the number of joules of work it can do in 1 second. Power is the *rate* of doing work. The unit of power is the **watt (W)**.

- One joule per second (J/s) = 1 watt of power.

G–E

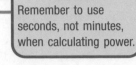

Top Tip!

Remember to use seconds, not minutes, when calculating power.

Calculating energy and power

- An object gains **energy** when lifted to a higher position:

  energy gained = work done

- The work done by a **force** is given by: **work done = force × distance moved**

- Raising an object of mass 50 kg by 1 m requires a **force** to lift the object's **weight** (measured in newtons):

  force to overcome weight = mass (kg) × 10 (force of gravity on 1 kg mass)

  potential energy gained = work done = force × distance moved
  = 50 × 10 × 1 = 500 J

- We calculate the power of a machine from:

  power (W) = work done (J) ÷ time taken (s)

- A crane lifts 2000 kg of bricks to a height of 2 m in 2 min:

  work done = force × distance moved = weight × distance moved
  = 2000 × 10 × 2 = 40 000 J
  power = work done ÷ time taken
  = 40 000 ÷ 120 = 333 W

D–C

Could you calculate the work you do to climb up a flight of stairs?

Efficiency

Efficiency

- An **efficient** machine does the maximum work with the minimum of energy input.

- **Efficiency =** $\dfrac{\text{useful energy output}}{\text{total energy input}}$

  – Useful energy output is the work a machine does or the energy that a device transforms into useful forms.

- A crane has an energy input of 2000 J and does 800 J of useful work. Its efficiency = 400 J per 1000 J.

- The greater the **percentage** of energy that is usefully transformed in a machine, the more efficient a machine is.

- For the petrol engine shown:

  efficiency = $\dfrac{\text{useful energy output}}{\text{total energy input}} = \dfrac{300}{1000} = 0.3 = 30\%$

How Science Works

You should be able to: calculate the efficiency of a machine using:

efficiency = $\dfrac{\text{useful energy transferred by the machine}}{\text{total energy supplied to the machine}}$

G–E

```
1000 J
chemical
energy
input          850 J kinetic
               energy in engine          300 J useful
                                          kinetic energy
                                          in car


       150 J wasted heat   550 J wasted heat
       in exhaust gases    energy in moving
                           engine parts
```

A Sankey energy transfer diagram for a petrol engine.

D–C

Questions

Grades G-E

1 A girl lifts twenty 2 kg bags of sugar from the ground onto a shelf 3 m above the ground.
How much work has she done? (Hint: remember to calculate the *weight* of the bags first.)

Grades D-C

2 It took the girl 1 min. Calculate her power output.

Grades G-E

3 Machine A has an input of 1000 J and a useful energy output of 500 J. Machine B has an input of 2000 J and a useful energy output of 900 J. Which machine is more efficient?

Grades D-C

4 Calculate the percentage efficiency of machine A.

Using energy effectively

Energy saving

How Science Works

You should be able to: evaluate the effectiveness and cost effectiveness of methods used to reduce energy consumption.

G–E

- If you cut down on energy use in the home, it saves money and is good for the environment:
  - turn off lights and turn down the temperature of central heating
  - replace old, energy-wasting devices with new, more efficient ones
  - replace incandescent light bulbs with **energy-saver light bulbs**.
- Energy-saver light bulbs cost more than incandescent bulbs but use less than one-fifth of the energy.

Home insulation

D–C

- Installing insulation reduces the heat energy lost through ceilings, walls, doors and windows. The table compares savings and costs. **Payback time** is how long it takes for savings to equal installation costs.

Types of insulation.

type of insulation	installation cost (£)	annual saving (£)	payback time (years)
loft insulation	240	60	4
cavity wall insulation	360	60	6
draught-proofing doors and windows	45	15	3
double glazing	2500	25	100

Savings as a result of different types of insulation.

Why use electricity?

Using electricity

G–E

- As an energy source in the home, **mains electricity** is instant, convenient, clean, safe and reliable.
- **Electrical energy** is **transformed** into:
  - heat and kinetic energy by a hairdryer
  - light and sound energy by a television.

heat energy   light energy

electrical energy   electrical energy

kinetic energy   sound energy

A hairdryer and a television transform electrical energy.

- **Batteries** store and supply a *small* amount of portable **chemical energy** that can be transformed into electrical energy, e.g. for a radio or a mobile phone.

Alternatives to electricity

D–C

- The following transform the chemical energy stored in either food or fuel:
  - human, animal and steam power all provide **kinetic energy**
  - various types of biomass (plant material and animal waste) can be burnt to supply **heat** energy for cooking; wax candles or gas or oil lamps can provide **light**.
- **Solar energy** and the **potential energy** in clockwork springs can charge rechargeable batteries. This means that people can use phones, computers and other devices even where there is no electrical supply.

Questions

(Grades G-E)

1 How does putting draught excluders against doors save energy?

(Grades D-C)

2 An ordinary light bulb loses 95% of its energy input as heat. A fluorescent bulb is 20% efficient. Which of these two kinds of bulb gives the most light for the same input of electrical energy?

(Grades G-E)

3 Name the kind of energy that is stored in a battery.

(Grades D-C)

4 Some illuminated signs on country roads are powered by solar and wind energy. Why is this done?

Electricity and heat

The correct fuse should stop a keyboard from catching fire.

Electrical energy and heat

- **Electrical current** from the mains carries **electrical energy** which is **transferred** to **electrical appliances**.

- When an appliance **transforms** electrical energy, some **heat energy** always results. It is often wasted.

- Sometimes, we want the heat energy. For example, the fine wire of a **fuse** gets hot and melts if too much current flows through it. This breaks the circuit and protects other **components**.

Heat, resistance and current

- A wire heats up because it has **resistance** to the current flowing through it. Thin wires have a greater resistance than thick wires.

- Different metals have different resistances. Copper wire has a relatively low resistance, so current flows easily through it, little electrical energy is transformed into heat, and more is available for circuit components.

- Thickness of wire used in an appliance is chosen according to the current through it. A large current needs a thicker wire. For a telephone it is thin. For an electric cooker it is very thick.

The cost of electricity

How Science Works

You should be able to:

- compare and contrast the particular advantages and disadvantages of using different electrical devices for a particular application

- use the correct equation to calculate:
  - the total *amount* of energy transferred from the mains
  - the total *cost* of energy transferred from the mains.

Electrical energy and power rating

- The electric current in the **circuit** of a house *does not get used up*: current in = current out.

- We pay for the **electrical energy** that the current *transfers* to **electrical appliances**.

- **Power rating** is the rate at which an appliance transforms electrical energy. With a power rating of 1 **watt (W)**, an appliance transforms 1 **joule (J)** of electrical energy per second. 1 **kilowatt (kW)** = 1000 W. (See also page 52.)

How to calculate energy transfer and its cost

- **i** A 1.2 kW microwave oven is used for 5 min.

  total amount of energy transferred (in J) = power rating (in W) × time (in *seconds*)

  Total energy transferred from mains supply to microwave oven = 1200 (W) × 300 (s) = 360 000 J

  ii Electricity bills charge for kilowatt hours (kWh):

  number of kilowatt hours used = power rating (in kW) × time (in *hours*)

  Number of kilowatt hours used = 1.2 (kW) × 5/60 (h) = 0.1 kWh (or 0.1 Unit on the electricity bill)

  iii The cost of using an appliance is given by:

  total cost = number of kWh used × cost per kWh

  At 8p per kilowatt hour, the cost of using the microwave: = 0.1 × 8 = 0.8p

Questions

(Grades G-E)

1 Explain why a fuse breaks if too much current flows through it.

(Grades D-C)

2 Explain why a thin wire gets hotter than a thick wire when the same current flows through both of them.

(Grades G-E)

3 A small electric desk fan has a power rating of 50 W. How many joules of electrical energy does it transform every second?

(Grades D-C)

4 If electricity costs 8p per kWh, calculate the cost of running the fan for 30 minutes.

The National Grid

Voltage in the National Grid

G–E

- The **National Grid** is a nationwide network of power stations and cables carrying electricity to homes and businesses.

- Electricity is transmitted through the National Grid at a very **high voltage** (400 000 V) and very **high energy**.

- At **substations, step-down transformers** reduce the voltage to 132 000 V. This is repeated by step-down transformers at smaller substations, so that electricity is at 230 V when it reaches homes and businesses.

Top Tip!

Very high voltages transmit very large amounts of energy at low current and therefore with relatively small energy losses.

Why the National Grid uses high voltages

D–C

- **Power** (in watts) is the energy that an electric current transfers per second:
  power = current × voltage, or **$P = IV$**

- Using very high voltage, the National Grid transmits very high power electricity with a small current.

- Small current means low loss of energy wasted as heat. Thinner, cheaper, copper wires can be used.

- At small substations, the current is split into **parallel**, smaller currents for supply to homes.

Generating electricity

Electric motors and dynamos

G–E

- The blades of a motor-driven fan spin. The **electric motor** transforms **electrical energy** from a battery into **kinetic energy**.

- When the blades are spun by hand and an ammeter replaces the battery, the ammeter records an electric current. Now, kinetic energy is being transformed into electrical energy. The set-up is now a **dynamo**.

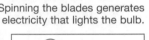

Spinning the blades generates electricity that lights the bulb.

When the magnet moves in or out of the coil, an electric current is generated in the circuit.

Magnets, current and movement

D–C

- An electric current flowing through a wire coil will generate a **magnetic field** that attracts (or repels) a magnet. Either the coil or the magnet will *move*.

  – An *electric current* and a *magnet* together cause *movement*.

When current flows through the coil, the magnet is attracted to the coil.

- Moving a magnet into or out of a coil generates an electric current.

- The direction of current as the magnet moves into the coil is reversed when the magnet moves out.

  – A *magnet* and *movement* together cause an *electric current*.

- Wind rotates the windmill blades attached to the wire coil between the magnet poles. This generates a current that the ammeter records. This is how an **electricity generator** works.

magnet

windmill blades fixed to axle

axle

magnet

wooden block with wire coiled round it

When the blades and coil turn, an electric current is generated.

Questions

(Grades G-E)

1 Name the device that changes high-voltage electricity into lower voltage electricity before it reaches your home.

(Grades D-C)

2 Explain why electricity is transmitted at very high voltages in the National Grid.

(Grades G-E)

3 Explain the difference between an electric motor and a dynamo.

(Grades D-C)

4 Explain how a magnet and a coil of wire can be used to produce electricity.

Power stations

How a power station generates electricity

- Most British **power stations** burn the **non-renewable**, **fossil fuels** coal, natural gas or oil:
  - heated water becomes steam, which rotates a **turbine**
  - the turbine turns a **generator**, which generates electricity
  - a **step-up transformer** increases the **voltage** to 400 000 V for transmission in the **National Grid**.

- A power station may serve several hundred thousand homes and businesses.

Stages in electricity generation of a coal power station.

G–E

Energy changes, efficiency and nuclear power

- Energy transformations in a power station inevitably cause **wasted energy**.

- Combined cycle gas turbine power stations make use of the 'waste' heat to produce steam and drive steam turbines. Efficiency can rise to 50%.

- A **combined heat and power** station reaches 70–80% efficiency if waste energy heats local houses and businesses.

- **Nuclear power stations** are 30% efficient. Uranium-235 undergoes **nuclear fission**, producing vast amounts of heat that power steam turbines. Radioactive plutonium-239 in the waste must be disposed of safely.

boiler	turbine	generator	transformer
chemical to heat (heat wasted)	heat to kinetic (heat wasted)	kinetic to electrical (heat wasted)	electrical transferred (heat wasted)

Energy wasted as heat in a power station. In a coal-burning power station, 65% of energy can be wasted.

Top Tip!

Make sure you know the sequence of stages in a power station and the energy changes at each stage.

D–C

Renewable energy

Renewable energy sources

- Some power stations burn household rubbish, and renewable **biomass** (including willow saplings and grass).

- **Wind turbines** transform the **kinetic energy** of moving air to **generate** electricity. They can be grouped into **wind farms**.

- **Hydroelectric** power stations transform the energy of falling water into electrical energy. The water directly turns the turbines as it flows through them. This is very clean, renewable energy.

- A **tidal power** station uses the energy of water as it moves with the rise and fall of tides. The incoming tide flows through a turbine. The outgoing tide turns the turbine the other way.

- A **wave power** generator uses the energy of wave water inside a tube. A turbine at the top turns when the air above the rising and falling water is forced or sucked through the turbine.

- In a **geothermal power** station, water is pumped down to hot rocks and returns as steam to drive a turbine.

A hydroelectric power station. The transformer steps up the voltage of electricity before transmission along power lines.

G–E

D–C

Questions

(Grades G-E)

1 What is the difference between a turbine and a generator?

(Grades D-C)

2 If 65% of the energy in the fuel a power station uses is wasted, what is the efficiency of the power station?

(Grades G-E)

3 Explain why hydroelectric power stations are said to use a 'renewable energy source'.

(Grades D-C)

4 In volcanic areas, such as Iceland, hot water and steam come up to the surface of the ground. How can this be used to generate electricity, and what is the name for this kind of energy source?

Electricity and the environment

Comparing means of electricity generation

- Most means of generating electricity have harmful effects and benefits.

source of electricity	harmful effects	benefits
fossil fuel-burning power station	gases add to global warming, and climate change	currently fuel is cheap and abundant
waste burning power station	toxic gases may cause cancer, birth defects	reduces need for landfill sites
hydroelectric power station	dams flood valleys, habitats, land, homes	cheap, clean, renewable energy
tidal and wave power station	change water flow patterns; disrupt shipping; destroy wildlife habitats	very low running costs
geothermal power station	might release dangerous gases from deep under the Earth's surface	small station, no deliveries required, so low environmental impact
wind turbine	may kill birds; some people think they are an eyesore	cheaper in remote areas than installing National Grid
nuclear power station	danger of radioactive pollution; problem of nuclear waste disposal	fuel available; no greenhouse gases produced

G–E

D–C

- To **limit costs**, power companies **locate**:
  - all stations as close to users as possible
  - fossil fuel power stations near roads or rail for fuel access, and a river for cooling water
  - wind turbines in windy areas
  - hydroelectric power stations in hilly landscapes
  - geothermal power stations on areas of easily drilled rock.

How Science Works

You should be able to: compare and contrast the particular advantages and disadvantages of using different energy sources to generate electricity.

Making comparisons

Which type of power generation?

- Fossil fuels produce carbon dioxide which promotes global warming and climate change. They will eventually run out.
- Scientists design **new electricity power stations** that:
  - use fossil fuels more **efficiently** and use a **reliable** energy source
  - use **renewable energy** sources; efficiency is less important
  - are **flexible** enough to meet changing demand.
- **Solar cells** transform the Sun's light energy into electrical energy that can be used for heating, cooling and running many electrical devices and tools.
- Power stations cannot store electrical energy. They have to match output to **demand**.
- Power station planners consider: **capital** (building) **costs**; **operating** (fuel and running) **costs**.
- Most **renewable** energy power stations have low operating costs. Capital costs are high, but improved technologies will reduce them.
- Fossil fuel power stations are cheaper to build, but fuel costs are rising.
- Wind turbines are only **reliable** where winds are regularly over 50 km per hour.
- Hydroelectric power needs reliable rainfall to give a regular flow of water.
- Solar power is practicable where there is year-round sunshine.

G–E

D–C

Top Tip!

Remember at least one really clear advantage and one really clear disadvantage for each type of power station.

Questions

(Grades G-E)

1 Explain why people may not want wind turbines near their houses.

(Grades D-C)

2 Why are oil-fuelled power stations often sited near to a major road?

(Grades G-E)

3 List **two** reasons why we need to use less fossil fuels to generate electricity.

(Grades D-C)

4 Explain why hydroelectric power stations are usually situated in mountainous areas of the country.

P1a summary

Heat is energy that flows from a hot object to a colder one.
Temperature indicates how hot something is.

Heat energy can be transferred by **conduction**, **convection** and **thermal radiation**.

Heat energy

Thermal (infrared) radiation is the transfer of heat energy by **electromagnetic waves**.

All materials **absorb** and **emit** thermal radiation.
Dull, dark surfaces do so more than shiny, light surfaces.

Energy **transfer** occurs when the *same* form of energy *moves* from one place to another.
Energy **transformation** occurs when *one* form of energy is *changed* to *another*.

Energy transfer and transformation

Thermal conductors (e.g. metals) transfer heat energy easily. **Thermal insulators** (e.g. plastic, glass) do not.

Free electrons in metals help to transfer energy between the fixed particles.

Types of energy include: **heat**, **light**, **sound**, **chemical**, **microwaves**, **radio waves**, **kinetic** and **potential**.

The **Law of Conservation of Energy** states:
Energy cannot be created or destroyed. It can only be transferred or transformed from one form to another. The total energy always remains constant.

A **Sankey diagram** shows the **energy input** and **output** of a device.
Wasted energy is usually heat energy.

Energy types and diagrams

Work is measured in joules (J):
work done = force × distance moved

Power is measured in watts (W):
power = work done ÷ time taken

$$\text{efficiency} = \frac{\text{useful energy output}}{\text{total energy input}}$$

Electrical current from the mains carries **electrical energy** which is **transferred** to **electrical appliances**.

The **power rating** of an appliance is the rate at which it transforms electrical energy.

The **cost** of the energy that current transfers to an electrical appliance can be calculated:
- **total amount of energy transferred (J) = power rating (W) × time (s)**
- **number of kilowatt hours (kWh) used = power rating (kW) × time (h)**
- **cost of electricity = number of kWh used × cost per kWh**

The **National Grid** transmits **electricity** around the country at **high voltage** and **low current** to reduce energy losses.

Electricity

A **dynamo** produces electricity when coils of wire rotate inside a **magnetic field**.

Fossil fuels and **biomass** are burned to produce heat.
Nuclear fuels release energy as heat.
The heat is used to produce **steam**:
- the steam turns a **turbine**
- the turbine turns a **generator**
- the generator produces **electricity**.

Renewable energy sources include **wind**, **hydroelectric**, **tidal**, **wave** and **geothermal power**.

All types of **electricity generation** have some **harmful effects** on people or the environment. There are also **limitations** on where they can be used.

Uses of electromagnetic radiation

What is electromagnetic radiation?

- **Electromagnetic radiation** is **energy** carried by **waves** that travel from one place to another. The waves:
  - move in **straight lines**
  - can travel through a **vacuum** (space); some can also travel through matter
  - have a **range** of energies, representing different types of radiation
  - all travel at the **same speed**, the speed of light.

- We group radiations into 'types' according to their effects and uses. We call some types of radiation **rays**.

- The different types of waves have different energies. The **electromagnetic spectrum** shows this.

Types of radiation and their uses

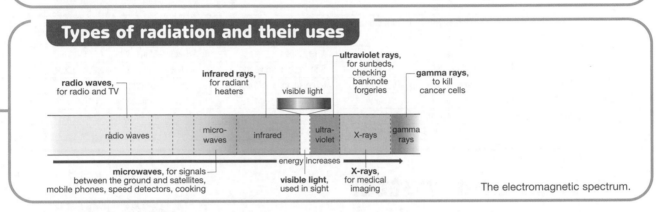

The electromagnetic spectrum.

Electromagnetic spectrum 1

Waves travel while material stays put

- An electromagnetic wave carrying energy moves through material at the speed of light but the material stays where it is.

Top Tip!

The shorter the wavelength, the higher the frequency. The longer the wavelength, the lower the frequency. The higher the frequency, the higher the energy.

Wavelength, frequency and energy

- A wave has a **wavelength**, measured from one crest to the next.

- The shorter the wavelength, the more **energy** it carries.

- The shorter the wave, the greater the number of waves passing a point in one second. This number is the **frequency** of the wave, measured in **hertz** (**Hz**). One Hz is one wave per second.

- The diagram shows the wavelengths and frequencies of the waves in the **electromagnetic spectrum**.

lower energy
wavelength

higher energy
wavelength

Wavelength, frequency and energy are linked.

The electromagnetic spectrum with wavelengths and frequencies.

Questions

1 What is carried by an electromagnetic wave?

2 Give **one** use of microwaves, other than cooking.

3 Which electromagnetic waves have the longest wavelength?

4 Which carries more energy – a wave with a short wavelength, or a wave with a long wavelength?

Electromagnetic spectrum 2

Colour, radiation types and sources

- We can see some wavelengths in the electromagnetic spectrum. They are in the **visible range** of wavelengths called **light**. Our eyes absorb energies in this range and send messages to the brain. The brain interprets messages about different groups of wavelengths as different colours.

- Sun, bulbs and lamps **transmit** radiation in the visible range.

- White light contains all colours in the visible spectrum.

- An object that: **reflects** all light radiation looks white; **absorbs** all but red, and reflects red only, looks red; is smooth and shiny and reflects all light is a mirror; absorbs all light radiation looks black.

- Radiation carries energy. A white object absorbs no light energy and a black object absorbs a lot. This is why an illuminated black object is hotter than a white object.

- Types of radiation do not have clear boundaries, even though diagrams show them this way. The table describes radiation types and sources.

wave type	wavelength	sources	detectors
gamma rays	10^{-12} m	radioactive nuclei	Geiger–Müller tube
X-rays	10^{-10} m	X-ray tubes	Geiger–Müller tube
ultraviolet	10^{-7} m	Sun, very hot objects	photographic film, fluorescent chemical
visible light	0.0005 mm	hot objects, Sun, lasers, LEDs	eyes, photographic film
infrared	0.1 mm	warm or hot objects	skin, thermometer
microwaves	1–10 cm	radar, microwave ovens	aerial, mobile phone
radio waves	10–1000+ m	radio transmitters	aerial, TV, radio

Waves and matter

How electromagnetic waves behave

- A mirror **reflects** the infrared radiation of a TV remote control.

- Food in a microwave oven **absorbs** microwave radiation.

- Soft tissue **transmits** X-rays – they pass through it.

- Dark, dull surfaces tend to absorb a large range of radiation.

The mirror reflects the infrared wave and the TV absorbs it.

- The energy from different wavelengths of electromagnetic radiation can be reflected, absorbed or transmitted by an object, depending on its substance and surface.

- To be absorbed, a wave's energy must match the energies that particles of a particular substance can absorb.

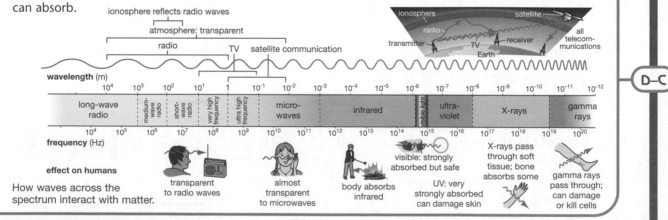

How waves across the spectrum interact with matter.

Questions

1 What do we call the part of the electromagnetic spectrum that our eyes can detect?

2 List the different types of electromagnetic radiation, in order of their wavelengths (smallest first).

3 Why does food inside a microwave oven get hot?

4 Your mobile phone produces microwaves. Are they absorbed in your head?

Dangers of radiation

Protection from radiation dangers

G–E

- Long wavelength radiations (low frequency, low energy) pass harmlessly through humans (see bottom diagram page 60).

- Microwaves cook food and also damage our cells, so ovens start only when the door is closed.

- Strong **infrared** radiation can cause burns if you are too close to a fire.

- Sun-block protects skin from the Sun's **ultraviolet** radiation, which can cause skin cancer.

- Lead aprons and lead-glass screens protect hospital staff from **X-rays** used for imaging and **gamma rays** used to kill cancer cells.

Effect of radiation on living cells

D–C

- Radiation can transfer energy to cells and damage them. Effects depend on the type of radiation and size of dose.

- Microwaves in a cooker transfer energy to water molecules, which vibrate more vigorously and heat up.

- Human bodies transmit and absorb infrared radiation. Small doses are safe.

- Molecules in the retina of the eye detect visible light. Too strong a light can damage the retina.

- Ultraviolet radiation, X-rays and gamma radiation are all **ionising** radiations. They can ionise atoms and molecules, especially DNA, in living cells. A cell with damaged DNA can become cancerous.

- Limits for safe doses are set for sun-bed users and hospital patients.

Telecommunications

Fibre optics and digital signals

G–E

- The bottom diagram on page 60 shows that radiations from long-wave radio to microwave are used for **telecommunications** through the atmosphere.

- Infrared and visible (**laser**) rays transmit information along **fibre optic cables**, which: replace expensive copper cables used for electrical transmission; transmit signals at the speed of light more clearly (no interference) than electricity; can carry many more messages than wires; allow information to travel curved paths.

- Messages travel as pulses of radiation, called a **digital signal**.

Communications technology

D–C

- **Telecommunications** use **geostationary satellites** that stay over the same ground position. The satellites receive and transmit ultra-high frequency and microwave radiations, such as TV signals.

How a satellite TV programme travels.

Questions

Grades G-E
1 What kind of radiation does sun-block protect you from?

Grades D-C
2 How can ionising radiation damage living cells?

Grades G-E
3 What is meant by a 'digital signal'?

Grades D-C
4 Explain what is meant by a 'geostationary satellite'.

Fibre optics: digital signals

How light travels along optical fibres

- Long-distance cables for phones, TVs and computers contain **optical fibres**.

- Laser light (infrared and visible) can travel inside a glass fibre. The light is **pulsed** (turned on/off) to produce a **digital signal**.

- An optical fibre is a very fine strand of pure silica glass. Black cladding prevents light from escaping.

- A light pulse in a fibre hits the side of the glass at an **angle of incidence** over 42°, so that the light undergoes **total internal reflection**.

- One fibre can carry several wavelengths of light simultaneously.

- Up to 10 billion pulses travel per second. One fibre can carry over 10 000 phone calls simultaneously.

- Between 12 and 200 optical fibres are bundled together in a cable.

- Light does not heat the cable, and signals in adjacent fibres do not interfere.

optical fibre

plastic coating: protects fibre

black cladding: stops light from emerging through wall of fibre

pure glass fibre: much thinner than a hair

light in an optical fibre

two light rays enter here

rays reflected from fibre wall: internal reflection

Fibre optic cables carry information for phones, cable TV and computers.

at 90°, light goes straight through

at an angle of incidence over 42°, light is reflected back at the boundary: **total internal reflection**

air: less dense than glass

glass/air boundary

glass: denser than air

light ray

angle of incidence

angle of reflection

Behaviour of light at a glass/air boundary.

G–E

D–C

Radioactivity

Atoms, isotopes and radiation

- All atoms of the same element have the same number of **protons** in their nucleus. But some of the atoms may have a different number of **neutrons**.

- Different forms of an element are called **isotopes**.

- If an atom is **unstable**, it breaks down and emits **radiation** from its nucleus. It is a **radioactive isotope** of the element.

- You cannot predict when an unstable atom will break down. Breakdown is random.

electron: negative charge, –

neutron: neutral

proton: positive charge, +

A helium atom contains 2 protons, 2 neutrons and 2 electrons.

G–E

element	hydrogen	helium	lithium	beryllium	boron	carbon
atomic number	1	2	3	4	5	6
full symbol	^1_1H	^4_2He	^7_3Li	^9_4Be	$^{11}_5B$	$^{12}_6C$

- For an element: all atoms have the same number of **protons**, called the **atomic number**; the number of **neutrons** can vary. Different forms are **isotopes** of the element, with different **mass numbers**. Some are unstable.

- When the unstable **nucleus** of an atom breaks down, it emits **energy**. This breakdown is called **radioactive decay**.

- The energy is emitted as alpha particles, beta particles or gamma rays, depending on the element. It is called **nuclear radiation**.

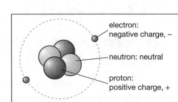

mass number = protons + neutrons

atomic number = protons

^7_3Li

D–C

Questions

Grades G-E

1 What is an optical fibre made of?

Grades D-C

2 Explain why light rays do not escape from optical fibres.

Grades G-E

3 An atom has a nucleus containing one proton. Another atom has a nucleus containing one proton and one neutron. Are they atoms of the same element?

Grades D-C

4 Explain why 'nuclear radiation' is given that name.

Alpha, beta and gamma rays 1

Three types of nuclear radiation

- An atom of a radioactive isotope emits **nuclear radiation** as its nucleus breaks up. This happens in one of three ways.

- An **alpha particle** is a helium nucleus. It is large, heavy and cannot go through paper.

- A **beta particle** is an electron. It leaves a nucleus when a neutron changes into a proton. It penetrates aluminium to about 3 mm.

- **Gamma rays** are rays of electromagnetic radiation travelling at the speed of light. Some penetrate lead to a depth of 10 cm.

- The energy carried by types of nuclear radiation varies according to the radiation source. The values in the table are averages.

The three radiations and their ability to penetrate materials.

How Science Works

You should be able to: evaluate the possible hazards associated with the use of different types of nuclear radiation.

radiation	alpha (α) particle	beta (β) particle	gamma (γ) rays
what it is; charge	helium nucleus, $^4_2$He (large, heavy); 2+	high energy, fast moving electron; e⁻	very high energy/frequency electromagnetic radiation; no charge
range in air	1–5 cm	10–100 cm	very long range
ionising effect	very high	high	low
depth in skin	fraction of a millimetre	0.1 mm	centimetres
damage done to cells	severe, but only if it gets inside the body	moderate	little damage done at low doses, because it is not strongly ionising
uses	smoke detectors, killing cancer cells	radiotracers, medical images	high doses used to destroy cancer cells; as radiotracer to produce images

Background radiation 1

Sources of background radiation

- Very low level, **background radiation** is all around us, in the air and the ground. It includes man-made radiation.

- The Earth's magnetic field and the atmosphere shield us from harmful solar **cosmic rays**.

- Living things have evolved to repair damage to their cells caused by average levels of background radiation.

- Minerals in granite rocks contain atoms that emit **gamma rays** and form **radon** gas, also radioactive. This raises background radiation levels in some parts of the UK. In granite areas, ventilation equipment is installed to clear radon gas from buildings.

- Aeroplanes fly in a weaker magnetic field and thinner atmosphere than at ground level, so there is more cosmic radiation. So aircrew monitor their flight schedules to maintain safe radiation levels.

- Anyone working with man-made radiation (e.g. radiographers and nuclear power station workers) wears a **radiation badge** to monitor exposure.

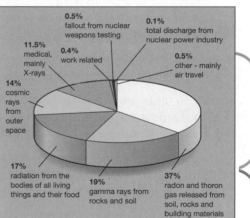

Background radiation chart.

Questions

Grades G–E
1 List the **three** kinds of nuclear radiation.

Grades D–C
2 Explain why an isotope emitting alpha radiation is very dangerous if you swallow it, but not if you just touch it.

Grades G–E
3 Name the **two** sources that produce the greatest amount of background radiation.

Grades D–C
4 Explain why people who work near radiation must wear radiation badges.

Half-life

Decay of radioisotopes

$\frac{1}{1}$	$\frac{1}{2}$	$\frac{1}{4}$	$\frac{1}{8}$
radioactive material			
at start	after 1 half-life	after 2 half-lives	after 3 half-lives

Amount of radioactive isotope left after each half-life.

- Another name for a **radioactive isotope** is **radioisotope**.

- Nuclei of the atoms of radioisotopes undergo the process of **radioactive decay**.

- We cannot predict when this will happen. But we do know how many atoms in a radioisotope sample will decay in a particular time. This is the **activity (count) rate**. Each radioisotope has its unique rate.

- We measure activity as **half-life**.
  The half-life of a radioisotope is the average time it takes for half of its atoms to decay.

- When radioisotopes decay, their nuclei can emit alpha, beta or gamma radiation. A Geiger counter detects and records this radiation.

- As decay continues, fewer unstable nuclei remain and activity rate decreases.

- Scientists identify radioisotopes from their unique rate of decay.

isotope	radiation	half-life
uranium-238	alpha	4.5 billion years
cobalt-60	gamma	5 years
barium-143	beta	12 seconds

- For a sample, the **half-life** is the average time taken:
  - for half the radioactive nuclei to decay, and therefore
  - for the activity (count rate) to decrease to half its starting rate.

How Science Works

You should be able to: evaluate measures that can be taken to reduce exposure to nuclear radiations.

Uses of nuclear radiation

Uses of nuclear radiation

- Gamma rays are used to **irradiate** surgical instruments sealed in plastic, killing bacteria and sterilising them.

- Gamma rays can penetrate the body. In large doses they kill cancer cells. This is called **radiotherapy**.

- Leaks in underground water pipes are located by adding a small amount of radioisotope to the water and detecting a build-up of gamma radiation with a Geiger counter.

- The radioisotope americium-241 in a **smoke detector** emits alpha radiation which ionises air particles. The alarm goes off when smoke particles prevent the ions from reaching the electrodes.

- Paper mills control **paper thickness**. A counter measures the beta radiation that passes through the paper from a radioactive source. If the paper is too thin, the counter reads high and the rollers controlling thickness open a little. If too thick, the counter reads low and the rollers move closer.

alarm   battery

3. Current detector: when smoke prevents alpha particles from ionising air, the current stops and the detector triggers the alarm.

1. Radioactive source, americium-241, emits alpha particles. They collide with air and ionise its particles.

2. Electrons move to positive electrode. Positive ions move to negative electrode.

air enters

How a smoke detector works.

counter (thickness detector)

source of beta radiation

paper sheet

Controlling paper thickness.

How Science Works

You should be able to: evaluate the appropriateness of radioactive sources for particular uses, including as tracers, in terms of the types of radiation emitted and their half-lives.

Questions

Grades G-E
1 A piece of a radioactive isotope contains 10 million atoms. After 12 years, 5 million of these atoms have decayed. What is the half-life of the isotope?

Grades D-C
2 What can you say about the count rates you could detect from the isotope in question 1, using a Geiger counter, at the start and then 12 years later?

Grades G-E
3 How can radiation help us to find leaks in underground pipes?

Grades D-C
4 Suggest why beta radiation, not alpha or gamma, is the best to use for checking the thickness of paper.

Safety first

Using radiation safely

G–E

- Alpha, beta or gamma radiations can damage cells inside our bodies (e.g. in our lungs).
- Workers using radiation sources follow strict health and safety procedures. They: handle radiation sources carefully, not touching them directly; wear protective gloves and other clothing; wear a badge that monitors the radiation they receive.
- The **energy** of different types of radiation can damage or kill cells by knocking electrons from atoms. This **ionises** the atoms and can disrupt the cells' chemical processes.
- If a cell's DNA is damaged, it can **mutate**. The cell may divide uncontrollably, forming a cancerous tumour.

D–C

- Long or intense radiation exposure causes **radiation sickness** and often death, because more cells die than the body can replace.
- There are debates on how to dispose safely of **spent nuclear fuel** which contains radioactive substances.

> **Top Tip!**
> Do not say that radiation ionises 'cells'. You cannot ionise cells, only atoms and molecules.

category of waste	typical waste materials	method of disposal
low-level hazard	protective clothing and used packaging materials	buried in land or at sea
intermediate hazard	irradiated equipment from reactors, cladding materials and reprocessing fluids	storage in concrete warehouses, or deep burial underground
high-level hazard	spent nuclear fuel and other radioactive sources from the reactor	vitrification (sealing in glass blocks), or deep burial

Searching space

Looking into space

> **How Science Works**
> You should be able to: compare and contrast the particular advantages and disadvantages of using different types of telescope on Earth and in space to make observations on and deductions about the Universe.

G–E

- **Optical telescopes** use light to form images of objects in space, such as stars and planets.
- The Earth's **atmosphere** makes it difficult to see clear images because: air density and humidity varies and distorts the image; dust pollution reduces the light reaching the telescope. Large ground-based telescopes are therefore built on high mountains in remote, dry places.
- Objects in space transmit radiation in all parts of the **electromagnetic spectrum** (see page 59).
- Ground and satellite telescopes form images of these objects from radio waves through to gamma rays.
- The **Hubble space telescope** was the first to make excellent images in the visible spectrum of objects in space such as the Eagle nebula on page 66.

D–C

- A simple optical telescope forms a poor image because it is too short and narrow to gather much light.
- A **reflecting optical telescope** produces far better images.
- The **Hubble space telescope**, a reflecting optical telescope 13.2 m long and 2.4 m in diameter, has formed images further and deeper into space than ever before.

The reflecting optical telescope has a large diameter, and mirrors extend its length.

Questions

Grades G-E

1 How can nuclear radiation cause cancer?

Grades D-C

2 Explain why disposing safely of spent nuclear fuel is extremely important.

Grades G-E

3 Why can a telescope in space produce clearer images of stars and planets than a telescope on the ground?

Grades D-C

4 Why did the invention of the reflecting optical telescope allow better images to be seen?

Gravity

Gravity, movement and orbits

- **Gravity** is a 'pull' force acting between *all* objects. A larger **mass** has greater gravity.

- Earth's gravity pulls the smaller Moon, keeping it moving in **orbit**:
  - without gravity, the Moon and Earth would drift apart
  - if the Moon did not orbit, the Moon and Earth would fall together.

- Distant objects have less pull than near objects. The further the **distance** between two objects, the weaker the gravity.

- Gravity and movement keep all planets and other bodies of the **Solar System** in their orbits.

G–E

Spacecraft, gravity and space travel

- Launched in 1977, Voyagers 1 and 2 spacecraft have investigated the giant planets and are travelling beyond the Sun's gravity to explore outer space.

- To reduce the fuel load, the spacecraft flew close to Jupiter and Saturn and accelerated past these planets, gaining momentum from their huge gravitational pull.

Voyager 2's route across the Solar System.

D–C

Birth of a star

Stars from dust

- Space is not empty, but contains gas and dust particles. In places, these form huge gas clouds called **nebulae**.

- Over millions of years, **gravity** pulls the particles closer, forming a dense ball of gas with a central core.

- Particles forced together can heat up to 15 million °C. Nuclei of atoms fuse together. This is called **nuclear fusion**. It becomes a **chain reaction**, giving out huge amounts of energy. A **star** is formed.

Spectacular gas clouds in the Eagle nebula, taken by the Hubble space telescope.

G–E

- The diagram shows the stages as:

  - mass and hence gravity increases, attracting particles that collide and heat up

  - core temperature increases; infrared radiation escapes

  - a protostar is formed; surrounding gas and dust are driven off

  - the protostar collapses; nuclear fusion starts:

nuclear fusion

hydrogen ⟶ helium + ENERGY

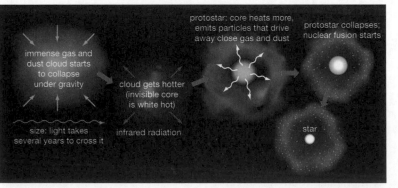

Steps in the formation of a star.

D–C

Questions

1 Write a sentence to say what 'gravity' is.

2 How did gravity help the two Voyager craft to save fuel?

3 What is 'nuclear fusion'?

4 What happens to hydrogen atoms when nuclear fusion takes place?

Formation of the Solar System

Birth and shape of the Solar System

G–E

- **Gravity** pulled gas and dust into a huge cloud. A central star heated up, exploded and flung gas outwards.

- The Sun formed and surrounding gases, dust particles and rocks clumped together, forming the planets.

- The inner **rocky planets** are Mercury, Venus, Earth and Mars.

- The outer **gas giants**, Jupiter, Saturn, Uranus and Neptune, are made of gases with a solid core.

How the Solar System was formed.

D–C

- The contracting mass of gas and dust spun into a disc. Dust particles clumped to form rocks.

- The exploding early star shot the gases further out than the rocks.

- The rocks formed the rocky planets, and beyond them, the **asteroid belt**.

- The gases (mostly hydrogen) cooled and condensed into the gas giants.

- On the edge of the Solar System is the Oort cloud of ice and particles, from where many **meteorites** come.

Life and death of a star

Events in a star's life and death

G–E

- Astronomers have pieced together the life of stars from their observations.

- A star gradually loses **mass**, as energy and particles in **solar wind**.

- As it loses mass, the star's gravity also gets less, so it expands. It cools as hydrogen fuel runs out.

- A Sun-sized star eventually becomes a **red giant** which collapses under gravity to form a **white dwarf**, then a **black dwarf**.

- A larger star becomes a **red supergiant** which collapses and explodes in a **supernova**. The small core becomes a **neutron star**. This may become a **pulsar**, and, if large enough, a **black hole**.

D–C

- In a stable star, two factors are in balance:
  - huge energy and streams of particles from **nuclear fusion** reactions produce an *outward* force called **radiation pressure**
  - the mass of gas exerts an *inward* **force of gravity**.

- These become unbalanced when hydrogen in the core runs out. Hydrogen fusion continues further out and the star expands. It continues to lose mass. The cooling core contracts, and the star collapses.
  - The star's **mass** (see diagram) determines the **speed** of collapse and the objects that may then form.

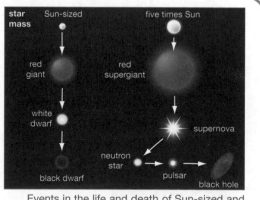

Events in the life and death of Sun-sized and larger stars.

Questions

Grades G-E

1 Which are the **four** gas giants?

Grades D-C

2 Explain why the gas giants are the planets furthest from the Sun.

Grades G-E

3 What will eventually happen to our Sun?

Grades D-C

4 Explain why a star collapses when its hydrogen fuel runs out.

In the beginning

The Big Bang

- **Galaxies** are moving apart at tremendous speeds. It is thought that:
  - there was once a huge amount of energy at a single point
  - the energy expanded violently in the **Big Bang**
  - within seconds, the temperature cooled, and energy began changing into **protons**, **neutrons** and **electrons**.
- Over billions of years, galaxies containing stars and planets have formed.

G–E

Energy becomes matter

- Following the Big Bang, the Universe expanded and cooled. Energy became matter, including quarks. Cooling quarks formed protons, neutrons and electrons.

- Hydrogen, the first element, formed. It made up the first gas clouds. These were the birthplace of stars, in which nuclear fusion produced the other elements.

a fraction of a second after   three minutes after   300 000 years after   3.8 billion years after

quarks, electrons   protons, neutrons, nuclei   hydrogen

The different stages of the Big Bang.

D–C

The expanding Universe

Red shift and an expanding Universe

- Light from a stationary source is at a particular **wavelength**.

- If the light source is moving away from an observer, the same wavelength *appears* longer, shifting towards the red part of the spectrum. The difference between actual and apparent wavelengths is the **red shift**.

The redder the galaxy, the faster it is travelling away, and therefore the greater its distance from Earth.

- Galaxies are moving apart. Their red shift tells us that the further away, the faster they recede.

- A **spectroscope** splits light into bands of different colours (wavelengths), forming a **spectral pattern**.

- When heated, every element emits light which forms a unique spectral pattern.

- Edwin Hubble looked at the spectral pattern for hydrogen in different galaxies. From the red shifts, he worked out how fast each galaxy was receding, and how far away it was.

The spectral pattern for hydrogen.

- George Gamow worked back to a single point as the place where the galaxies started from. This and the red shift support the Big Bang theory.

G–E

D–C

Questions

(Grades G-E)

1 According to the Big Bang theory, what happened at the very beginning of the Universe?

(Grades D-C)

2 Which was the first element that was formed by the Big Bang?

(Grades G-E)

3 If something is moving away from us, the wavelength of the light coming to us appears to be longer. Rewrite this sentence, stating how the appearance of the *frequency* of the light will change.

(Grades D-C)

4 Explain how the red shift provides evidence for the Big Bang.

P1b summary

Electromagnetic radiation is **energy** carried by **waves** that:
- travel in **straight lines**
- can travel through a **vacuum**
- have a range of energies
- travel at the **speed of light**.

Radiation can be **reflected**, **absorbed** or **transmitted** by an object depending on its substance and surface.

Wavelength is measured from one **crest** to another.
Frequency is the number of waves passing a point in a second, measured in **hertz** (Hz).

Optical fibres use pulsed laser light to produce a **digital** signal.

Electromagnetic radiation

The **electromagnetic spectrum** is divided into regions to show the energies of each particular type of wave.
From low frequency and energy (longest wavelength) to high frequency and energy (shortest wavelength) they are:
- radio waves
- microwaves
- infrared
- visible light
- ultraviolet
- X-rays
- gamma rays.

Radiation can **damage** living cells. **Ionising** radiation (ultraviolet, X-rays and gamma rays) can ionise atoms, including in **DNA**, and can cause cancer.

Radiation has many uses in **communication** systems including: radio, TV, satellites, cable and mobile phone networks.

Radioactive substances have **unstable atoms** that emit radiation.
Three forms are: **alpha particles**, **beta particles** and **gamma rays**.
Each has its own uses and hazards.

Radioactivity

Background radiation is all around us. **Granite** rocks can emit **gamma rays** and form radioactive **radon** gas.

The **half-life** of a **radioisotope** is the average time it takes for half of its atoms to **decay**.

Reflecting optical telescopes built on high ground in dry places and **space telescopes** (e.g. the **Hubble** telescope) form the best images of space as they reduce or avoid **atmospheric distortion** of the image.

Gravity and **movement** keep all planets and other bodies in the **Solar System** in orbit.

A **star** is born when the particles in a **nebula** are pulled together by **gravity**, heat up and form a **protostar**. This collapses and **nuclear fusion** starts:

nuclear fusion

hydrogen ⟶ **helium + ENERGY**

The gases, dust particles and rocks surrounding the **Sun** formed the inner **rocky planets**, the outer **gas planets**, the **asteroid belt** and the outer **Oort cloud** of the **Solar System**.

The Universe

A star's **mass** determines the **speed** of its **collapse** and the objects that may form.
A sun-sized star becomes:
    red giant → white dwarf → black dwarf
A larger star becomes:
    red supergiant → supernova → neutron star → pulsar → black hole.

Current evidence suggests that: the **Universe** is **expanding**; a huge amount of **energy** at a single point expanded violently in the **Big Bang**; energy became **matter**; **hydrogen** was formed into gas clouds; **stars** were born; and **nuclear fusion** produced the other **elements**.

Red shift indicates that:
- **galaxies** are moving apart
- the **further** away a galaxy is, the **faster** it recedes.

Cells

Animal and plant cells

- The top diagram shows the structure of the **cells** of **animals**, including humans.

- A typical **plant** cell includes the parts shown in the animal cell. It is also enclosed in a cell wall and may contain chloroplasts and a permanent vacuole.

Top Tip!

Be very careful to use the terms cell *wall* and cell *membrane* correctly.

cell membrane, which controls the passage of substances in and out of the cell

cytoplasm, where metabolic reactions controlled by enzymes take place

mitochondrion, where energy is released in respiration

ribosomes, where protein synthesis takes place

nucleus, which controls the activities of the cell

Detail of an animal cell.

cell wall, which strengthens the cell

ribosome

mitochondrion

cytoplasm, where metabolic reactions take place

cell membrane, which controls the passage of substances in and out of the cell

chloroplast, which absorbs light energy to make food

permanent vacuole, filled with cell sap

nucleus, which controls the activities of the cell

A plant cell.

Cell organelles

- The parts inside a cell are called **organelles**. Each organelle has a particular function.

- Energy is released in respiration in the **mitochondria**.

- Protein synthesis occurs in **ribosomes**.

How Science Works

You should be able to: relate the structure of different types of cells to their function in a tissue or an organ.

Specialised cells

Different cells perform different functions

- All the functions of a single-celled organism are carried out by the one **cell**.

- A larger, multicelled organism contains different tissues and organs. The cells of each tissue or organ have a structure that is specialised for their **function**.

- The diagram on the right shows how a sperm cell is adapted for its function.

- Here are two more examples of specialised cells in the human body.

vesicle containing enzymes, to digest a way into the egg

nucleus containing one set of chromosomes

cell membrane

cytoplasm

mitochondria, to provide energy for swimming

tail for swimming

30 μm

Note: 1 μm is 1 millionth of a metre

A sperm cell.

Goblet cells are found in the lining of the alimentary canal, and in the tubes leading down to the lungs. They make mucus, which helps food to slide easily through the alimentary canal, and helps to stop bacteria getting down into your lungs.

mucus that has been made by the cell

cell membrane

cytoplasm

nucleus

20 μm

A goblet cell.

cilia, which move in time with each other like a Mexican wave

cell membrane

cytoplasm

many mitochondria, to supply energy for the movement of the cilia

nucleus

Ciliated cells are found alongside goblet cells in the trachea and other tubes leading into the lungs. The sweeping movements of their cilia move mucus up the tubes and into the back of the throat, where you swallow it.

20 μm

A ciliated cell.

Questions

Grades G-E

1 Name **two** structures in a plant cell that are not in animal cells.

Grades D-C

2 Name the organelle where protein synthesis takes place.

Grades G-E

3 State **two** ways in which a sperm cell is adapted for its function.

Grades D-C

4 Where are goblet cells found and what do they do?

Diffusion 1

Moving particles

Top Tip!
Particles do not go in a particular direction on purpose. They just move around randomly.

- **Diffusion** is the spreading of particles of a gas, or of a substance dissolved in a solution, from a region where the particles are at a higher **concentration** to where they are at a lower concentration.

Diffusion and temperature

- Water and sugar molecules in the solution move around in all directions. The sugar **concentration** is higher at the bottom of the beaker than at the top.

- By diffusion, there is a **net movement** of sugar molecules from a more concentrated to a less concentrated region since, on average, there are more to move to the less concentrated region than in the other direction. Eventually, sugar and water molecules are evenly spread throughout the solution.

- Particles move faster at higher **temperatures**. Therefore, a temperature rise speeds up diffusion.

dilute — water molecule
sugar molecule
concentrated

Diffusion in a beaker of sugar solution.

Diffusion 2

Diffusion in cells and jelly

- Cells need **oxygen** for **respiration** according to the equation:

glucose + oxygen → carbon dioxide + water

- Respiration uses up oxygen, so oxygen concentration inside the cell is lower than outside. This causes oxygen to diffuse into the cell through the cell membrane.

- The faster the cell respires, the faster it uses up oxygen, increasing the difference between oxygen concentrations inside and out. The greater the concentration difference, the faster oxygen diffuses into the cell.

- **Alveoli** in the lungs allow oxygen to diffuse rapidly into the bloodstream for transport to cells around the body.

- **Antibiotics** kill bacteria and can diffuse through **agar jelly**.

- The agar contains bacteria, and each disc contains a different concentration of the same antibiotic. The table shows results for bacterial growth after two days.

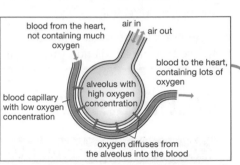

blood from the heart, not containing much oxygen — air in — air out
blood to the heart, containing lots of oxygen
alveolus with high oxygen concentration
blood capillary with low oxygen concentration
oxygen diffuses from the alveolus into the blood

Gas exchange in an alveolus. Why does each lung have millions of alveoli?

disc	concentration of antibiotic (arbitrary units)	diameter of clear area (mm)
A	0	0
B	1	4
C	5	6
D	10	9

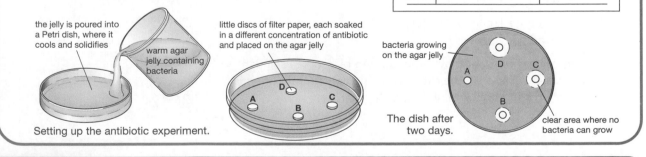

the jelly is poured into a Petri dish, where it cools and solidifies
warm agar jelly containing bacteria

little discs of filter paper, each soaked in a different concentration of antibiotic and placed on the agar jelly

bacteria growing on the agar jelly

clear area where no bacteria can grow

Setting up the antibiotic experiment.

The dish after two days.

Questions

1 Which of these contain particles that can diffuse: solid, solution, gas?

2 How does a decrease in temperature affect the rate of diffusion?

3 Which gas diffuses into a cell when it is respiring?

4 What do the results of the experiment above tell you about how a difference in concentration affects the rate of diffusion?

Osmosis 1

Demonstrating osmosis

- A **partially permeable membrane** allows small molecules like water to **diffuse** through it, but blocks larger molecules.

- If a partially permeable membrane separates a dilute sugar solution from a concentrated solution, water molecules diffuse from the dilute solution through the membrane into the concentrated solution (see top diagram). This diffusion of water molecules only is called **osmosis**.

- In the bottom diagram, *water* molecules are at a *higher* concentration in the dilute sugar solution, and move to where water molecules are at a *lower* concentration in the concentrated sugar solution.

- In the experiment, all the water and sugar molecules move and bump into each other and the membrane.

- Only water molecules can pass through the holes in the membrane.

- More water molecules from the dilute sugar solution will cross the membrane into the concentrated sugar solution than in the reverse direction.

- Now there is less water in the dilute sugar solution and more in the concentrated solution than at the start of the experiment.

water rises in glass tubing

dilute sugar solution

Visking tubing – a partially permeable membrane

concentrated sugar solution

Using Visking tubing to demonstrate osmosis.

dilute sugar solution ('concentrated' water)

partially permeable membrane

concentrated sugar solution ('dilute' water)

Key ○ water molecule   ● sugar molecule

Explaining osmosis.

Osmosis 2

Osmosis and animal cells

- The **cytoplasm** of an animal cell is a fairly concentrated solution. The **cell membrane** is **partially permeable**: water can diffuse through it.

- The diagrams show the movement of water by osmosis: *into* an animal cell from distilled water (highest water concentration), the cell bursts; *out* of an animal cell into a more concentrated solution, the cell collapses.

distilled water

cytoplasm – a fairly concentrated solution

cell membrane – a partially permeable membrane

Water moves into this animal cell by osmosis.

concentrated solution

water moves out of the cell by osmosis

Water moving out.

Osmosis and plant cells

- As well as a partially permeable cell membrane, a plant cell has a strong, fully permeable **cell wall** that limits expansion of the cell.

- Therefore, only a limited amount of water can enter the cell by osmosis, and the cell does not burst.

- A plant cell will lose water by osmosis when placed in a solution more concentrated than the cytoplasm. The **cell membrane** may pull away from the cell wall. If it tears, the cell dies.

How could you try to revive this cell?

distilled water

cytoplasm – a fairly concentrated solution

partially permeable cell membrane

cell wall made of cellulose

Water moves into this plant cell by osmosis.

concentrated solution

cell membrane is pulled away from the cell wall

strong cell wall stays the same

the cell shrinks, and pulls away from the cell wall

Questions

1 Name the substance that diffuses across a partially permeable membrane during osmosis.

2 Which contains more water molecules – a dilute solution or a concentrated solution?

3 Which part of a cell is a partially permeable membrane?

4 Name the structure that stops a plant cell bursting when it absorbs a lot of water.

Photosynthesis

Making food and using energy

- In **photosynthesis**, the green plant pigment, **chlorophyll**, traps energy from sunlight. The energy is used to convert **carbon dioxide** and water into **glucose**:

  carbon dioxide + water (+ light energy) → glucose + oxygen

- Glucose is a **sugar**, a kind of **carbohydrate**. Plants can use glucose to make other kinds of food, such as **starch**, fats and proteins.

- The by-product of photosynthesis is **oxygen**, which plants release.

- In photosynthesis, the **energy** from light becomes **chemical energy** stored in glucose molecules.

- This energy can be unlocked again when organisms break down glucose in respiration:

  glucose + oxygen → carbon dioxide + water + energy

- Our cells use this energy for warmth, growth and movement.

- Plants also respire, and use the energy to grow and produce flowers and fruit.

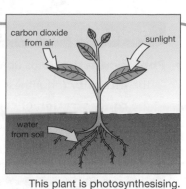

This plant is photosynthesising.

> **Top Tip!**
>
> Remember that plants respire and photosynthesise during the day, and just respire at night.

Leaves

Leaves: adaptations for photosynthesis

- The diagram shows how a **leaf** is **adapted for photosynthesis**.

- Photosynthesis uses up carbon dioxide and its concentration in the leaf becomes less than in air, so more carbon dioxide **diffuses** into the leaf through the **stomata**.

- Photosynthesis produces oxygen which becomes more concentrated in the leaf than outside it, so this gas diffuses out through the stomata into the air.

there are tiny pores (**stomata**) on the underside of the leaf which allow carbon dioxide to diffuse into the leaf

the broad, flat surface of the leaf gives it a very large surface area, so lots of sunlight and carbon dioxide reach it

the leaf is very thin, so sunlight and carbon dioxide can easily reach the cells where photosynthesis happens

the veins in the leaf contain tubes which run all the way up from the roots, carrying water and minerals

How a leaf is adapted for photosynthesis.

- Plant cells adapted for photosynthesis have chloroplasts, which contain chlorophyll to absorb the light energy used to make glucose.

- Glucose can be used immediately to release energy in respiration. Cells store excess glucose as starch.

- The **tissue** best adapted for photosynthesis is the palisade mesophyll.

- Spongy mesophyll and the cells of stomata are adapted for gas exchange.

chloroplast

tissues
upper epidermis
palisade mesophyll
spongy mesophyll
lower epidermis

air space   guard cell   stoma (small hole)

The tissues of a leaf in section.

Questions

Grades G-E

1 Write down the word equation for photosynthesis.

Grades D-C

2 Name the process in which cells get energy from glucose.

Grades G-E

3 Give **two** ways in which being thin helps a leaf to photosynthesise.

Grades D-C

4 How does the position of palisade cells in a leaf help them to photosynthesise?

Limiting factors

G–E

Speeding up photosynthesis

- **Photosynthesis** speeds up when:
  - the **intensity** (brightness) of the **light** increases
  - the **concentration** of **carbon dioxide** increases
  - the **temperature** increases, as happens through spring and summer.

- These factors can be controlled in glasshouses.

Limiting factors for photosynthesis

- The upper graph shows that no photosynthesis occurs at zero **light intensity**. As light intensity gradually increases, the **rate** of photosynthesis increases. In region A, the low light intensity is **limiting** the rate of photosynthesis – it is a **limiting factor**.

- Beyond point B, any further increase in light intensity does not affect the rate of photosynthesis.

- The lower graph shows that, in region A, **carbon dioxide concentration** is a limiting factor for photosynthesis.

- Within certain values, **temperature** also limits the rate of photosynthesis.

- Light intensity, carbon dioxide concentration and temperature are all limiting factors for the rate of photosynthesis, which they can affect separately or together.

How light intensity affects the rate of photosynthesis.

How carbon dioxide concentration affects the rate of photosynthesis.

D–C

Healthy plants

Healthy plants need mineral salts

- Using carbon dioxide and water, plants make glucose and other sugars, starch, fats and cellulose. All these are **carbohydrates** – they contain the elements carbon, hydrogen and oxygen only.

- To make proteins and chlorophyll from carbohydrates, plants need other elements. They exist in ions of **mineral salts** in the soil – **nitrogen** as **nitrate ions** and **magnesium** as **magnesium ions**. Plants absorb these ions through their roots.

- For healthy growth, plants need mineral ions.

- If a plant is **deficient** in (lacks) an ion it needs, it will not grow properly.

- A plant deficient in **nitrate ions** cannot make enough amino acids and therefore proteins. Its growth is stunted.

- Each chlorophyll molecule contains a **magnesium** atom. A plant deficient in **magnesium ions** cannot produce enough chlorophyll and its leaves are yellow.

substance	elements it contains				
	C	**H**	**O**	**N**	**Mg**
carbohydrate	✓	✓	✓		
fat	✓	✓	✓		
protein	✓	✓	✓	✓	
chlorophyll	✓	✓	✓	✓	✓

A healthy potato plant.

A potato plant short of nitrate ions.

A potato plant short of magnesium ions.

G–E

D–C

Questions

(Grades G-E)

1 List **three** factors that could increase the rate of photosynthesis.

(Grades D-C)

2 In the light intensity graph, why does the line stay flat in region C?

(Grades G-E)

3 Name a compound found in plants that contains magnesium ions.

(Grades D-C)

4 How do nitrate ions help to make proteins?

Food chains

Energy flow

A lion with the carcass of an antelope, and vultures in the background.

- A **food chain** shows how energy (originally from sunlight) passes from one organism to another.

- In the food chain: grass → antelope → lion

  grass is a **producer**. It uses light energy to grow, providing nutrients that the next organism consumes. Food chains always start with a producer.

  - the antelope is the **primary** (first) **consumer** and is a **herbivore**: it eats plants.

  - the lion is a **secondary consumer** and is a **carnivore**: it eats other animals. It is also a **predator**, and the animals it kills are its **prey**.

Food webs and energy wastage

- Food chains can interact to form a **food web**.

- A leaf uses only a little of the solar energy reaching it to store in cell substances. The rest is reflected or passes through the leaf without reaching the **chlorophyll** or it is the wrong wavelength (colour). This energy is wasted.

This food web shows that both vultures and lions feed on antelope.

Biomass

How Science Works

You should be able to: interpret pyramids of biomass and construct them from appropriate information.

Pyramids of biomass

- A mass of living material is called **biomass**.

- The biomass of grass that an antelope eats greatly exceeds the antelope's biomass. The biomass of the antelopes a lion eats greatly exceeds its biomass. Also, antelopes do not eat entire plants or lions entire antelopes, yet biomass includes the uneaten parts.

- A **pyramid of biomass** represents the mass at each step in a food chain. In each step, the biomass is less than in the step before.

mass of lions

mass of antelopes

mass of grasses

A pyramid of biomass.

Energy losses

- The biomass reduces along a food chain because not all the energy from one organism is transferred to the next.

energy lost as heat from respiration

energy in grass → energy in antelope → energy in lion

energy lost as waste

How energy is lost in this food chain.

- The diagram shows that energy:

  - is transferred to tissues and organs as the antelope and lion grow

  - is lost as heat in respiration and as waste.

- To survive in cold surroundings, mammals and birds use **respiration** to maintain a constant body temperature. They are **homeotherms** (meaning 'same temperature').

- Snakes, frogs and fish stay at the temperature of their surroundings. The biomass of the food they need is less than for homeotherms.

Questions

Grades G-E
1 What do the arrows in a food chain show?

Grades D-C
2 List **three** reasons why plants do not use all the energy in the sunlight where they grow.

Grades G-E
3 What is biomass?

Grades D-C
4 Why do mammals need more food than reptiles of the same size?

Food production

Shortening food chains

- Biomass gets less along **food chains** because **energy** is lost at each stage.

- The shorter the food chain, the smaller the energy loss.

- In farming, reducing energy loss increases the **efficiency** of **food production**.

- An inefficient food to produce is salmon: plankton → small fish → bigger fish → salmon → humans

- An efficient food to produce is maize: maize plants → humans

It would be more energy-efficient for us to eat the maize rather than eat the cattle.

G–E

Energy-saving methods of food production

- Farmers of poultry and livestock find ways to reduce energy losses:
  - birds and animals are housed in barns to reduce their movement
  - barns are warmed to reduce heat loss in respiration.

- The energy in animal feed is then converted efficiently into eggs, meat and milk.

- In other cultures, animals have other uses that affect their living conditions:
  - pigs wander around eating discarded food scraps
  - cattle pull ploughs in fields.

D–C

The cost of good food

Food miles and animal welfare

- Food production costs include transport costs.

- To save costs, supermarket food is carried in container lorries from huge central distribution warehouses.

- Transport costs may be less for locally produced food.

- Cheap food may result from poor **animal welfare**, including bad housing and poor quality feed.

- The minimum welfare standards for rearing animals in Britain may not apply to farming overseas. Therefore, food from overseas is often cheaper.

- Humans need food, but cheap food can mean poor animal welfare.

G–E

Three means of egg production

- **Battery hens** are cramped in small cages in heated buildings. Automated systems provide their food and water.

- **Barn hens** are kept in heated buildings, but are free to walk around and have perches.

- **Free-range hens** are indoors at night but are free to walk around a fenced area outside during the day.

Battery hens.

D–C

Questions

1 Why is it more energy-efficient to eat bread than to eat beef?

2 How can keeping animals warm reduce a farmer's costs?

3 How do supermarkets reduce transport costs for the food they sell?

4 Why are free-range eggs more expensive than eggs from battery hens?

Death and decay

Decay

G–E

- All plants and animals eventually die.

- **Microorganisms**, including bacteria and fungi, use **enzymes** to break down (**digest**) dead and waste plant and animal materials. This is the **decay process**.

- In order to break down these materials, microorganisms need oxygen for **aerobic respiration**.

Speeding or slowing decay

D–C

- The growth and decay action of microorganisms are affected by temperature, moisture levels and oxygen levels.

- Microorganisms function most rapidly at **warm temperatures**, typically between 25 and 45 °C. (High temperatures kill microorganisms, but the spores of some survive.)

- To grow and cause decay, microorganisms need moisture.

- To be active, many microorganisms need oxygen for aerobic respiration.

freezer (–4 to –10 °C)	fridge (0 to 4 °C)	boiling water (100 °C)	pressure cooker (120 °C)
no growth	slow growth	fast growth / no growth – cells killed in 10 minutes (but not spores)	cells and spores killed in 15 minutes

How temperature affects the activity of microorganisms.

There is no water in this dried food (left), and there is no air in this vacuum-packed food (right), so microorganisms cannot grow.

Cycles

Recycling

G–E

- Without **decay**:
  - dead and waste plant and animal materials would pile up
  - nutrients for plants would be locked up in these materials and the soil would be less fertile.

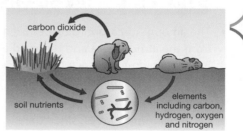

A food chain with decay microorganisms.

- Decay by **microorganisms** enables substances taken from the environment to be returned to it.

- **Detritus feeders** are animals that eat dead bodies and waste material, for example earthworms.

- Within a **community** of organisms, materials are removed from the environment and also returned to it – they are **cycled**. If removal processes and return processes are balanced, the community is stable.

D–C

- Microorganisms continually break down dead parts of plants and the waste products of animals.

- The diagram shows that microorganisms break down all organisms in a **food chain** when they die.

- The processes of removing materials from the environment and returning them continue all the time, as the diagram shows.

This diagram shows how materials are recycled in an ecosystem.

Questions

Grades G-E
1 How do microorganisms cause decay?

Grades D-C
2 Why does decay happen more quickly when it is warmer?

Grades G-E
3 Give an example of a detritus feeder.

Grades D-C
4 What kind of organisms are fed on by decay microorganisms?

The carbon cycle 1

The importance of carbon

A simple carbon cycle.

- Carbon starts in the **carbon dioxide** of the air. In **photosynthesis**, green plants use carbon dioxide to make glucose:

carbon dioxide + water $\xrightarrow{\text{light energy}}$ **glucose (a carbohydrate) + oxygen**

- For growth, plant cells make fats, proteins and carbohydrates from glucose. All these contain carbon.

- Plants are eaten by animals, and some animals eat each other.

- Plant and animal cells also use glucose in **respiration** to produce carbon dioxide which enters the air, and to release usable energy:

glucose + oxygen → carbon dioxide + water + energy

- This constant cycling of carbon is called the **carbon cycle**.

A Martian greenhouse

- Life on Earth is adapted to live in Earth's atmosphere. About the same amount of carbon dioxide that plants remove from the atmosphere is returned when organisms respire.

- Mars is cold and has a very thin atmosphere, mostly of carbon dioxide.

Why would plants not be able to grow in the Martian atmosphere?

- Scientists suggest that they could grow plants on Mars inside huge transparent glasshouses. There would be natural light and carbon dioxide. As plants photosynthesised, oxygen would gradually build up to a level in which visitors from Earth could survive.

The carbon cycle 2

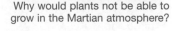

The carbon cycle

- Microorganisms that are **decomposers**, and the **detritus feeders**, return the plant nutrients in dead organisms to the soil. When decomposers respire, they release carbon dioxide. In this way, all the energy originally captured by green plants has been transferred.

The carbon cycle.

- The diagram shows how animals, plants and decomposers all interact in the carbon cycle.

Do trees store carbon?

The numbers = mass of carbon taken in or given out per year, in each square metre of forest.

- The amount of carbon dioxide taken from the atmosphere and returned to it is becoming imbalanced: its concentration is increasing in air.

- A way to remove some carbon may be to plant more trees, since they take in carbon dioxide in photosynthesis.

Questions

Grades G-E

1 Name **three** types of compound in a plant that contain carbon.

Grades D-C

2 Why would we need plants if we lived on Mars?

Grades G-E

3 How do decomposers return carbon dioxide to the air?

Grades D-C

4 How do trees help to remove carbon dioxide from the air?

B2a summary

Cells

Animal cells have a cell membrane, cytoplasm and a nucleus.
In addition, **plant cells** may also have a permanent vacuole, chloroplasts and a cell wall.

In multicellular organisms, different cells are **specialised** to perform different functions.

Particles of gas, or substances dissolved in a solution, move by **diffusion**. Oxygen and carbon dioxide diffuse into and out of cells.

Osmosis is the special diffusion of water molecules through a partially permeable membrane.

Photosynthesis

Green plants use **chlorophyll** to trap **energy from sunlight** to perform **photosynthesis**:
carbon dioxide + water → glucose + oxygen

Leaves are **adapted** for photosynthesis by being very thin and having stomata, a broad, flat surface and veins.

Glucose can be used in **respiration**:
glucose + oxygen → carbon dioxide + water + ENERGY

The **rate** of photosynthesis is affected by:
– light intensity
– carbon dioxide concentration
– temperature.
These can all be **limiting factors**.

Ions of **mineral salts** in the soil can be used by plants to make proteins or chlorophyll.

Lack of a particular mineral ion results in a **deficiency symptom** in a plant.

Food chains and cycles

Food chains show the feeding relationships of organisms.

Energy passes along food chains, but some energy is lost at each transfer.

The shorter the food chain, the smaller the energy loss.

Reducing energy loss increases the **efficiency of food production**.

In **decay**, **microorganisms** use **enzymes** to break down waste materials.

Decay is affected by the levels of:
– temperature
– moisture
– oxygen.

Decomposers and **detritus feeders** feed on the organisms in the food chain, recycling materials, e.g. carbon.

In the **carbon cycle**, carbon is removed from the air as carbon dioxide during photosynthesis and returned to it by respiration.

Enzymes – biological catalysts

Enzyme action, heat and pH

- **Enzymes** are **biological catalysts** that speed up chemical reactions and are not used up. They are **protein molecules**, affected by **temperature** and **pH**. Each controls one type of reaction.

- Enzymes catalyse the reactions in **respiration**, **protein synthesis** and **photosynthesis**.

- An enzyme molecule is a long, folded chain of **amino acids**. It has a hollow **active site** where its **substrate** molecule fits and reacts.

- High **temperatures** and extreme **pHs** (beyond a narrow range) change the *shape* of the active site. Then the substrate cannot fit or react.

- Some enzymes (e.g. digestive enzymes) *break large* molecules into smaller ones. Others *join small* molecules to make larger ones (e.g. enzymes joining amino acids to form proteins).

- The temperature at which an enzyme works best is its **optimum temperature**. In humans, this is 37 °C. Cooling enzymes **inactivates** them. Overheating **denatures** (destroys) them.

- Each enzyme has an optimum **pH**. Extreme pHs denature enzymes.

How an enzyme breaks molecules down.

Graphs showing the effect on enzyme activity of temperature (left) and of pH (right).

Top Tip!

Do not say that enzymes are 'killed' by high temperatures. They are just chemicals, not living things. They are *denatured*.

Enzymes and digestion

Digestive enzymes

- **Digestive enzymes** are secreted into the gut cavity by specialised cells in glands. They catalyse the splitting of large, insoluble food molecules into smaller ones that pass through villi in the small intestine, into the bloodstream and reach all parts of the body.

- The pancreas produces all three groups of digestive enzymes: amylase, protease enzymes and lipase enzymes.

Where enzymes of the human digestive system are produced.

mouth
salivary glands produce amylase
oesophagus
liver produces bile
stomach produces protease (pepsin) and hydrochloric acid
gall bladder stores bile
pancreas produces amylase, protease and lipase
small intestine produces amylase, protease and lipase
large intestine
appendix
rectum
anus

starch – a long molecule made of many glucose molecules joined together
glucose molecules
amylase enzymes cut glucose molecules off
these **glucose** molecules are soluble and can be absorbed into the blood

Digestion by salivary amylase of starch in the mouth at pH6.5–7.5.

protein molecule - made of many amino acid molecules joined together
protease enzymes cut amino acid molecules off
amino acid molecules are soluble - they can be absorbed into the blood

Digestion by a protease (pepsin) in the stomach at pH2.0.

glycerol
fatty acid
fatty acid
fatty acid
lipid
lipase enzymes cut the **glycerol** from the **fatty acids** - they can be absorbed into the blood

Digestion by lipase of fat and oil in the small intestine at pH7.0.

- Cells in the stomach produce hydrochloric acid which kills bacteria in food, unravels proteins to allow their digestion, supplies optimal acid conditions for the stomach's digestive enzymes, and activates pepsin (inactive before food arrives).

- **Bile** is a strong alkali produced by the **liver** and stored in the **gall bladder**. Bile joins partly digested food arriving in the small intestine. It neutralises the acid and provides the alkaline conditions the small intestinal enzymes need, and **emulsifies** (breaks down) fats.

Questions

Grades G-E
1 What is an 'active site'?

Grades D-C
2 Explain why enzymes do not work at extreme pHs.

Grades G-E
3 How does amylase help in digestion?

Grades D-C
4 How does bile help in digestion?

Enzymes at home

Biological washing powders

G–E

- Protein and fat stains on clothes can be difficult to wash out, requiring nearly boiling water. This can damage the clothes.

- Some microorganisms secrete protein- and fat-digesting enzymes from their cells. These enzymes are purified and added to **biological detergents** (washing powders). They work best at about 35–40 °C.

    - **Protease** enzymes digest protein (e.g. in blood and egg).

    - **Lipase** enzymes digest fat (e.g. butter and oil).

D–C

- Insoluble stains are difficult to remove from fabrics. In biological washing powders, **detergent** molecules make greasy stains **soluble**, lifting them from the fabric, and **enzymes** break them down.

- **Proteases** digest protein in stains.

- **Lipases** digest fat droplets when detergent frees them from fabric fibres and they pass into the water.

Removing a protein stain.

Enzymes and industry

Enzymes and the food industry

G–E

- Small babies do not digest proteins easily. Manufacturers use **proteases** to pre-digest the protein in baby foods.

- **Carbohydrases** are used to convert **starch** from potatoes and maize into **sugar syrups** (saturated sugar solutions) that are used in sports drinks.

- To reduce the sugar content of sugars in slimming foods, the enzyme **isomerase** is used to convert **glucose** into the sugar **fructose**, which tastes much sweeter.

How Science Works

You should be able to: evaluate the advantages and disadvantages of using enzymes in the home and in industry.

These are made with the help of enzymes.

D–C

- Small babies cannot break up protein molecules in their food into amino acids. **Baby food** manufacturers add **proteases** to do this. Babies can then absorb the amino acids into their bloodstream for transport to all their body tissues. Older babies produce all the digestive enzymes they need.

Questions

Grades G-E

1 How do proteases in washing powders help to remove blood stains?

Grades D-C

2 Explain how detergents make it easier for enzymes in washing powders to work.

Grades G-E

3 Why is fructose used in 'slimming' foods?

Grades D-C

4 Why are proteases sometimes used in the production of baby foods?

Respiration and energy

Energy: getting it and using it

- Energy for cells is supplied by **aerobic respiration**, a process that happens in cell organelles called **mitochondria**. The equation that summarises aerobic respiration is:

glucose + oxygen → carbon dioxide + water (+ energy)

- The energy is used:
  - to build large molecules from small ones
  - by plants to build sugars, nitrates and other nutrients into amino acids, then proteins
  - by mammals and birds to maintain a steady body temperature
  - by animals to make muscles **contract**.

- Energy from respiration allows:
  - muscles to contract, enabling movement
  - mammals and birds to maintain their body temperature in a colder environment
  - all living things to make larger molecules from smaller ones. In animals, this includes glycogen (storage carbohydrate) and proteins – see diagram.

Small molecules join to make big ones.

Top Tip!

Do not confuse respiration and breathing. Respiration happens inside cells. Breathing is movement of air in and out of the lungs.

Removing waste: lungs

The lungs and gas exchange

- **Carbon dioxide** is a waste product of **respiration** in cells and must be removed. It **diffuses** from cells into the blood, and then diffuses from blood into the air sacs (alveoli) of our lungs.

- Breathing out (**expiration**) causes a pressure that forces the air, high in carbon dioxide, out of the lungs. Breathing in (**inspiration**) lowers the pressure and draws air into the lungs.

- Muscles for breathing include **intercostal muscles** between the ribs, which raise and lower them, and the **diaphragm**, which flattens as it contracts.

Structure of the lungs and muscles of breathing.

- To maximise carbon dioxide **diffusion** from bloodstream to lung cavity, **alveoli** have a rich blood supply, thin walls and a very large total surface area.

- In the opposite direction, **oxygen** from incoming air diffuses through alveolar walls into the bloodstream.

- The gases diffuse towards a region of lower concentration – they diffuse down a **concentration gradient**.

- The brain detects excess carbon dioxide in the blood. It sends nerve impulses to the intercostal and diaphragm muscles to make us breathe faster and more deeply.

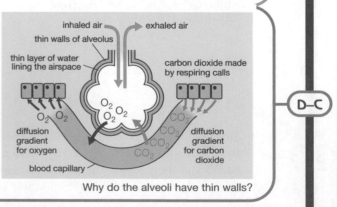

Why do the alveoli have thin walls?

Questions

(Grades G-E)

1 Without checking back, write the word equation for aerobic respiration.

(Grades D-C)

2 Why do mammals and birds need more energy than fish?

(Grades G-E)

3 Name the **two** sets of muscles involved in breathing.

(Grades D-C)

4 List **three** ways in which alveoli are adapted for quick gas exchange.

Removing waste: liver and kidneys

Producing urea and excreting urine

- During digestion, **protein** in the diet is broken down into **amino acids**. Cells then use these to build their own proteins.

- The **liver** converts any surplus amino acids into **urea**, a soluble substance which the blood transports to the **kidneys**.

- The kidneys filter waste substances from the blood, including urea, producing **urine** which flows to the **bladder** before being excreted.

- The body's largest organ, the **liver**, breaks down many surplus and toxic substances. It makes urea, which then travels to the kidneys.

- In the kidneys, tiny tubes called **nephrons** collect waste substances from the blood, including urea and water. The waste (urine) flows down the ureters and into the bladder.

Top Tip!

Remember: urea is made in the liver, not in the kidneys. The kidneys make *urine*, which contains urea dissolved in water.

How our bodies remove urea.

How urea is made.

The kidneys and bladder.

Homeostasis

Blood sugar, temperature and water

- To remain healthy, the body's **internal conditions** must be kept constant. This is called **homeostasis**.

- **Insulin**, secreted by the pancreas, enables cells to take in **blood sugar**, glucose, for respiration. Excess blood sugar can be fatal, and insulin also causes it to convert into glycogen.

- Enzymes in cells need a **temperature** of about 37 °C to work properly. When the body overheats, **sweating** cools it. When it is too cold, blood is re-routed deeper in the body.

- Cells gain **water** and **mineral ions** from food. Excess water could burst cells, and excess ions could be toxic, so the body regulates their concentration.

- The diagram on the right shows how the body balances its **water** gains and losses to avoid cell dehydration or damage by excess water.

- Water loss in urine differs in hot and cold weather, as the table shows.

How homeostasis works.

Water in, water out – it must balance.

Summer and winter mean daily urine output.

day	1	2	3	4	5
mean summer output (cm³)	995	1000	975	1005	990
mean winter output (cm³)	1325	1295	1310	1300	1305

Questions

Grades G-E
1 Which organ changes surplus amino acids into urea?

Grades D-C
2 What is a nephron?

Grades G-E
3 List **three** internal conditions that are kept constant.

Grades D-C
4 Why is more urine often produced on a cold day than on a hot day?

Keeping warm, staying cool

Keeping close to 37 °C

- Normal **core body temperature** is between 35.5 and 37.5 °C.

- The **thermoregulatory centre** is in the brain. It contains **temperature receptors** that monitor blood temperature.

- Also, temperature receptors in the **skin** send nerve impulses to the thermoregulatory centre.

- If the blood is too hot or too cold, the centre sends impulses to parts of the body to make them restore the temperature to normal.

- If a person gets much too cold, the thermoregulatory centre stops working and **hypothermia** results: they cannot warm up. At 25 °C, the person will die.

Temperature control is a homeostatic mechanism.

G–E

D–C

Treating diabetes

Glucose balance and insulin

- The **pancreas** monitors blood **glucose** and controls its level by producing the **hormone insulin**.

- Insulin allows cells, particularly muscle and liver cells, to absorb glucose and convert it to the storage material **glycogen**.

- It is life-threatening for blood glucose concentration to rise (or fall) too far.

- People with **diabetes** produce too little or no insulin. They may have to inject insulin regularly and follow a diet to control sugar intake.

- Diabetes symptoms include: frequent urinating and glucose in the urine, thirst, tiredness, weight loss and blurred vision.

Blood glucose, insulin and cell membranes.

G–E

The history of controlling diabetes

- Scientists linked the islets of Langerhans, groups of cells in the pancreas, to diabetes.

- In 1921, Canadians Banting and Best collected fluid from the islets of Langerhans in the pancreas of a healthy dog. They injected it into a dog with diabetes, which was restored to health.

- The effective substance was named **insulin**, and since then has kept countless people with diabetes alive.

How Science Works

You should be able to:

- evaluate the data from the experiments by Banting and Best which led to the discovery of insulin

- evaluate modern methods of treating diabetes.

D–C

Questions

Grades G-E
1 Where is the body's thermoregulatory centre?

Grades D-C
2 Explain why a person with hypothermia cannot get warm.

Grades G-E
3 Which organ secretes the hormone insulin?

Grades D-C
4 How did we discover what insulin does?

Cell division – mitosis

Making new body cells

- Cells divide in two to produce more cells, so that the body grows and replaces worn and dead cells.

- A human **body cell** has two sets of 23 **chromosomes**, one set from each parent, therefore 23 pairs totalling 46. (Sex cells, gametes, have only one set.) Chromosomes contain **genes** that control inherited characteristics.

- In cell division, chromosomes must be duplicated so that each new cell contains the identical two sets of chromosomes.

Stages in mitosis

- The process of making new body cells is **mitosis**, shown in the diagram.

- Some organisms reproduce **asexually**. They produce new cells by mitosis. The new cells grow into a new organism. It is genetically identical to its parent.

Top Tip!

Cells produced by mitosis have identical genes.

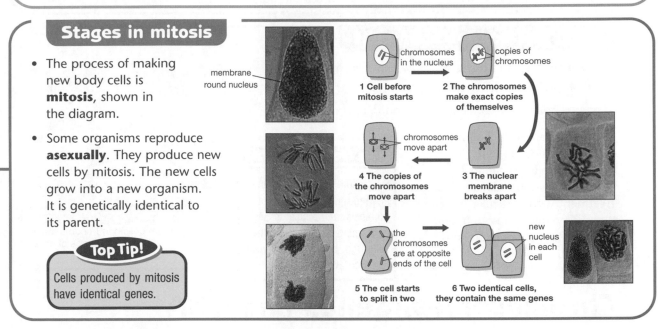

membrane round nucleus

1 Cell before mitosis starts

chromosomes in the nucleus

2 The chromosomes make exact copies of themselves

copies of chromosomes

chromosomes move apart

4 The copies of the chromosomes move apart

3 The nuclear membrane breaks apart

the chromosomes are at opposite ends of the cell

5 The cell starts to split in two

new nucleus in each cell

6 Two identical cells, they contain the same genes

Gametes and fertilisation

How gametes are formed

- **Gametes** are formed in the reproductive organs (**ovaries** and **testes**).

- A gamete must have only *half* the body cell number of chromosomes so that, when two gametes fuse in **fertilisation**, the cell formed contains the normal number.

- In humans, the new cell formed at fertilisation contains two sets of chromosomes, 23 from the sperm and 23 from the egg. The cell divides over and over again by mitosis, to form a new organism.

Importance of fertilisation

- Genes determine inherited characteristics. **Alleles** are different varieties of a gene. A person has two copies of each gene, one from each parent. Different alleles determine different variations of a particular characteristic.

- The alleles in sperm and egg are brought together at fertilisation. So an offspring inherits alleles, and hence characteristics, from both parents. The mixture of alleles is different from the mixture in either of the parents, so each offspring has a unique set of genes. This is one cause of **variation**.

Questions

(Grades G-E)
1 How many sets of chromosomes are there in a body cell?

(Grades D-C)
2 Name the kind of cell division that makes new cells for growth, repair and asexual reproduction.

(Grades G-E)
3 Why must sperm and eggs contain only one set of chromosomes each?

(Grades D-C)
4 Why are offspring produced by sexual reproduction not exactly like their parents?

Stem cells

Unspecialised, specialised and stem cells

- Some cells start by being **unspecialised**: they could develop into any type of cell. Then in **animals** they **differentiate** to do just one job – they become **specialised**.

- **Stem cells**, in embryos and certain parts of adults, are undifferentiated. They can divide by mitosis and develop into different cell types.

- Scientists hope to use stem cells to repair damaged body parts (e.g. nervous tissue).

- Many **plant** cells can differentiate into any type of cell throughout their life.

Adult and embryo stem cells

- Different stem cells in **bone marrow** can develop:
  - either into red or white blood cells
  - or into bone cells, cartilage cells, tendon cells or fat cells.

- Stem cells from 4- to 5-day-old **embryos** placed in **cell growth medium** can divide for months without differentiating. Treatment with particular proteins turns some genes on or off, and the cells then differentiate into different tissues.

How Science Works

You should be able to: make informed judgements about the social and ethical issues concerning the use of stem cells from embryos in medical research and treatments.

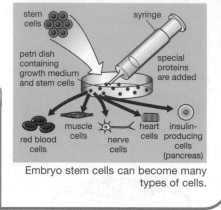

Embryo stem cells can become many types of cells.

Chromosomes, genes and DNA

Chromosomes, DNA and genes

- A **chromosome** is a very long molecule of **deoxyribonucleic acid** (**DNA**). The structure of DNA was discovered by Francis Crick and James Watson.

- There are 46 chromosomes in the **nucleus** of each human body cell.

- A gene is a tiny section of a chromosome. The genes contain the **genetic code**, the instructions for joining amino acids together to make proteins.

- Many proteins are **enzymes**. Thousands of enzymes control the chemical reactions of all the cells in our body, and thereby determine our characteristics.

A cell — cell membrane — nucleus

gene

chromosome

Expanding diagram to illustrate DNA.

How DNA copies itself in mitosis

- The diagram shows part of the DNA double helix with its complementary bases.

- Before **mitosis** begins (see page 85), the base pairs separate, unzipping the two strands. Complementary bases move to each single strand and attach themselves, making two double strands.

- The strands coil up to form two identical chromosomes, ready to separate into two new cells.

■ A
□ T
■ C
□ G

triplet of bases — single strand of unwinding double helix

The DNA double helix.

Questions

Grades G-E

1 What is a stem cell?

Grades D-C

2 Name **two** different kinds of cells that stem cells in bone marrow can produce.

Grades G-E

3 What is the chemical that chromosomes are made of?

Grades D-C

4 Why must DNA duplicate itself before mitosis begins?

Inheritance

Mendel's experiments on breeding pea plants

- Mendel repeated same-colour crosses of pea plants until the offspring had either red only flowers or white only flowers. These were **pure breeding** plants, and Mendel concluded that a 'factor' from each parent determined flower colour.

- The 'factor' is a **gene**. The **characteristic** of flower colour is controlled by a *single gene* which has two forms or **alleles** – red and white. A plant inherits one allele from each parent. Where both alleles are the same, the plant is **homozygous**. Where they are different, it is **heterozygous**. These rules apply to all characteristics controlled by a single gene with two alleles.

- Mendel then crossed homozygous red plants with homozygous white plants and all the offspring had red flowers.

This is a Punnett square. It shows how the genes in this cross are inherited.

Parents:	Rr		Rr	
Gametes:	ⓡ and ⓡ		ⓡ and ⓡ	

Offspring:

		ⓡ	ⓡ
	ⓡ	**RR** red flowers	**Rr** red flowers
	ⓡ	**Rr** red flowers	**rr** white flowers

- All these offspring inherited a white and a red allele. But red was **dominant**. White was **recessive** and would show only if red was absent.

- Mendel then crossed these red offspring with each other. Three-quarters of the offspring had red flowers and one-quarter had white flowers. He concluded that a characteristic may not show but can still pass to the next generation. The genetic diagram on the right shows what happened. **R** is the dominant allele for red flowers and **r** is the recessive allele for white flowers.

Top Tip!

Genetic diagrams do not tell you how many offspring two parents will have. They just tell you the chances of getting different combinations of alleles.

How is sex inherited?

Sex inheritance and sex ratio

- Human **body cells** contain 23 chromosome pairs. The pair of **sex chromosomes**, **X** and **Y**, carry the genes which determine sex.

- An egg or sperm contains one sex chromosome only.

- As the table shows, an embryo always inherits X from the egg. If the sperm supplies: an X, it will be a girl (XX); a Y, it will be a boy (XY).

How sex is determined.

sex chromosome in egg	sex chromosomes in sperm	sex chromosomes in embryo	sex of embryo
X	X	XX	female
X	Y	XY	male

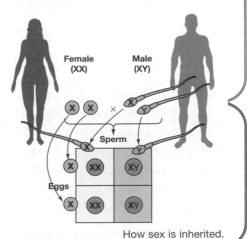

How sex is inherited.

- As the diagram shows, a sperm has an equal chance of containing the X or the Y chromosome. The sperm determines a child's sex, and all eggs have an equal chance of fusing with either X or Y.

- The same number of males as females would be expected. This is roughly the case.

Questions

Grades D-C

1 What does 'pure-breeding' mean?

Grades D-C

2 Explain what is meant by a 'recessive allele'.

Grades G-E

3 Which determines the sex of a baby – the sperm or the egg?

Grades D-C

4 Why are roughly equal numbers of boy babies and girl babies born?

Inherited disorders

Inherited disorders

- An **inherited disorder** prevents the body from working properly. The disorder lasts for life because chromosomes in all body cells carry the gene.

- A **recessive allele** of a certain gene causes **cystic fibrosis**, which affects cell membranes. Two healthy parents may be **carriers** and may each pass on a recessive allele to their child.

- A **dominant allele** causes **Huntington's disease** which affects the nervous system. A child will inherit Huntington's disease if it gets this allele from just one parent.

father (carrier) — mother (carrier)

child 1 (carrier) — child 2 (unaffected) — child 3 (CF) — child 4 (carrier)

Inheritance of cystic fibrosis.

father (with Huntington's disease) — mother (healthy)

child 1 — child 2 — child 3 — child 4 — child 5 — child 6

Inheritance of Huntington's disease.

G–E

Cystic fibrosis and its treatment

- A child with cystic fibrosis (CF) but who has healthy parents inherits the recessive allele from both parents. They are protected by the dominant (healthy) allele, but half their gametes will carry the recessive allele.

- A person with CF cannot digest food properly, has frequent chest infections, fertility problems and a shortened life.

- There is no cure, but treatments can minimise the symptoms and tackle infections.

- Embryos can be **screened** to check for this disorder.

How Science Works

You should be able to: make informed judgements about the economic, social and ethical issues concerning embryo screening that you have studied, or from information you have received.

D–C

DNA fingerprinting

What is a DNA fingerprint?

- All your body cells contain the same **DNA**. It differs from anyone else's except an identical twin – it is **unique** and cannot be changed.

- DNA (from e.g. saliva, hair, blood, semen) displayed on a sheet of film is called a **DNA** (or **genetic**) **fingerprint** or a **DNA profile**. It:
  - requires only a tiny sample
  - identifies one person only, alive or dead
  - can be used to identify close relations.

A good DNA fingerprint identifies just one person.

G–E

Making and using DNA fingerprints

- In **DNA fingerprinting**, DNA is cut into different lengths by enzymes. The bits are separated into bands using a technique like chromatography. The bands are treated with radioactive chemicals to make an image of them on X-ray film.

- In **solving crime**, DNA from the crime scene can be matched to DNA from a suspect.

- In **paternity disputes**, the number of bands shared by a man and a child shows whether he is the child's father.

D–C

Questions

(Grades G-E)
1 Give **two** examples of inherited disorders.

(Grades D-C)
2 How can two parents without CF have a child with CF?

(Grades G-E)
3 Why can a DNA fingerprint identify a single person?

(Grades D-C)
4 Name **two** uses of DNA fingerprinting.

B2b summary

Enzymes

Enzymes are **biological catalysts** that speed up chemical reactions and are not used up.

Enzymes are involved in **respiration**, **photosynthesis** and **protein synthesis**.

Each enzyme works at an **optimum pH** and **temperature**.

The digestive enzymes **amylases**, **lipases** and **proteases** break down food.

Enzymes are used in industry in **food production** and in the home in **biological detergents**.

Respiration and waste removal

Aerobic respiration is carried out in **mitochondria** in our cells to provide energy:
glucose + oxygen → carbon dioxide + water + ENERGY

Oxygen for respiration **diffuses** into the blood from the lungs. Carbon dioxide from respiration diffuses in the opposite direction, and is removed from the lungs when we breathe out. The dissolved gases are carried round the body in the blood.

Waste proteins are broken down by the **liver** into **urea**.

The **kidneys** make **urine**, which contains dissolved urea.
The urine is stored in the **bladder** then excreted.

Homeostasis

Blood sugar levels are controlled by **insulin** produced by the **pancreas**.
People with **diabetes** produce too little insulin or do not respond to it.

Human body **temperature** is kept constant at about 37 °C.
The **thermoregulatory** centre in the brain contains **temperature receptors**.

Sweating cools the body down.
Rerouting blood deeper in the body reduces heat loss.

Water inputs from drinking and cell respiration are balanced with water outputs from tears, evaporation from the lungs, sweat, urine and faeces.

Inheritance

Chromosomes are long molecules of **DNA** made up of **genes**.
Genes contain the **genetic code**, which contains the instructions for making proteins, including **enzymes**.

Body cells divide by **mitosis**. Each new cell has the same number of identical chromosomes as the original cell.

Gametes (sperm and egg) fuse at **fertilisation** to produce a new organism.
The mixture of **alleles** from the parents causes **variation** in the offspring.

Sex chromosomes determine the sex of the offspring (male XY, female XX).

Some inherited **characteristics** are controlled by a single gene.
Different forms of a gene are called **alleles**.

In **homozygous** individuals, the alleles inherited from each parent are the same.
In **heterozygous** individuals, they are different.

Alleles can be **recessive** or **dominant**.

Undifferentiated **stem cells** can **specialise** into many types of cells.

Atomic structure 2

Top Tip!
Remember the relative mass and charge of a proton, a neutron and an electron.

The structure of an atom

- An **atom** contains **sub-atomic particles** (see also page 26). Its small central **nucleus** contains **protons** and **neutrons**. Round the nucleus are **electrons**.
  - For the **relative masses** and **charges** of sub-atomic particles, see the table on page 26.

- In an atom, the number of electrons = the number of protons. It has no overall electrical charge.

G–E

- All atoms of an **element** have the same number of protons. Atoms of different elements have different numbers of protons.

- For information on the **symbol**, **mass number** and **atomic number** of an element, see page 26.

- The elements are arranged in the periodic table (see page 26) in order of atomic number.

D–C

23 **Na** mass number
11 atomic number
The element sodium.

Electronic structure

Top Tip!
Electrons fill the electron shells in order, starting with the shell closest to the nucleus.

The arrangement of electrons

- Electrons are arranged round a nucleus in **shells**. Each shell represents a different **energy level**. The shells closest to the nucleus have the lowest energy levels.

- An electron always takes up the lowest available energy level, in the innermost shell available.

- The first shell holds up to two electrons. The second shell holds up to eight electrons.

- Atoms with a full outer electron shell are stable and unreactive.

- When an atom reacts, it gains, loses or shares one or more electrons to achieve a full outer electron shell.

How Science Works
You should be able to: represent the electronic structure of the first 20 elements of the periodic table in the following forms.

For sodium:
and 2, 8, 1

one electron in the third shell
two electrons in the first shell
Na
eight electrons in the second shell

Arrangement of electrons in a sodium atom. Sodium reacts by losing **one** electron and so is in Group **1** of the periodic table.

G–E

Electron notation

- A **magnesium** atom has 12 electrons.

- Shells 1 and 2 contain the maximum two and eight electrons. The remaining two electrons occupy the third shell.

- The **electron notation** for a magnesium atom is 2, 8, 2.

- Magnesium reacts by losing the **two** electrons in its outermost shell, and so is in Group **2** of the periodic table.

- A **calcium** atom has 20 electrons. Its electron notation is 2, 8, 8, 2.

- Calcium reacts by losing its outermost **two** electrons, placing it in Group **2**.

two electrons in the third shell
two electrons in the first shell
Mg
eight electrons in the second shell

Arrangement of electrons in a magnesium atom.

D–C

Arrangement of electrons in a calcium atom.

eight electrons in the second shell
two electrons in the first shell
Ca
eight electrons in the third shell
two electrons in the fourth shell

Questions

Grades G-E
1 State the charges on each of these sub-atomic particles: a proton, a neutron, an electron.

Grades D-C
2 Look at the periodic table on page 26. What is the mass number of potassium? What does this tell you about the structure of a potassium atom?

Grades G-E
3 A potassium atom has 19 electrons. The inner shell contains two electrons, the next shell contains eight and the next shell also contains eight. How many electrons are there in the outer shell of a potassium atom?

Grades D-C
4 How can a potassium atom become stable?

Mass number and isotopes

Elements, groups and relative atomic mass

G–E

- Atoms of an **element** all have the same number of **protons** and **electrons**. An atom has no net charge. **Isotopes** of an element have different numbers of neutrons (see also page 127).

- The **periodic table** arranges the elements in order of **atomic number**.

- A **group** contains elements with the same number of electrons in their outermost shell (the highest energy level). This determines the reactions an element can undergo.

D–C

- The **mass number** of an element is the sum of protons and neutrons.

- For most elements, the mass number is same as the element's **relative atomic mass** or A_r for short.

Ionic bonding

Ion formation and ionic compounds

G–E

- To form **ionic compounds**, the atoms of reacting elements either lose electrons and become **positively** charged **ions**, or gain electrons and become **negatively** charged ions.

- An **ionic bond** is the force of attraction between oppositely charged ions. Ionic bonds hold ionic compounds together strongly, making them solids at room temperature.

- All ions achieve a full outermost electron shell, giving the electronic structure of a noble gas (Group 0).

- **Metal** atoms *lose* electrons to form positive ions.

- **Non-metal** atoms *gain* electrons to form negative ions.

D–C

- Diagram 1 shows the formation of ions when magnesium metal reacts with oxygen gas to form the solid, magnesium oxide.

- Diagram 2 shows the formation of ions when calcium metal is heated in chlorine gas to form the solid, calcium chloride.

When a sodium atom becomes a sodium ion, its electron arrangement changes from 2, 8, 1 to $[2, 8]^+$. When a chlorine atom becomes a chloride ion, its electron arrangement changes from 2, 8, 7 to $[2, 8, 8]^-$.

How Science Works

You should be able to: represent the electronic structure of the ions in sodium chloride, magnesium oxide and calcium chloride in the following forms.

For sodium ion (Na^+): and $[2, 8]^+$

1 When magnesium and oxygen react, a magnesium atom loses two electrons to form an ion with a 2+ charge, and an oxygen atom gains two electrons to form an ion with a 2– charge.

2 When calcium and chlorine react, a calcium atom loses two electrons to form an ion with a 2+ charge, and *each of two* chlorine atoms gains one electron to form an ion with a 1– charge.

Questions

Grades G-E

1 A lithium atom has three protons. How many electrons does it have? What is its charge?

Grades D-C

2 What do all the elements in the same group in the periodic table have in common?

Grades G-E

3 A potassium atom has one electron in its outer shell. A chlorine atom has seven electrons in its outer shell. How can each of these atoms become stable?

Grades D-C

4 Explain why the *overall* charge of the ions in an ionic compound is zero.

Ionic compounds

The structure and properties of ionic compounds

- An **ionic compound** contains **positive metal ions** and **negative non-metal ions**.

- Electrostatic **forces of attraction** form strong bonds between oppositely charged ions. (See also page 27.)

- Ionic compounds:
  - are solids at normal temperatures
  - do not conduct electricity as solids
  - have high melting points and boiling points.
  - conduct electricity when molten
  - conduct electricity when dissolved (in water)

G–E

- Ionic compounds are **giant ionic structures**.

- At normal temperatures the strong electrostatic forces of attraction act in all directions, fixing the positions of oppositely charged ions in an ordered **lattice**.

- An electric current is the movement of charged particles. Ionic compounds can **conduct electricity** when **molten** because positive and negative ions are then free to move independently.

- Many ionic compounds are soluble in water. Then, the ions can move independently. This enables the **solution** to conduct electricity.

positive ion

negative ion

Lattice structure of an ionic compound. Ionic bonds fix each ion in position next to oppositely charged ions only.

D–C

How Science Works

You should be able to: suggest the type of structure of a substance, given its properties.

Covalent bonding

Non-metal atoms in compounds

- A **covalent bond** is formed when two atoms **share** electrons. Covalent bonds are very strong.

- Covalent bonds hold atoms together in **simple molecules** such as hydrogen, chlorine, oxygen, hydrogen chloride, water and ammonia.

The atoms in a hydrogen molecule are held together by a covalent bond. The molecule is represented as H_2 or H–H.

G–E

Covalently bonded simple molecules

- The first electron shell is **stable** when it contains two electrons, and the second when it contains eight electrons.

- Atoms in covalent compounds share electrons to achieve stable outermost shells of electrons.

An oxygen atom has 6 electrons in its outermost shell. To make 8, each oxygen atom shares 2 electrons with the other oxygen atom, making a **double** bond.

A molecule of oxygen, O_2, O=O.

An oxygen atom shares 1 electron from each hydrogen atom to make 8, and each hydrogen atom shares 1 electron from oxygen to make 2.

A molecule of water, H_2O, H–O–H.

How Science Works

You should be able to: represent the covalent bonds in molecules such as water, ammonia, hydrogen, hydrogen chloride, chlorine, methane and oxygen.

For ammonia (NH_3):

and/or

D–C

Questions

Grades G-E

1 Name **two** states in which an ionic compound can conduct electricity.

Grades D-C

2 Explain what is meant by a 'giant ionic structure'.

Grades G-E

3 Name the kind of bond that is formed when two atoms share a pair of electrons.

Grades D-C

4 Explain why two oxygen atoms can both become stable if they share two pairs of electrons.

Simple molecules

Simple molecules and their properties

- **Simple molecules** contain a few atoms joined by strong **covalent bonds** (see page 92).

- **Elements** include:
  - hydrogen, H_2, H—H
  - oxygen, O_2, O=O
  - chlorine, Cl_2, Cl—Cl

- **Compounds** include:
  - water, H_2O, H—O—H
  - ammonia, NH_3, H—N—H with H below
  - hydrogen chloride, HCl, H—Cl
  - methane, CH_4, H—C—H with H above and below

Top Tip!

Simple molecules consist of small numbers of atoms held together by shared pairs of electrons forming covalent bonds.

- **Properties** of simple molecular compounds:
  - they do not conduct electricity
  - they have low melting and boiling points
  - they are often liquids or gases at room temperature.
  - when solid, they are brittle and often soft or waxy
  - they are often insoluble in water but may dissolve in other liquids

- Covalently bonded simple molecules have no charged particles. Therefore they do not conduct electricity.

Giant covalent structures

Giant covalent structures and their properties

- **Giant covalent structures** are macromolecules. Strong covalent bonds join many atoms arranged in an orderly, three-dimensional **lattice**.

- These macromolecules have high melting points. Many are very hard. They do not conduct electricity or dissolve in water.

- Examples include the element carbon in the form of **diamond** and **graphite** and the compound **silicon dioxide**.

Top Tip!

In general, covalent bonds are just as strong as ionic bonds.

Diamond and silicon dioxide

How Science Works

You should be able to: represent the covalent bonds in giant structures, e.g. diamond and silicon dioxide.

- **Diamond** is a form of the **element** carbon and is the hardest **mineral**. Very strong covalent bonds join each carbon to four others, which is what makes diamond so hard.

A small part of the diamond lattice. All carbon atoms are strongly bonded to four others.

- **Graphite** is also a form of carbon. In each atom, three electrons form covalent bonds with other carbon atoms, giving a layered arrangement. The layers slide over each other easily, so graphite is soft and slippery.

Graphite has free electrons between layers.

- The **compound silicon dioxide** is sand. It is very hard and has a high melting point. It resists weathering because its silicon-oxygen covalent bonds are very strong.

A small part of the silicon dioxide lattice.

oxygen atom   silicon atom

Questions

Grades G-E

1 Name **two** elements that exist as simple molecules.

Grades D-C

2 Explain why compounds made of molecules do not conduct electricity.

Grades G-E

3 Give **one** example of an element that can form a giant covalent structure.

Grades D-C

4 Explain why graphite is soft.

Metals

How the structure of metals relates to their properties

- **Metals** have few electrons in their outer shells. They are lost when metals form **positive** ions.

- Metals can be bent and shaped because the layers of atoms easily slide over each other.

- Heavier metals have high densities and strength, and high melting and boiling points.

- The properties of metals determine the many uses that metals are put to.

- When a potential difference (voltage) is placed across some metal, the delocalised electrons move, carrying the electrical **charge** through the structure.

- Heating increases the kinetic energy of the delocalised electrons in the metal. They move faster, transferring **thermal energy** to positive ions. In this way, heat moves rapidly through the metal by conduction.

An atom of the metal element magnesium. It loses the two outer electrons when it reacts.

delocalised electrons move through the metal

In a circuit, delocalised electrons move in one direction to carry charge to the positive terminal of a cell.

Alkali metals

Group 1 – the alkali metals

- The elements of **Group 1** are the **alkali metals**.

- Alkali metals are very reactive. All react vigorously with water. Reactivity increases down the group.

- In reactions, they lose one outer electron to form **ions** with a 1+ charge.

- They react with non-metals to form **salts** which are **ionic compounds**.

The Group 1 metals are on the far left of the periodic table.

good electrical conductors — good thermal conductors — they are soft — properties of Group 1 metals — they have low melting points — although shiny when freshly cut the surfaces quickly **tarnish** in air

Properties of Group 1 metals.

How alkali metals react down Group 1

- When an alkali metal loses one electron, the ion is left with the full outer shell of a noble gas.

A lithium ion has the configuration of the noble gas helium and a charge of 1+.

loses an electron

lithium atom

to form a lithium ion

Questions

1 How do metals form ions?

2 Explain why metals can be bent and shaped.

3 How many electrons are there in the outer shell of the Group 1 metals?

4 Which is more reactive – sodium or rubidium? Explain your answer.

Halogens

Group 7 – the halogens

- The elements of Group 7 are the halogens:
  - **fluorine**, pale yellow gas
  - **chlorine**, green gas
  - **bromine**, dark red liquid
  - **iodine**, dark grey crystalline solid
  - **astatine**, highly radioactive solid.

- All exist as molecules with two covalently bonded atoms. (See also page 92.)

- On heating, solid **iodine** sublimes to give a purple vapour.

- With alkali metals, halogens form **salts** which are **ionic compounds**.

- Halogens have seven electrons in their outermost shell. They react by gaining one electron to acquire the electron structure of a noble gas.

- Sodium metal burnt in chlorine gas forms the **ionic compound** sodium chloride: **sodium + chlorine → sodium chloride**

- Halogen ions are called **halide** ions. They have a 1– charge.

- Fluorine is the halogen with the smallest atoms and is the most reactive. Reactivity decreases *down* Group 7.

When chlorine atoms react they gain an electron to form a chloride ion which has a 1– charge and the electron structure of noble gas argon.

Nanoparticles

What is a nanoparticle?

- A **nanoparticle** is a tiny manufactured structure of a few hundred atoms, from 1 to about 100 nanometres long. (A human hair is 100 000 nanometres wide.)

- The atoms in nanoparticles are regularly arranged in hollow structures such as tubes and spheres one atom thick.

- Being so small, nanoparticles have a very high **surface area to volume** ratio.

- Nanoparticles are likely to find uses in computers, as catalysts, sensors and coatings, and as new materials.

Scientists have used test tubes like this nanoparticle to catalyse new reactions.

Nanoscience using nanoparticles

- Scientists expect to make **materials** from nanoparticles that are harder, lighter and stronger than any known materials.

- Nanoparticles may make good **biosensors**, detecting very low levels of chemicals and biological agents.

- Nanoparticles could be designed to **process information** in computers millions of times faster than current components.

- Nanoparticle **catalysts** are expected to speed up manufacturing processes.

Questions

Grades G-E
1 What are the halogens?

Grades D-C
2 How do the halogens form ions?

Grades G-E
3 Explain what is meant by 'nanoparticles'.

Grades D-C
4 How might nanoparticles be used to speed up chemical reactions?

Smart materials

Smart materials

- A **smart material** has a particular **property**: when there is a change in the **environment**, it **responds** by changing in some way that is **reversible**.

- Smart materials include alloys that have shape memory (see page 33), photochromic materials, thermochromic materials and electroluminescent materials.

G–E

Types and uses of smart materials

- **Photochromic** materials darken on exposure to strong light and are used in spectacles and sunglasses.

- **Thermochromic** materials respond to changes in **temperature** by changing **colour**. These materials are black at most temperatures but change colour over a particular range of temperatures. They are used in paper thermometers for use on children's foreheads.

- **Electroluminescent** materials emit light of different colours when an alternating current passes through them. These materials form a layer on the plastic sheeting of some advertising signs.

How Science Works

You should be able to:
- relate the properties of substances to their uses
- evaluate the developments and applications of new materials, e.g. nanomaterials and smart materials.

D–C

Compounds

Compounds and ratios of elements in them

- A **compound** contains atoms of more than one **element**.

- Atoms of two or more elements react to form a **compound**. The elements are **chemically combined** and are joined by **chemical bonds**.

- **Glucose** has a large **covalently bonded** molecule with the **chemical formula** $C_6H_{12}O_6$. The letters indicate the elements in glucose, and the numbers show how many atoms of each element there are. The **ratio** of elements is: $6:12:6 = 1:2:1$.

- Salt, **sodium chloride**, has a **giant ionic structure** of formula NaCl. Its sodium ions (Na^+) and chloride ions (Cl^-) are in the **ratio** 1:1.

This shows the way that atoms are arranged in one molecule of glucose. The lines between the atoms represent chemical bonds.

Each grain of salt contains a huge number of sodium and chloride ions.

G–E

Mixtures and compounds

- A **mixture** is composed of different substances that are *not* chemically combined. The substances can be mixed in any proportion. They are not in fixed ratios. They can be separated.

- In a **compound**, elements are chemically joined in particular, fixed ratios.

In water, hydrogen and oxygen are chemically combined in the ratio of 2:1.

Substances in this mixture are not chemically combined or in fixed ratios.

D–C

Questions

Grades G-E
1 Explain what is meant by a 'smart material'.

Grades D-C
2 Explain how a photochromic material behaves when exposed to light.

Grades G-E
3 The chemical formula of methane is CH_4. What does this tell you about the numbers and types of atoms in a methane molecule?

Grades D-C
4 Explain the difference between a 'compound' and a 'mixture'.

Percentage composition

Relative formula mass and percentage of elements in compounds

G–E

- Each element has a **relative atomic mass** (see page 91).
- To find the **relative formula mass** (M_r) of a molecule or compound, the relative atomic masses of all the atoms are added together in the **numbers** shown by the formula (see page 96).

The formula of an oxygen **molecule** = O_2
Relative atomic mass of an oxygen atom = 16

Relative formula mass of O_2 = 2 × 16 = 32

- The formula used to calculate the **percentage mass** of an **element** in a **compound** is:

$$\text{% mass of an element in a compound} = \frac{\text{relative atomic mass of the element} \times \text{number of atoms of element in the formula}}{\text{relative formula mass}} \times 100$$

D–C

- The formula of the **compound** potassium nitrate is KNO_3.

Relative atomic masses: K = 39, N = 14, O = 16
Relative formula mass of KNO_3 = 39 + 14 + (3 × 16) = 101
% mass of potassium in potassium nitrate = $\frac{39 \times 1}{101} \times 100 = 38.6\%$

How Science Works

You should be able to: calculate chemical quantities involving formula mass (M_r) and percentages of elements in compounds.

Moles

What is a mole?

G–E

- Atoms and molecules are so tiny that their masses cannot be measured individually. Instead, we use an amount called the mole:

a mole of a substance is the relative formula mass in grams of that substance

- There is the same number of particles in 1 mole of any substance.

- Formula of the compound water = H_2O

Relative formula mass of H_2O = (2 × 1) + (1 × 16) = 18
So the mass of 1 mole of water molecules is 18 g

The mass of 1 mole of water molecules is 18 g.

$$H_2O$$
(2 x 1) + (1 x 16)
$$= 18$$

- Water is formed when hydrogen is burnt in air (oxygen):

hydrogen + oxygen → water
The formula for hydrogen is H_2 and for oxygen it is O_2. Water is H_2O.
The balanced equation is: $2H_2 + O_2 \rightarrow 2H_2O$
This shows that 2 moles of hydrogen molecules react with 1 mole of oxygen molecules.

Top Tip!

The particles in a mole of a substance can be atoms or molecules (or ions).

D–C

- Magnesium oxide is formed when magnesium metal burns:

magnesium + oxygen → magnesium oxide
Magnesium oxide is MgO.
The balanced equation for the reaction of magnesium and oxygen is: $2Mg + O_2 \rightarrow 2MgO$
This shows that 1 mole of oxygen molecules reacts with 2 moles of magnesium atoms.

Top Tip!

The percentage composition is calculated by mass.

Questions

Grades G-E

1 The formula of methane is CH_4. The relative atomic mass of carbon is 12. The relative atomic mass of hydrogen is 1. Calculate the relative formula mass of methane.

Grades D-C

2 Calculate the per cent by mass of carbon in methane.

Grades G-E

3 What is the mass of 1 mole of methane? (Hint – look at your answer to question 1.)

Grades D-C

4 The equation for the reaction between carbon and oxygen is: $C + O_2 \rightarrow CO_2$. How many moles of carbon atoms react with each mole of oxygen molecules?

Yield of product

The amount of product obtained in a reaction

- No atoms are gained or lost in a chemical **reaction**. Yet the **reactants** that take part in the reaction do not always give as much **product** as expected. In other words, the **actual** amount can be less than the **theoretical** amount. The diagram shows some reasons for this.

Why the actual amount of product in a reaction may be less than the theoretical amount.

the reaction is reversible and does not go to completion

why the actual amount of product may be less than the theoretical amount

some of the product may be lost when it is separated

some of the reactants may react in an unexpected way

G–E

Theoretical yield and actual yield

- The maximum amount of product from a reaction is the **theoretical yield**. The amount that actually forms is the **actual yield**.

- To work out the cost of making a chemical, a manufacturer first calculates how much product to expect from a reaction process.

- A reaction is said to have a high **atom economy** if a high proportion of the reactants end up as useful products.

- It is important for **sustainable development** and to limit costs to use reactions with a high atom economy.

D–C

Reversible reactions

Examples of reversible reactions

- In some chemical reactions, the **products** can themselves react to produce the original **reactants**. Such reactions go **forwards** to give the products, and in **reverse** to give the reactants again. They are **reversible reactions**, represented as:

A + B ⇌ C + D

where A and B are reactants and C and D are products.

solid ammonium chloride — funnel — beaker — solid ammonium chloride — HEAT

The reversible reaction of ammonium chloride.

- When heated, the white solid **ammonium chloride** forms the gases **ammonia** and **hydrogen chloride**:

ammonium chloride ⇌ ammonia + hydrogen chloride

When the gases cool down, they react to re-form ammonium chloride.

Top Tip!

You need to be able to explain what a reversible reaction is.

- Copper sulfate reacts with water to give blue **hydrated copper sulfate**. When it is heated, the water is lost and it becomes white **anhydrous copper sulfate**. This reaction is reversible:

hydrated copper sulfate ⇌ anhydrous copper sulfate + water

- **Anhydrous cobalt chloride** is blue. It reacts with water to form pink **hydrated cobalt chloride**. This reaction is used to test whether water is present, and can be reversed by heating:

hydrated cobalt chloride ⇌ anhydrous cobalt chloride + water

G–E

D–C

Questions

Grades G-E

1 What can you say about the total number of atoms in the reactants and the products of a chemical reaction?

Grades D-C

2 State **two** reasons why the quantity of a product from a reaction may not be as great as expected.

Grades G-E

3 The breakdown of ammonium chloride to ammonia and hydrogen chloride is a reversible reaction. How can you show this when you write the equation for the reaction?

Grades D-C

4 Nitrogen and hydrogen can react to produce ammonia. This is a reversible reaction. Write the word equation for this reaction.

Equilibrium 1

Reversible reactions

- The **reversible reaction** between nitrogen and hydrogen forms ammonia:
  nitrogen + hydrogen ⇌ ammonia

- A reversible reaction does not go to **completion**. The **forward reaction** and the **reverse** reaction take place at the same time.

The effects of reaction conditions on reversible reactions

- **Reaction conditions**, especially pressure and temperature, affect the yield of product in a reversible reaction.

- **Removing products** increases the forwards reaction.

- **Catalysts** increase the rate of a reversible reaction. They do *not* affect the amount of reactants and products, but will shorten the time taken for the reaction to reach a balance.

Haber process

The Haber process for making ammonia

- Ammonia is used to make fertilisers. Ammonia is produced by the **Haber process** in the reaction:
  nitrogen + hydrogen ⇌ ammonia

- Nitrogen is obtained from the air and hydrogen comes from natural gas.

- The **conditions** used in the Haber process are:
  – an iron catalyst
  – a pressure of about 200 atmospheres
  – a temperature of about 450 °C.

- The process is continuous. The ammonia is **liquefied** by cooling and then removed. Removing the product drives the reaction forwards.

An ammonia molecule contains three hydrogen atoms and one nitrogen atom.

The Haber process.

Questions

1 When hydrogen and nitrogen react to form ammonia, we say that the reaction 'does not go to completion'. What does this mean?

2 What effect does a catalyst have on the amount of product formed in a reversible reaction?

3 What is produced by the Haber process?

4 Name the catalyst that is used in the Haber process.

C2a summary

An **atom** contains **sub-atomic particles**:
- a central **nucleus** of **protons** and **neutrons**
- **electrons**, which orbit the nucleus.

An element's **mass number** = total number of **protons** + total number of **neutrons** per atom. An element's **atomic number** = number of **protons** in an atom.

Sub-atomic particles

Electrons are arranged around the nucleus in **shells**, which represent **energy levels**: shell 1 has 2 electrons; shell 2 has 8 electrons.

Electron notation shows the **arrangement** of electrons in the shells in an element, e.g. the notation for magnesium is 2, 8, 2.

Ionic bonding is the attraction between **oppositely charged ions**.

Metal atoms **lose** electrons to form **positive** ions. **Non-metal** atoms **gain** electrons to form **negative** ions.

Ionic compounds:
- are giant ionic **lattices**
- are solids at room temperature (high m.p.)
- conduct electricity when molten or dissolved.

An **alkali metal** (Group 1 of the periodic table) reacts losing one electron to form a positive ion. With non-metals (e.g. halogens), alkali metals form **salts** (ionic compounds).

A **nanoparticle** is a tiny structure made with special properties due to the precise way in which its atoms are arranged.
Nanoparticles are used as **biosensors** and may process information, act as catalysts and provide new, very hard, light materials.

Structure, properties and uses

Non-metal atoms can **share** pairs of electrons to form **covalent bonds**.

Simple molecular compounds (e.g. water, methane) with covalent bonds:
- do not conduct electricity
- have low melting and boiling points
- are often gases at room temperature
- when solid, are brittle or waxy
- are often insoluble in water.

A **halogen** (Group 7 of the periodic table) exists as a **molecule** of two covalently bonded atoms.

Giant covalent structures (e.g. diamond, silicon dioxide) are macromolecules that:
- have very high melting points
- are very hard
- do not conduct electricity.

Smart materials have particular properties:
- **photochromic** materials react to light
- **thermochromic** materials respond to changes in temperature
- **electroluminescent** materials emit light when an alternating current passes through them.

The **relative formula mass** of a compound or molecule = sum of **relative atomic masses** of all atoms in its **formula**.

A **mole** of a substance is the **relative formula mass** in **grams** of that substance.

Percentage yield can be used to compare the **actual yield** of a **chemical reaction** with the **theoretical yield**.

Composition, yield and equilibrium

High **atom economy** is achieved if a high proportion of the reactants end up as useful products. This is important for **sustainable development** and to limit costs.

Reversible reactions can proceed in both directions. They are affected by: **pressure** and **temperature**, removal of products and presence of a **catalyst**.
Reversible reactions carried out in a '**closed system**' will eventually reach **equilibrium**.

The **Haber process** is a reversible reaction:
nitrogen + hydrogen ⇌ ammonia

The **percentage mass** of an element is given by:

$$\% \text{ mass of an element in a compound} = \frac{\text{relative atomic mass of the element} \times \text{no. of atoms of element in the formula}}{\text{relative formula mass}} \times 100$$

Rates of reactions

Measuring the rates of reactions

- Magnesium ribbon reacts with hydrochloric acid, forming magnesium chloride and hydrogen gas.

- The speed of this reaction can be measured by finding:
  - the time taken for the magnesium to dissolve
  - the speed at which hydrogen bubbles form.

- To speed up the reaction, we could:
  - increase the **temperature** or **concentration** of the acid
  - use **powdered** metal
  - or use a **catalyst**.

Magnesium + hydrochloric acid → magnesium chloride + hydrogen.

- The **rate** (speed) **of a chemical reaction** can be measured as the amount of:
  - **reactant** used up in a set period of time:

    rate of reaction = $\dfrac{\text{amount of reactant used up}}{\text{time}}$

  - **product** formed in a set period of time:

    rate of reaction = $\dfrac{\text{amount of product formed}}{\text{time}}$

- The rate of reaction between magnesium and an acid can be measured by finding the volume of gas formed (cm^3) in the time taken (seconds). The faster the reaction, the faster the gas fills the syringe.

gas syringe

acid
magnesium ribbon

Measuring the volume of gas formed using a gas syringe.

Following the rate of reaction

Methods of measuring reaction rate

- To follow the reaction between magnesium ribbon and hydrochloric acid, we can measure:
  - how quickly it makes hydrogen gas
  - how quickly magnesium loses mass.
  - how long magnesium takes to dissolve

- If a **product** of a reaction is a **gas**,

  (**1**) measure its **volume** by:
  - collecting the gas in an upturned burette or measuring cylinder
  - collecting the gas in a syringe
  - counting bubbles of gas fed into a test tube of water

  (**2**) measure the **mass** of gas lost by finding the mass of the reacting mixture.

- In all these methods, several measurements can be made at timed intervals.

- If a **reactant** is a **solid** that **dissolves**, we can make just one measurement, of the time it takes to disappear.

- For two solutions reacting to form a solid that clouds the solution, measure the time for a cross beneath a flask to disappear. Alternatively measure several times using a sensor and **datalogger**.

Questions

(Grades G-E)

1 State **three** ways in which you could *slow down* the rate of reaction between a piece of magnesium and hydrochloric acid.

(Grades D-C)

2 When following reaction rates, what is a gas syringe used for?

(Grades G-E)

3 A conical flask containing a piece of magnesium and some hydrochloric acid was placed on a top pan balance. The reading on the balance gradually got less. Why did this happen?

(Grades D-C)

4 To find out the rate at which this magnesium ribbon was reacting with the hydrochloric acid in the conical flask, what else would you need to measure?

Collision theory

Increasing the rate of collisions

- A chemical reaction occurs when reacting particles **collide** with **sufficient energy**.

- The minimum energy needed for a reaction to take place is called **activation energy**.

- **Successful collisions** are more likely when there is an *increase* in:
  - the **concentration** of solutions
  - the **pressure** on reacting gases
  - the **temperature** of reactants
  - the **surface area** of reacting solids

  and when there is a **catalyst** for the reaction.

G–E

key
- ○ reactant 1
- ▪ reactant 2
- ⌕ successful reaction

- In a solution of two reactants, doubling the concentration of *one* reactant will double the **frequency** of **successful collisions** with the other reactant.

- The frequency of successful collisions increases when:
  - **pressure** is increased on reactants that are **gases**, because this brings particles closer
  - **temperature** rises, increasing the **kinetic energy** of particles, so they collide more frequently, with greater energy
  - **solids** are broken up (e.g. crushed or powdered), to increase the **surface area** for collisions
  - a **catalyst** is added that weakens bonds in reacting particles.

Beaker A   Beaker B

The concentration of reactant 2 (▪) has been doubled in Beaker B.

D–C

Heating things up

Temperature and reaction rate

conical flask
sodium thiosulfate and hydrochloric acid
piece of paper with an X on it

Reacting sodium thiosulfate with hydrochloric acid.

- Sodium thiosulfate and hydrochloric acid react to form products that include **solid** sulfur:

  sodium thiosulfate + hydrochloric acid →
  sulfur + sulfur dioxide + sodium chloride + water

  Sulfur clouds the water and it eventually looks opaque, obscuring the cross under the flask, as shown in the diagram.

G–E

- The reaction is repeated with exactly the same reactants but at different temperatures, at 10 °C intervals. The time taken to obscure the cross is recorded for each temperature.

- **Trial experiments** for this reaction show:
  - volumes and concentrations that give a sensible reaction time
  - how to ensure that all **factors** except *temperature* are constant.

- The **graph** for this reaction shows that:
  - temperature increase causes reaction *time* to decrease
  - each 10 °C increase halves reaction time
  - the temperature change 20–30 °C has more effect than 30–40 °C
  - results form a perfect curve, so the results are reliable.

Graph to show the time for a cross to disappear at different temperatures when sodium thiosulfate and hydrochloric acid react.

D–C

Questions

1 Why does increasing the concentration of a reactant increase the rate of a reaction?

2 If you double the concentration of hydrochloric acid in which magnesium is reacting, by how much would this increase the frequency of collisions between particles of reactants?

3 Explain why a cross under a flask becomes invisible when sodium thiosulfate reacts with hydrochloric acid.

4 Use the graph to calculate how much faster the reaction was at 30 °C than at 20 °C.

Grind it up, speed it up

Surface area and reaction rate

G–E

- Two sizes of marble chips (calcium carbonate) and hydrochloric acid react to form products that include carbon dioxide **gas**:

calcium carbonate + hydrochloric acid → carbon dioxide + calcium chloride + water

As carbon dioxide bubbles off, the mass of the reaction mixture falls. Mass is measured at 30-second intervals.

D–C

- You can use **trial experiments** to find out:
  - a mass of marble and a volume and concentration of acid for a reaction lasting about 10 minutes
  - how much acid to use so that all the magnesium reacts
  - how to ensure all **factors** except the *size* of the marble chips are constant.

- The **graph** shows that: the total volume of gas is the same in both experiments; the reaction rate is highest at the start.

- We can use the 1-minute reading to find the **initial reaction rate** as *mass* of gas per minute:

rate of reaction = $\dfrac{\text{amount of product formed}}{\text{time}}$

- The rate of reaction is greater for the smaller chips than the larger chips.

cotton wool bung — conical flask

hydrochloric acid and large marble chips

hydrochloric acid and small marble chips

Finding the rate of loss in mass when marble chips of two different sizes react with acid.

Graph to show the mass of carbon dioxide gas lost when acid reacts with marble chips of two different sizes.

Concentrate now

Concentration and reaction rate

G–E

- Magnesium reacts with sulfuric acid to form products that include hydrogen gas which can be collected:

magnesium + sulfuric acid → hydrogen + magnesium sulfate

An upturned measuring cylinder collects the hydrogen.

- Gas volume is recorded at 1-minute intervals. The reaction is repeated for five concentrations of acid, with all other conditions constant.

D–C

- **Trial experiments** can be used to show:
  - masses of magnesium and concentrations of sulfuric acid that give sensible reaction times
  - how to ensure all **factors** except acid *concentration* are constant.

- The **graph** shows that: increase in acid concentration causes rate of reaction to increase; the reaction rate is highest at the start.

upturned measuring cylinder

delivery tube

HCl

Mg

Measuring the volume of gas formed when magnesium reacts with different concentrations of acid.

Graph to show the volume of hydrogen formed with five different concentrations of sulfuric acid.

- 0.1 mol/dm³
- 0.2 mol/dm³
- 0.3 mol/dm³
- 0.4 mol/dm³
- 0.5 mol/dm³

Questions

Grades G-E

1 Marble reacts with hydrochloric acid to produce hydrogen. Which would produce hydrogen faster – lumps of marble or tiny marble chips (assuming the total mass of marble is the same)?

Grades D-C

2 Explain why measuring the mass of gas lost after 5 min would not give you useful information about the rate of the reaction between hydrochloric acid and marble chips.

Grades G-E

3 Name the gas produced when magnesium reacts with hydrochloric acid.

Grades D-C

4 Explain why using 2 mol/dm³ acid rather than 1 mol/dm³ of acid would double the rate of reaction.

Catalysts

Effect and behaviour of catalysts

- Catalysts speed up reaction rates, or allow reactions to take place that would otherwise not happen. Catalysts are not used up in reactions. Different reactions have different catalysts.

- Hydrogen peroxide (H_2O_2) decomposes slowly to water and oxygen, but very quickly with manganese(IV) oxide used as a catalyst.

  hydrogen peroxide → water + oxygen

- Burning petrol produces the poisonous gas carbon monoxide, and oxides of nitrogen that contribute to **acid rain**. **Catalytic converters** make them react to produce less harmful gases:

  carbon monoxide + nitrogen dioxide → carbon dioxide + nitrogen

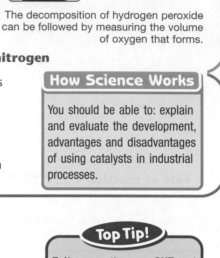

gas syringe

hydrogen peroxide

manganese(IV) oxide

The decomposition of hydrogen peroxide can be followed by measuring the volume of oxygen that forms.

- Weighing manganese(IV) oxide before and after the reaction proves that it is not used up when hydrogen peroxide decomposes.

- Catalysts include **transition metals** (e.g. manganese) and their compounds.

- In the **Haber process**, an **iron** catalyst helps nitrogen and hydrogen gases to react to form ammonia gas.

How Science Works

You should be able to: explain and evaluate the development, advantages and disadvantages of using catalysts in industrial processes.

Energy changes

Exothermic and endothermic reactions

Top Tip!

Exit means the way **OUT** and Exothermic reactions give **OUT** energy. Endothermic is the opposite.

- In a reaction, **energy** is either given out (transferred *to* the surroundings), or taken in (transferred *from* the surroundings).

- During **exothermic** reactions, energy is given out, often as heat. Examples include all **combustion** reactions, many **oxidations**, and **neutralisations** (e.g. hydrochloric acid + sodium hydroxide → sodium chloride + water).

- During **endothermic** reactions, energy is taken in, often as heat. Examples are **thermal decompositions** (e.g. calcium carbonate → calcium oxide + carbon dioxide).

Energy changes in reversible reactions

- A **reversible reaction** that is **exothermic** in one direction is **endothermic** in the opposite direction. In both directions the same amount of energy is transferred.

- When water is added to **anhydrous copper sulfate** (see page 98), the white powder turns blue and *gives out* a lot of heat. This reaction (a test for water) is exothermic.

  anhydrous copper sulfate + water $\underset{\text{endothermic}}{\overset{\text{exothermic}}{\rightleftharpoons}}$ **hydrated copper sulfate**

- When hydrated copper sulfate is heated, the crystals *take in* an equal amount of heat that drives off the water. This reaction is endothermic.

Questions

Grades G-E

1 Name a catalyst that can speed up the breakdown of hydrogen peroxide to water and oxygen.

Grades D-C

2 Name the group of metals that make good catalysts.

Grades G-E

3 When hydrochloric acid reacts with sodium hydroxide in a test tube, the tube gets hot. Is this an exothermic reaction, or an endothermic reaction?

Grades D-C

4 What can you say about the amount of energy taken in or given out in a reversible reaction?

Equilibrium 2

Reaching a balance point

G–E

- When heat breaks down calcium carbonate in a closed system (no products can enter or leave), there is a **reversible reaction**:

  calcium carbonate $\rightleftharpoons$ **calcium oxide + carbon dioxide** (See page 98.)

- In time, the reverse reaction balances the forward reaction. This balance point is called the **equilibrium**.

D–C

- Increasing the heat breaks down more calcium carbonate, and less of the calcium oxide and carbon dioxide re-form calcium carbonate. **Changing a condition** (temperature) has moved the equilibrium forwards. (See page 99.)

Industrial processes

The Haber process

G–E

- **Ammonia** is required to manufacture **nitrate** fertilisers. Ammonia is made from nitrogen and hydrogen in a reversible reaction (see also page 99):

$$\text{nitrogen} \quad + \quad \text{hydrogen} \quad \rightleftharpoons \quad \text{ammonia}$$
$$N_2(g) \quad + \quad 3H_2(g) \quad \rightleftharpoons \quad 2NH_3(g)$$

  number of molecules: 1 nitrogen   3 hydrogen   2 ammonia

- Conditions are:
  - temperature 450 °C
  - pressure 200 atmospheres
  - iron catalyst.

D–C

- **Yield** is 30%. This means that 30% of the starting materials are converted to product.

- This is acceptable because the **reaction rate** (see page 101) is good.

- Also, costs are relatively low. Reducing energy costs contributes to **sustainable development**.

Conditions for the Haber reaction

D–C

- For a good **yield**:
  - the reaction is **exothermic**, so low temperature increases yield; however, this lengthens the time to reach equilibrium
  - high pressure drives the reaction forwards.

- For a good **rate** but lower yield, higher temperatures are used to produce ammonia faster.

How Science Works

You should be able to: describe the effects of changing the conditions of temperature and pressure on a given reaction or process.

Questions

Grades G-E

1 A reversible reaction eventually settles down so that the reverse reaction balances the forward reaction. What name is given to this point?

Grades D-C

2 Look at the equation for the thermal decomposition of calcium carbonate. How could we move the equilibrium *backwards*?

Grades G-E

3 Ammonia is made in the Haber process. Give **one** use of this ammonia.

Grades D-C

4 State the conditions under which the reactions in the Haber process take place.

Free ions

Freeing ions in a solid

When a metal and non-metal react together they form an ionic solid which is made of positive and negative ions held tightly together, forming a giant lattice.

- Ionic compounds are solids. Their **ions** are arranged in a **giant lattice** (see page 92).

- Ions are free to move when the solid is **dissolved** in water to form a solution, or is **melted**.

- Heat melts solid lead bromide:

  $PbBr_2(s) \rightarrow PbBr_2(l)$

- Passing an electrical current through the molten lead bromide splits it into its **elements**, molten lead and bromine gas:

  $PbBr_2(l) \rightarrow Pb(l) + Br_2(g)$

  This process is **electrolysis** (see also page 31).

- (s) = solid, (l) = liquid, (g) = gas. Also, (aq) = aqueous (dissolved in water).

positive electrode   negative electrode
crucible
containing
molten lead
bromide

To free the ions, an ionic compound is kept molten. Then, it can be electrolysed to separate its elements.

G–E

Electrolysis of ionic compounds

- Ionic compounds always contain **positive metal** ions and **negative non-metal** ions. They are solids at normal temperatures, so the ions are not free to move.

- They can be electrolysed when either molten or in solution.

- In electrolysis, the ions move towards the *oppositely* charged electrode. Positive metal ions move to the **negative** electrode, and negative non-metal ions move to the **positive** electrode.

You should use your periodic table to help you with questions that ask you about the charges on ions.

positive   negative
electrode   electrode

The ions in molten sodium chloride move to the electrode with a charge opposite to theirs.

D–C

Electrolysis equations

Electrolysis of molten sodium chloride

- Sodium ions are positively charged, so they are attracted to the negative electrode. Chloride ions are negatively charged, so they are attracted to the positive electrode.

- The diagram shows what happens to the sodium ions and chloride ions at the electrodes.

- At the negative electrode, a positively charged sodium **ion** *gains* an electron: it is **reduced**.

  – It becomes an uncharged atom of sodium **metal**.

- At the positive electrode, a negatively charged chloride ion *loses* an electron: it is **oxidised**.

  – It becomes an uncharged chlorine atom. It shares electrons with another chlorine atom to form a molecule of chlorine **gas**.

- Page 107 describes the uses made of the products from the electrolysis of sodium chloride.

Na^+ sodium ions
Cl^- chloride ions
• electrons

The sodium ions (Na^+) and chloride ions (Cl^-) move to the oppositely charged electrodes.

Na^+ + • ➡ Na

Cl^- + Cl^- ➡ Cl Cl + ⦂

Each sodium ion gains an electron and each chloride ion loses an electron.

G–E

D–C

Questions

Grades G-E

1 In which **two** states can lead bromide be electrolysed?

Grades D-C

2 Explain why lead ions go to the negative electrode when lead bromide is electrolysed.

Grades G-E

3 A student wrote this sentence: 'Sodium is positive so it goes to the negative electrode.' Decide what is wrong, and write the sentence correctly.

Grades D-C

4 Describe what happens to chloride ions at the positive electrode. Are they reduced or oxidised?

Uses for electrolysis

Electrolysing sodium chloride (salt) solution.

Sodium chloride and copper solutions

- A **solution** of sodium chloride contains Na^+ and Cl^- ions from salt, and H^+ and OH^- ions from water.

- The **reactivity series** (see below) arranges metals and hydrogen in order of their reactivity.

- In **electrolysis**, H^+ and Na^+ ions go to the negative electrode. Hydrogen forms, but not sodium; it is more reactive than hydrogen.

- The electrolysis **products** are the gases chlorine and hydrogen. Na^+ and OH^- remain in solution as ions of **sodium hydroxide**.

- **Chlorine** is used to make bleach, chlorinated solvents and PVC. **Hydrogen** is a fuel, and is used to make ammonia and to change vegetable oils to margarine. Sodium hydroxide is used to make soap, detergents, paper and to purify bauxite (aluminium ore).

- To **purify copper** metal, a piece of the impure copper is used as the positive electrode in a solution containing copper ions, Cu^{2+}.

- Copper ions leave the positive electrode and move through the solution to the **pure** copper *negative* electrode. They gain two electrons and become Cu atoms, building up on the electrode.

- More reactive impurities, including zinc, stay in solution. Unreactive impurities collect as sludge.

Copper transfers from the positive to the negative electrode.

How Science Works

You should be able to: predict the products of electrolysing solutions of ions.

Acids and metals

Which metals react with acids to make salts?

- Any **metal** *higher* than hydrogen in the **reactivity series** can react with an acid. The reactive metal **displaces** the less reactive hydrogen in the acid. A soluble **salt** is formed, and hydrogen is given off.

- Potassium and sodium are high in the series. They are too reactive to react safely with sulfuric acid. Copper and silver are lower than hydrogen and they will not react with dilute sulfuric acid.

- Salts from acids: sulfuric acid → sulfates; nitric acid → nitrates; hydrochloric acid → chlorides.

- A **salt** is formed when hydrogen in an acid is replaced by a metal or an **ammonium** ion.

- To make zinc sulfate, the **reagents** are zinc metal and sulfuric acid:

  zinc + sulfuric acid → zinc sulfate + hydrogen

  Zinc is added in **excess** to ensure that all the acid is used up and the solution contains only zinc sulfate. The solution is filtered to remove any remaining zinc. The water in the solution is evaporated to leave zinc sulfate crystals.

Top Tip!

If you are asked whether a metal reacts with acids, look at the reactivity series.

Reactivity series

K	**more reactive**
Na	
Mg	
Al	
Zn	
Fe	
Pb	
H	
Cu	
Ag	**less reactive**

Top Tip!

Learn the formulae of these acids:
sulfuric acid = H_2SO_4
nitric acid = HNO_3
hydrochloric acid = HCl

Questions

1 Name the gas that is formed at the negative electrode when sodium chloride solution is electrolysed.

2 Give **two** uses of this gas.

3 You have a piece of zinc. What acid would you add it to, to make zinc chloride?

4 Explain why you need to filter the mixture when the reaction finishes, if you want to get pure zinc chloride.

Making salts from bases

Bases, alkalis and reaction with acids

- **Bases** react with acids to form **salts**. Oxides and hydroxides of metals are bases. Soluble bases are called **alkalis**.

- **base + acid → salt + water**

- Solid zinc oxide reacts with warm acid:

  zinc oxide + hydrochloric acid → zinc chloride + water

- Unreactive metals will not react with acid, but their oxide (or hydroxide) *will* react.

BASES
CuO   ZnO
MnO₂   Ca(OH)₂
MgO   LiOH
NiO   NaOH
KOH   **ALKALIS**

All the substances are bases, but only some are alkalis.

G–E

Making a salt from an acid and an insoluble base

- Magnesium oxide and hydrochloric acid react to form magnesium chloride salt:

  magnesium oxide + hydrochloric acid → magnesium chloride + water

  $MgO(s)$ + $2HCl(aq)$ → $MgCl_2(aq)$ + $H_2O(l)$

- Adding **excess** oxide to warmed acid ensures all the acid reacts. So the solution contains only magnesium chloride. The excess magnesium oxide can be removed by filtering. Evaporating water in the solution leaves magnesium chloride powder.

D–C

Acids and alkalis

Using indicators to follow reactions of acids and alkalis

- The colour of an **indicator** shows how acid or alkaline a solution is – its **pH**.

- You can add indicator to an alkali and then gradually add acid to it. The indicator will change colour when just enough acid has been added to react with all the alkali. This is the point of **neutralisation**.

- At neutralisation, the solution contains a **salt** and water only.

- **Ammonia** dissolves in water to give an alkaline solution. This can be reacted with acids to produce ammonium salts, which are important as fertilisers.

G–E

Making a salt from an acid and an alkali

- **Phenolphthalein** is a good indicator to follow the reaction of sodium hydroxide with hydrochloric acid to form sodium chloride (common salt).

  sodium hydroxide + hydrochloric acid → sodium chloride + water

  pH   14   1   7

  At pH 14, phenolphthalein is pink.
  At pH 7 or below, it is colourless.

- The diagrams show the procedure.

- If pure salt is required, the procedure is repeated using the same concentrations and volumes but without the indicator. The water is then evaporated to produce solid salt.

pipette

conical flask

25 cm³ sodium hydroxide

1 A pipette is used to transfer exactly 25 cm³ of 2 mol/dm³ hydrochloric acid into a conical flask.

2 Two or three drops of phenolphthalein indicator are added to the flask. Because the solution is alkaline, it goes pink.

phenolphthalein

25 cm³ sodium hydroxide

3 Hydrochloric acid is added until one drop of the hydrochloric acid turns the indicator colourless.

burette

2 mol/dm³ hydrochloric acid

D–C

Questions

(Grades G-E)

1 What products would you get if you add magnesium oxide to hydrochloric acid?

(Grades D-C)

2 Name an oxide and an acid that would react to produce zinc sulfate.

(Grades G-E)

3 What does an indicator tell you about a liquid?

(Grades D-C)

4 Explain why, if you want to make a pure sample of sodium chloride, you would first react the base and acid using an indicator, and then do it again *without* the indicator.

Neutralisation

What are acids and alkalis?

Acids are substances that contain H⁺(aq).
Alkalis are substances that contain OH⁻(aq).

- All **acids** contain hydrogen. In water, it is released as hydrogen ions, **H⁺(aq)**.

- All **alkalis** contain hydroxide, released in water as hydroxide ions, **OH⁻(aq)**.

- The **strength** of an acid or alkali depends on the amount of H⁺ or OH⁻ ions it releases in water.

- **Universal indicator** goes red in a strong acid, indicating a pH of 1. In a weak acid (e.g. vinegar), pH is 3.

Universal indicator: the pH scale shows how strong an acid or alkali is.

- In **neutralisation** reactions, the hydrogen and hydroxide ions react to form water:

$$H^+(aq) + OH^-(aq) \rightarrow H_2O(l)$$

Most solutions around us are not neutral.

Precipitation

Forming a precipitate, and hard water

- **Insoluble salts** can be made by mixing two soluble salts that contain the appropriate ions. If the ions react to produce an insoluble salt, it forms a solid **precipitate** that can be separated.

- To form the insoluble yellow compound lead chromate, compounds containing lead ions and chromate ions are required, for example:

  lead nitrate + potassium chromate → potassium nitrate + lead chromate

- **Hard water** contains dissolved **calcium** ions. To remove them, adding soluble sodium **carbonate** forms insoluble calcium carbonate that precipitates out:

  soluble calcium ion + soluble carbonate ion → insoluble calcium carbonate

Water and effluent treatment

- Before treatment, the water supply contains **nitrate** ions from fertilisers and **phosphate** ions from washing powders.

- Phosphate ions are removed using calcium, iron or aluminium ions:

  soluble calcium ion + soluble phosphate ion → insoluble calcium phosphate

- **Heavy metals** are removed as insoluble **carbonates**.

- Several **metals** can be identified from their insoluble **hydroxides** which have characteristic colours. To form the precipitate, sodium hydroxide is added to water containing the metal ions.

How Science Works

You should be able to:

- suggest methods to make a named salt
- explain and evaluate processes that use the principles described in this unit.

Questions

Grades G-E
1 Write the formula of the ion that is present in all acids.

Grades D-C
2 Write an equation to show what happens when any acid reacts with any alkali.

Grades G-E
3 What is a 'precipitate'?

Grades D-C
4 Explain why adding calcium ions to water can remove phosphate ions from it.

C2b summary

Collision theory states that **particles** must collide with sufficient energy in order to react.

The **minimum energy** required for a reaction to take place (a **successful collision**) is the **activation energy**.

Rates of reaction

The **rate of a reaction** can be measured as:
– the amount of **reactant used up** in a set time
– or the amount of **product formed** in a set time.

Reactions can be **speeded up** by **increasing** the:
– **temperature**
– **concentration** of a solution
– **pressure** of a gas
– **surface area** of a solid
or by using a **catalyst**.

Exothermic reactions give OUT energy.
Endothermic reactions take IN energy.

If a **reversible reaction** is exothermic in one direction, it is endothermic in the opposite direction, e.g.

anhydrous copper sulfate + water

$$\text{anhydrous copper sulfate + water} \underset{\text{endothermic}}{\overset{\text{exothermic}}{\rightleftharpoons}} \text{hydrated copper sulfate}$$

Types of reaction

In reversible reactions, **equilibrium** is reached at a point when the rate of the reverse reaction **balances** the rate of the forward reaction.

The conditions used for the **Haber process** are:
– temperature 450 °C
– pressure 200 atmospheres
– iron catalyst.

When **molten** or **dissolved** in water, the **ions** in **ionic compounds** are **free** to move.

If ions are free to move, passing an **electric current** through an ionic compound breaks it down into its **elements**. This is called **electrolysis**.

Electrolysis of **sodium chloride** (common salt) solution makes hydrogen, chlorine and sodium hydroxide. **Copper** can be purified by electrolysis.

Ions

During **electrolysis**:
– **negative** non-metal ions **lose electrons** at the positive electrode to form atoms and molecules
– **positive** metal ions **gain electrons** at the negative electrode to form atoms.

OIL RIG:
Oxidation **I**s **L**oss
Reduction **I**s **G**ain – of electrons.

Metal oxides and hydroxides are **bases** and react with **acids** to form **salts**.

Soluble bases are **alkalis** and these react with acids to make salts.

In **neutralisation** reactions, hydrogen ions and hydroxide ions react to form water:
$$H^+(aq) + OH^-(aq) \rightarrow H_2O(l)$$

Some **metals** can be reacted with **acids** to make **salts**. The **reactivity series** of metals can be used to find out if a metal will react with an acid.

Making salts

Salts from acids:
– sulfuric acid → sulfates
– nitric acid → nitrates
– hydrochloric acid → chlorides.

Insoluble salts can be made as **precipitates** when two solutions are mixed together.

To remove **ion impurities** from the **water supply**, they can be made into **insoluble salts**, e.g. phosphate ions from washing powders can be removed using calcium ions to form insoluble calcium phosphate.

See how it moves!

Emma's journey to school.

Distance-time graphs

G–E

- A **distance-time graph** shows the **distance** an object moves in a period of **time**. The **slope** (distance/time) shows its **speed**.
- The graph shows Emma's journey to school:
  - A–B: walks between home and bus stop; *shallow* slope = *slow* speed
  - B–C: waits for bus; *horizontal* line = staying in the same place, **stationary**
  - C–D: on the bus; *steep* slope = *fast* speed
  - D–E: still on the bus; less steep slope = bus speed slower
  - E–F: off at the bus stop near the school, waits for a friend; horizontal line = stationary.
- A distance-time graph does not show whether a journey is straight or changes direction.

Speed and average speed

D–C

- Speed on a journey may vary. For the total journey:

$$\text{average speed} = \frac{\text{total distance travelled}}{\text{total time taken}}$$

> **Top Tip!**
>
> A distance-time graph that slopes downwards shows an object coming back towards its starting point, not one that is slowing down.

Speed isn't everything

> **Top Tip!**
>
> If a question does not tell you which way velocity is being measured, you can choose which way is positive, but keep it the same for the whole question.

Velocity involves speed and direction

G–E

- **The velocity, *v*, of an object is its speed in a given direction.**
- The diagrams show the velocities of three objects in an *upward* direction.

 $v = 12000$ m/s

 $v = -2$ m/s

 $v = 0$ m/s

The **positive velocity** of the rocket tells us that it is moving at 12 000 m/s upwards, away from the ground.

The parachute has a **negative velocity** because it is travelling **towards** the ground and velocity is measured **away** from the ground.

The aeroplane might have a very high speed, but its velocity away from the ground is 0 m/s. It is not moving towards or **away from the ground**.

> **How Science Works**
>
> You should be able to: construct distance-time graphs for a body moving in a straight line when the body is stationary or moving with a constant speed.

Acceleration and direction

D–C

- Any **change** in velocity, whether speeding up or slowing down, is called **acceleration**. The value is negative for slowing down.

$$\text{acceleration (in m/s}^2\text{)} = \frac{\text{change in velocity (in m/s)}}{\text{time taken for change (in s)}}$$

- Like velocity, acceleration must have a direction.
- The archer's arrow accelerates towards the right.

acceleration

The arrow is fastest when it leaves the bow, so its acceleration is negative.

Questions

Grades G-E

1 Look at the distance-time graph for Emma's journey to school. At which stage was she travelling fastest? How can you tell?

Grades D-C

2 Use the average speed equation to calculate Emma's average speed for her journey to school.

Grades G-E

3 Explain the difference between 'speed' and 'velocity'.

Grades D-C

4 A stationary car started to move, and reached a speed of 20 m/s in 20 seconds. Use the acceleration equation to calculate its acceleration in m/s².

Velocity-time graphs

Velocity-time graphs

- A **velocity-time** graph shows how the **velocity** of an object changes with time.

- The graph shows the changing velocity of a cyclist. A horizontal line indicates **constant velocity**.

- The **slope** (gradient) of a velocity-time graph represents **acceleration**.

- Steep slope A on the cyclist's graph (1) is a positive slope, showing a positive acceleration.

- Shallower slope C is a negative slope, indicating a negative acceleration.

- The area under a velocity-time graph (2) shows the **distance** an object travels.

- The graphs (3) are for a skydiver who:
  - A: leaves plane; velocity increases
  - B: reaches maximum velocity
  - C: opens parachute; slows rapidly
  - D: reaches constant, lower velocity
  - E: hits ground.

(1) Velocity-time graph for a cyclist.

(2) The area under the graph shows the distance a model plane travels.

(3) Velocity-time graph and distance-time graph for a skydiver.

Let's force it!

Adding forces

- All the different forces on an object act together as a single, **resultant force**. **The resultant force has the same effect as all the forces acting on the object.**

- Forces acting in the *same* direction are added. Forces acting in the *opposite* direction are subtracted. For the car:

  forward force – air resistance – friction = 1000 N – 400 N – 400 N

- If the resultant force is zero:
  - **stationary** objects remain stationary
  - moving objects maintain the same **speed**, in the same **direction**.

- If the resultant force is not zero:
  - stationary and moving objects start to **accelerate** in the direction of the resultant force.

The resultant force acting on the car moves it forwards.

Questions

1 Look at the velocity-time graph for the cyclist. Describe what the line on the graph looks like when he is cycling at a constant speed.

2 Sketch a velocity-time graph to show a car moving at a steady speed, slowing down and stopping at traffic lights, then accelerating back up to the same steady speed.

3 A parachute is falling through the air. The downwards force of gravity on it is exactly the same as the upwards force of air resistance. What can you say about its velocity? Explain your answer.

4 Tom is trying to push a box along the ground. He pushes with a force of 10 N. Friction between the box and the ground exerts a force of 10 N in the other direction. What happens to the box?

Force and acceleration

Unbalanced forces

- When a car travels at a **steady** (constant) **speed**, the forwards **driving force** exactly balances the backwards **frictional forces** of air and road.

- **Acceleration** occurs in the *direction* of the resultant force:
  - accelerating increases the driving force, increasing speed
  - braking increases the frictional force, reducing speed.

- For the same mass, a bigger force gives a bigger acceleration.

- For the same force, a bigger mass gives a smaller acceleration.

Forces on a car.

Measuring acceleration

- Different **forces** and different **masses** are used with the apparatus shown. In each trial, the **force** is made constant:

Apparatus to investigate how acceleration changes when force changes.

  - the **force meter** measures **force** (newtons)
  - the **ticker-timer** measures **velocity** (metres/second).

- Velocity-time graphs drawn from the results show that **acceleration** is:
  - given by the **slope** (velocity/time)
  - **directly proportional** to force, $a \propto F$
  - inversely proportional to mass, $a \propto 1/m$

- **resultant force (N) = mass (kg) × acceleration (m/s²): $F = ma$**

Top Tip!

The bigger the force, the bigger the acceleration. The bigger the mass, the smaller the acceleration.

Balanced forces

Balanced forces and stationary objects

- **For every force there is an equal and opposite force.**

- When two forces are balanced, they have the *same* value but act in *opposite* directions. Examples for **stationary** objects are:
  - the tug-of-war teams (see picture)
  - a person floating in water; the **upthrust** of the water balances the downwards force of the person's weight.

The two teams are stationary, pulling with equal but opposite force. The forces are balanced.

Forces and moving objects

- Objects moving at **constant velocity** also have balanced forces.

- When a gun is fired, the force of explosive drives:
  - the bullet forwards from the gun at high speed
  - the gun backwards a little, at lower speed.

air resistance

weight

When a parachutist is falling at constant velocity, his weight acting downwards balances the **air resistance** acting upwards.

Questions

Grades G-E

1 A small mass and a large mass are each pushed with the same force. Which one accelerates more?

Grades D-C

2 Complete this sentence: Acceleration is .... proportional to force.

Grades G-E

3 Complete this sentence: If equal and .... forces act on a stationary object, it does not move.

Grades D-C

4 Complete this sentence: If equal and opposite forces act on a moving object, it does not ....

Terminal velocity

Reaching terminal velocity

- To walk into a strong wind, a man exerts a **resultant force** greater than the forces of **wind resistance** and **friction** between shoes and ground. As he **accelerates**, the wind resistance increases. He reaches a steady slow top speed. Then, opposing forces balance.

- A falling object always reaches a top speed, its **terminal velocity**. At this speed, the force of **weight** downwards cancels the force of **air resistance** upwards.

Top Tip!

When a parachute opens it *reduces* the resultant downward force – but the resultant force is still downwards.

G–E

Terminal velocity graph

- As a skydiver falls through the air, she accelerates. As she gets faster, air resistance increases. Eventually, air resistance exactly balances the downwards force of her weight. She reaches her terminal velocity.

- The parachute opens. Air resistance increases, exerting a force greater than the skydiver's weight. The resultant force is upwards and she slows down until weight = air resistance at a lower terminal velocity than before.

Velocity-time graph for a skydiver.

B Weight = air resistance. No resultant force. She reaches a terminal velocity.

C Parachute opens, increasing air resistance. Now air resistance is greater than her weight, so the resultant force is upwards. She slows down.

A Skydiver's weight is greater than air resistance. Resultant force is downwards, so she accelerates.

D Weight = air resistance. No resultant force. She reaches a slower terminal velocity.

D–C

Stop!

Stopping distance depends on the speed of the car.

thinking distance   braking distance

stopping distance

Stopping distance

- A driver sees a cat in the road. Her car moves a **thinking distance** before she brakes. Then the car moves a **braking distance** before halting: **stopping distance = thinking distance + braking distance**

- A moving car has kinetic energy. To stop the car, this kinetic energy must be reduced to zero. This is done by the braking force.

- The greater the car's speed, the greater the braking force needed to stop it within a particular distance.

G–E

Thinking distance and braking distance

- A driver's **reaction time** and the car's **speed** both affect the thinking distance:

thinking distance = speed × reaction time

- Drinking, taking drugs or being tired prolongs the reaction time.

- **Braking distance** depends on:

  - **speed squared**; doubling the speed increases the braking distance four times

  - the car's **mass**; kinetic energy is proportional to mass

  - **friction** of brakes and tyres; greater friction shortens the braking distance, icy or wet roads lengthen it.

How Science Works

You should be able to:

- draw and interpret velocity-time graphs for bodies that reach terminal velocity, including a consideration of the forces acting on the body

- calculate the weight of a body using: **weight (N) = mass (kg) × gravitational field strength (N/kg)**

D–C

Questions

Grades G-E

1 What is meant by the 'terminal velocity' of a falling object?

Grades D-C

2 Explain how a parachute reduces the terminal velocity of a person falling through the air.

Grades G-E

3 Complete this equation: stopping distance = ... distance + ... distance.

Grades D-C

4 State **two** factors that might *increase* the thinking distance.

Moving through fluids

Forces in fluids

G–E

- Liquids and gases are **fluids** because they flow.

- An object moving through a fluid pushes fluid particles aside and rubs against them.

- **Frictional** forces between fluid and object include **air resistance** and **drag** in water. They increase as:
  - the **speed** of a moving object increases
  - the fluid gets **denser**, so an object displaces a greater **mass** of fluid to move the same distance.

- An object falling through a liquid accelerates due to the force of gravity, until it reaches a **terminal velocity**. Then, the resultant force is zero and:

  upthrust (liquid's resistance force) = object's weight due to gravity

The cyclist must apply greater force than on the road because the drag force of water is greater than air resistance.

Maximising speed through fluids

D–C

- Water resistance (drag) slows objects much more than air resistance does.

- At high speed, a jet ski lifts up higher in the water, reducing water resistance.

- Birds and fish have a **streamlined** shape to reduce air or water resistance and so minimise the energy they require to move.

Energy to move

Kinetic energy

G–E

- **Kinetic energy** is the energy of **movement**.

- Kinetic energy comes, for example, from: chemical energy in food and in petrol; electrical energy to a motor.

- Kinetic energy is **transformed** when: a hammer hits a horseshoe and **does work** flattening it; a gong is hit and produces **sound**; rubbing your hands produces **heat**.

- **Energy transfer diagrams** show how energy is transferred and transformed.

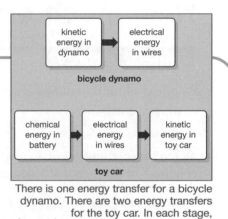

There is one energy transfer for a bicycle dynamo. There are two energy transfers for the toy car. In each stage, energy is also transformed.

Friction, heating and transformation

D–C

- A car's **kinetic energy** is transformed into **work** to overcome **frictional forces**.

- Much of the **work** against frictional forces is transformed into **heat**.

- A swing continually gains height and loses speed, then gains speed and loses height. **Kinetic energy** is transformed into **potential energy** and back again. The swing slows down because of friction between the swing and air.

How Science Works

You should be able to: discuss the transformation of kinetic energy to other forms of energy in particular situations.

Questions

Grades G-E

1 What is the difference between 'air resistance' and 'drag'?

Grades D-C

2 Why do fish need a more streamlined shape than birds?

Grades G-E

3 A builder drops a brick from the top of a wall. What happens to the kinetic energy in the brick when it hits the ground?

Grades D-C

4 A child plays on a swing. At which point on the swing's movement does it have: **a** most potential energy; **b** most kinetic energy?

Working hard

Work and energy

- When a **force** moves an object, **work** is done and **energy** is transferred:

  work done = energy transferred

- In the drawings:
  - the man does work against **air resistance**
  - the gardener does work against **friction**
  - the dog owner does work against **gravity**, transforming muscle energy into **potential energy** in the dog he lifts.

Each person is doing work by moving or lifting something.

G–E

How much work?

- Pushing a heavy trolley 10 metres requires greater **force**, transferring more energy, than pushing a light trolley.

- Pushing a trolley 20 metres requires more work than pushing it 10 metres.

- Amount of **work done** depends on the force and on the **distance**:

  work done = force × distance moved in the direction of the force
  (joules, J)   (newtons, N)   (metres, m)

Both men move their blocks the same amount. But the man below also does work against the friction between the ramp and the block.

D–C

How much energy?

Kinetic energy

- An object's **kinetic energy** increases when its:
  - **mass** increases
  - **speed** increases.

- All the **kinetic energy** of an object is transferred if it collides with another object and then stops. If it continues to move, but at a slower speed, then only some of its kinetic energy has been transferred.

G–E

Potential energy and elastic potential energy

- The **potential energy** of the top trolley is **transformed** into kinetic energy as it runs down the ramp.

- **Work done** to stretch the elastic band behind the trolley is stored as **elastic potential energy**. This is transformed into kinetic energy when the trolley moves forwards.

- Work done to blow up the balloon, stored as elastic potential energy, is transformed into kinetic energy when the trolley moves.

Three energy transformations.

D–C

Questions

(Grades G-E)
1 What are the units in which work is measured? Choose from: joules, newtons, grams.

(Grades D-C)
2 Jake lifts two parcels, each weighing 15 N, onto a desk 1 m above the ground. How much work does he do?

(Grades G-E)
3 A lorry and a motorbike are both moving at the same speed. Which one has more kinetic energy, and why?

(Grades D-C)
4 Kay uses an elastic band as a catapult to fire a paper pellet across the room. Where did the kinetic energy in the pellet come from?

Momentum

What is momentum?

The faster and more massive train has greater momentum than the bird.

- **Moving** objects have **momentum**. Momentum depends on an object's **mass** and its **velocity**:

  momentum = mass × velocity
  (kg m/s)   (kg)   (m/s)

- Momentum increases as mass and velocity increase.

- Momentum indicates how hard it is to halt a moving object.

- Momentum is a **vector** quantity – it has **size** (**magnitude**) and **direction**. Depending on direction, its value is positive or negative (see diagram opposite).

- Momentum is *not* the same as **kinetic energy**. Kinetic energy is a **scalar** quantity – it has size but not direction. Its value is always positive.

positive velocity +V m/s → positive momentum

negative velocity –V m/s ← negative momentum

The two identical masses move in opposite directions. One has a positive momentum, the other a negative momentum.

How Science Works

You should be able to: use the conservation of momentum to calculate the mass, velocity or momentum of a body involved in a collision or explosion.

Conservation of momentum

- To investigate **momentum**, two trolleys of known **masses** collide. Ticker tape gives their **velocities** immediately before and after collision.

before   (v = 0)   after

An inelastic collision.

- The **Law of Conservation of Momentum** states that:

  total momentum before collision = total momentum after collision

  – This applies as long as *no external forces* act on the bodies.

- In an inelastic collision, a 1 kg mass collides at 10 m/s with a stationary 4 kg mass:

before	after	rearranged
$(1 \times 10) + (4 \times 0)$ =	$(1 \times v) + (4 \times v)$	$10 + 0 = 5v$
		$v = 10/5 = 2\,\text{m/s}$

Off with a bang!

Explosions and jet engines

total momentum = 0

total momentum = +mv −mv = 0

momentum = −mv   momentum = +mv

before   after

Identical masses explode apart. Positive momentum = negative momentum, so total momentum is zero.

- Before an object **explodes**, nothing moves – momentum is zero. When it explodes, pieces fly in all directions. In any direction, the net momentum of the pieces will be zero. Momentum is conserved.

- In an aeroplane, the jet engine blasts gas backwards and the aeroplane moves forwards.

Examples of momentum

total momentum = 0
$MV = mv$

A lighter gun would recoil more.

- All beads in a Newton's cradle have the same **mass**. One bead hitting one end transfers just enough momentum to make one bead move from the other end.

Questions

Grades G-E

1 Complete this equation:
  momentum = mass × ....

Grades D-C

2 How does momentum differ from kinetic energy?

Grades G-E

3 Two inflated balloons, one small and one large, are resting on a table. What is the momentum of the balloons?

Grades D-C

4 One of the balloon bursts and shoots off the table. What can you say about the momentum of the moving balloon and the air that bursts out of it?

Keep it safe

Force causes a change in momentum

- **Momentum** is conserved as long as *no* **force** acts on an object.

- A force *changes* an object's momentum by changing its **velocity** or **direction**.

- The change in momentum depends on the **size** of the force and the **time** it acts.

- The diagram opposite shows how a small braking force changes the car's momentum slowly. A large force from the tree changes the momentum quickly.

Different ways of changing the car's momentum.

G–E

Changing momentum safely

- In a head-on collision, a car's seat belts let passengers move forward slightly before stopping them. They change momentum more slowly than the car, avoiding injury.

- On impact, the car bonnet crumples. This: slows the *rate of change* of the car's momentum; absorbs some of the car's **kinetic energy**; and transfers less energy to passengers.

How Science Works

You should be able to: use the ideas of momentum to explain safety features.

D–C

Static electricity

Static electricity

- Rubbing some materials together builds up **static electricity** on them.

- This happens because one material loses **electrons** and the other gains them. Electrons are **negative electrically** charged particles.

- Similar (like) charges **repel**. Opposite (unlike) charges **attract**.

- The charged ruler attracts the fine stream of water, because the water contains charged particles.

The charged plastic ruler attracts the stream of water, which curves towards the ruler.

G–E

Investigating static electrical charge

- In an atom, **protons** carry **positive charge** and **electrons** carry **negative charge**.

- An **electrical insulator** is a material that will hold static electrical charge and not let it flow away.

- When two insulators are rubbed together:
  - one gains electrons (e.g. polythene), becoming negatively charged
  - the other loses electrons (e.g. cellulose acetate) and is left with an equal positive charge.

- The suspended strip of cellulose acetate in the diagram carries a positive charge. If the strip approaching it:
  - is also charged cellulose acetate, the like charges repel
  - is charged polythene, the strips attract because they carry unlike charges.

A charged strip (left) is brought close to a positively charged cellulose acetate strip.

D–C

Questions

(Grades G-E)

1 An ice hockey puck slides across the ice. It hits a player's foot and moves off in a different direction, at the same speed. Has its momentum changed? Explain your answer.

(Grades D-C)

2 Use the idea of momentum to explain how wearing a seat belt helps to avoid head injuries in a car crash.

(Grades G-E)

3 A camper's shirt sleeve rubs against the fabric of her tent. The tent fabric becomes negatively charged. What is the name for this kind of electric charge?

(Grades D-C)

4 What has moved, and in which direction, to cause the tent fabric to become negatively charged?

Charge

Charge and force

G–E

- **Electrons** give an object a negative **static electrical charge**.
- Removing electrons gives an object a positive charge.
- Because charges exert a **force** on each other:
  - like charges (both positive or both negative) **repel**
  - unlike charges (positive and negative) **attract**.
- **Insulators** (e.g. plastics) hold electrical charge.
- **Electrical conductors** (e.g. metals, water) allow charge (electrons) to flow as an **electrical current**.
- A **gold leaf electroscope** detects if an object has an electrical charge. The gold leaf and rod are insulated.

a The gold leaf electroscope is **uncharged**. It has equal numbers of **positive** and **negative** charges all over it.
b The negative electrical charge on the strip repels the negative electrical charge on the electroscope. They move as far away as they can.
c The metal rod and the gold leaf both get a negative electrical charge. They move away from each other. The gold leaf rises.
d The more electrical charge the strip has, the more the gold leaf rises.

Identifying charges

D–C

- The electroscope can also tell whether a charge is positive or negative.

a Wipe a negatively charged polythene strip across the cap of the electroscope. The negative charges try to get as far apart as possible, so some of them will move down onto the electroscope.
b Remove the polythene strip. The electroscope now has a negative electrical charge. The gold leaf stays up.
c Holding a negatively charged object near the electroscope repels more negative charges down the metal rod of the electroscope. The gold leaf rises more.
d Holding a positively charged object near the electroscope attracts negative charges up to the cap. The gold leaf falls.

Van de Graaff generator

The Van de Graaff generator and lightning conductor

G–E

- In a **Van de Graaff generator**, two materials rub together and build up equal and opposite static electrical charges.
- The dome gains a large but safe **positive charge**. The **potential difference** (**voltage**) is high.
- When a girl standing on an insulating mat touches the dome, the positive charge redistributes over her and the dome. Her positively charged hairs repel each other. They move apart, standing on end.

Positive charge causes hairs to repel each other.

Moving charges

D–C

- The cloud's negative static charge induces a positive charge on the earthed lightning conductor. This charge streams up towards the cloud, discharging it and reducing the chance of a lightning strike.

The earthed conductor discharges the cloud.

Questions

Grades G-E
1 Why cannot static electricity build up on an electrical conductor that is earthed?

Grades D-C
2 Look at the diagrams of the electroscope in the second panel. Explain why the gold leaf goes up after the polythene strip is wiped across the cap.

Grades G-E
3 Why do the hairs on the girl's head repel each other, when she touches the dome of the Van de Graaff generator?

Grades D-C
4 What happens if you connect a charged object to earth with a conductor?

Sparks will fly!

Danger of electrostatic charge

- **Friction** builds up **static electrical** (**electrostatic**) **charge** on **insulating** materials.

- Charge can jump as a **spark** (flow of charge) between insulated objects, and set fire to flammable materials.

- **Earthing** prevents build-up of charge. Where this is not possible, charge can be redistributed evenly so that charge does not flow.

- In an operating theatre, fluids and gases are flammable. Plastics can become charged. They are **discharged** safely by keeping air humid, so that it conducts electricity, and installing a floor of material that conducts (earths) charge.

- Aircraft can build up static charge as they move through the air. Before mid-air refuelling, a wire connects the two aircraft to redistribute charge evenly between them.

- Friction in high pressure hoses charges the water droplets. Hoses used to clean oil tankers are earthed to prevent sparks from igniting the oil vapour.

Mid-air refuelling. Above, if charge is not evenly distributed, a spark may explode the fuel. Below, distributing charge evenly prevents sparks.

G–E

D–C

Making use of static electricity

- Paint droplets gain positive charge as they stream through the spray gun. The car bodywork is earthed, so a negative charge is **induced** near the spray. It attracts the paint droplets so they are deposited evenly over the bodywork.

Positively charged paint droplets are attracted to the car's bodywork.

- Flue gases from power stations must be cleaned to reduce pollution. **High voltage** wires in the chimney are negatively charged. Air particles are **ionised** between the wires. Positively charged particles stay near the wires. Negative charges are picked up by **ash** and **dust** particles which move to earthed metal plates lining the chimney. There they are collected.

- The drum inside a **photocopier** is positively charged. When you press Copy, light reflected from white areas of a sheet being photocopied illuminates the drum. Charge from those areas flows away, leaving positive charge only where there is colour on the sheet. These positively charged areas of the drum attract particles of black toner, which is then transferred to the copy.

Using static charge to remove ash and dust from flue gas in power stations.

G–E

D–C

How Science Works

You should be able to:
- explain why static electricity is dangerous in some situations and how precautions can be taken to ensure that the electrostatic charge is discharged safely
- explain how static electricity could be useful.

Questions

Grades G-E

1 Why could static electricity cause an explosion during mid-air refuelling of an aircraft?

Grades D-C

2 Explain why the high-pressure water hoses that are used to clean oil tanks are always made of metal and not plastic.

Grades G-E

3 How does static electricity help to stop paint being wasted when a car is sprayed?

Grades D-C

4 The toner in a photocopier has a negative charge. Why do particles of toner stick to some parts of the drum but not others?

P2a summary

Distance-time graphs help us to 'picture' how an object moves:
- the **slope** of the graph shows the **speed**.

$$\text{average speed} = \frac{\text{total distance travelled}}{\text{total time travelled}}$$

The **velocity** of an object tells us its **speed** in a given **direction**. Acceleration is the **rate of change** of velocity.

$$\text{acceleration (m/s}^2) = \frac{\text{change in velocity (m/s)}}{\text{time taken for change (s)}}$$

Velocity-time graphs show how the velocity changes:
- the **slope** shows the acceleration
- the **area** under the graph shows the **distance travelled**.

Forces can add up or cancel out to give a **resultant force**.

When the resultant force is *not* zero, an object accelerates.

The **acceleration** is:
- directly proportional to **force**, a ∝ F
- inversely proportional to **mass**, a ∝ 1/m

$$\text{resultant force (N) = mass (kg)} \times \text{acceleration (m/s}^2)$$
$$F = ma$$

For every force there is an equal and opposite force.

Balanced forces have the same value but act in opposite directions, e.g. a person floating in water has balanced forces.

Objects moving at **constant velocity** (including **terminal velocity**) have balanced forces.

stopping distance = thinking distance
+ braking distance
thinking distance = speed × reaction
time

Motion and forces

Stopping distance increases with:
- increasing **speed**
- a greater **mass**
- reduced **friction**
- increased **reaction time**.

Air resistance (in gases) and **drag** (in liquids), from frictional forces between the fluid and an object, increase as:
- the object's **speed** increases
- the fluid gets **denser**.

An object **falling** through a **fluid**, **accelerates** until it reaches **terminal velocity**, when the **resultant force** is **zero**:
upthrust (fluid's resistant force) = object's weight due to gravity

Every **moving** object has **kinetic energy** that can be transformed into other forms. An object's kinetic energy increases when its:
- **mass** increases
- **speed** increases.

Using a **force** to do **work** on an object gives it **energy**:
work done = energy transferred

The amount of work done depends on **force** and **distance**:
work done (J) = force (N) × distance moved in the direction of the force (m)

Every **moving** object has **momentum** that depends on its **mass** and its **velocity**:
momentum (kg m/s) = mass (kg) × velocity (m/s)

Momentum has **size** and **direction** and is **conserved** in collisions and explosions.

Forces change an object's momentum by changing its **velocity** or **direction**. The bigger the force on a moving object, the larger its **rate of change** of momentum.

Rubbing **electrical insulators** together can build up **static electricity**, because **electrons** transfer from one material to another.

If an object **gains** electrons, it has a **negative** charge. If it **loses** electrons, it has a **positive** charge.

Similar (like) charges **repel**. Opposite charges **attract**.

Static electricity

A **gold leaf electroscope** shows the **size** and **type** of charge on an object.

Static electricity can be dangerous if **sparks** cause flammable materials to **ignite** or **explode**. **Earthing** prevents a build-up of charge.

Paint sprayers, smoke precipitators and photocopiers are all useful **applications** of static electricity.

Circuit diagrams

Standard symbols and circuit diagrams

open switch

closed switch makes and breaks electrical circuits

cell a single unit producing electricity

battery several cells connected together

resistor reduces the flow of current in a circuit

variable resistor its resistance can be varied

lamp a device which gives out light

fuse protects equipment from electrical surges

voltmeter measures the p.d. across a component

ammeter measures current flowing in a circuit

diode allows current in one direction only. Has high resistance until a minimum p.d. reached

thermistor temperature affects its resistance

LDR (light dependent resistor) amount of light on it affects its resistance

series circuit ammeter measures current *in* circuit. Voltmeter measures potential difference (p.d.) *across* lamp. Current is *same* at any point in circuit

parallel circuit current at a junction splits. Current into junction = sum of currents in splits

variable resistor moving slider varies resistance. Current in circuit reduces as resistance increases, and increases as resistance reduces

fuse too high a current heats and melts fuse. Circuit breaks, protecting lamp

G–E

D–C

How Science Works

You should be able to: interpret and draw circuit diagrams using standard symbols.

Resistance 1

Electrons as energy carriers

- **Electrical current** is the flow of **electrons** carrying negative **charge**.

- **Free electrons** move between the atoms of a conductor.

- **Collisions** with atoms slow the free electrons down. This is **resistance**. The more collisions, the greater the resistance and the smaller the current.

- Copper has very low resistance, so allows for large currents.

copper wire

e⁻ → electron

copper atoms

Electrons travel through the spaces between the atoms of the copper wire.

G–E

Resistance

- A **potential difference** (**voltage**) exerts a **force** that pushes free electrons along in a circuit.

- Since all electrons have a like (negative) charge, electrons fixed in atoms repel free electrons. The **force of repulsion** slows down moving electrons. This causes resistance.

- Free electrons in **metals** make them good conductors with low resistances.

- **Insulators** do not have free electrons and so are poor conductors with very high resistances.

Top Tip!

Potential difference is the amount of 'push' a battery has to move current around a circuit.

Unit of potential difference = volt (V)

Symbol for potential difference = *V*

D–C

Questions

(Grades G-E)
1 Without looking, draw the symbols for a lamp and a resistor.

(Grades D-C)
2 Draw a series circuit containing two ammeters, one lamp and a voltmeter measuring the potential difference across the lamp.

(Grades G-E)
3 What are the particles that carry charge in an electrical circuit?

(Grades D-C)
4 Why are metals good conductors?

Resistance 2

Factors affecting resistance

G–E

- Long wires have greater **resistance** than short wires. Thin wires have greater resistance than thick wires.

- Electrical resistance is due to electrons colliding with atoms. Electrons in a longer wire have more collisions. Increasing **length** *increases* the wire's resistance.

- A thick wire has more electrons free to flow as a **current** than a thin wire. Increasing **thickness** *decreases* the wire's resistance.

D–C

- As **temperature** rises, atoms in a wire gain **kinetic energy** and **vibrate** more vigorously. They **repel** and **deflect** electrons more frequently and with greater energy. **Heating** *increases* a wire's resistance.

- **Different materials** have different resistances, depending on the size of atoms and the distance between them. High resistance can cause energy to be transformed into light and heat, as in filament light bulbs and electric bar heaters.

Ohm's Law

Ohm and resistance

G–E

- **Georg Ohm** investigated how **potential difference** relates to **current**, using the circuit shown. As he adjusted the variable resistor, he recorded potential difference across the **resistor** and the current through it.

- Ohm's straight-line graph shows that:
  - resistance of the **resistor** is *constant*

$$\text{resistance} = \frac{\text{potential difference}}{\text{current}} = \text{Ohm's Law}$$

  - current is *directly proportional* to potential difference across the resistor.

- The equation shows that, for a given potential difference across a component, the greater the resistance, the smaller the current.

The simple electrical circuit set up by Georg Ohm.

Ohm's graph for potential difference plotted against current.

Resistance of components

D–C

- To find the resistance of a **component**:
  - the component replaces the resistor in the circuit diagram above
  - current through it and potential difference across it are measured.

- The graph for a **filament lamp** shows that current levels off as potential difference increases. This is because resistance increases with heating.

Current begins to level off as potential difference increases.

Questions

1 Two series circuits are identical, except that circuit A has long, thin wires connecting the components, and circuit B has short, thick wires. In which circuit is resistance greater?

2 When a current flows through the metal wire in a filament lamp, it gets hot. How does this affect its resistance?

3 Write down the equation that links current, potential difference and resistance (Ohm's Law).

4 Why does a filament lamp not obey Ohm's Law?

More components

Thermistors, LDRs and how they work

- When the **temperature** of a **thermistor** is increased, its resistance *decreases*.

- So, as the thermistor heats up, the current through it *increases*.

- A thermistor is a semiconductor material (e.g. germanium, silicon). Heating frees more electrons, so increasing the current.

- Thermistors are used in equipment for detecting temperature changes, including electronic thermometers and fire alarms.

How Science Works

You should be able to: apply the principles of basic electrical circuits to practical situations.

- When **light** is shone on a **light dependent resistor** (**LDR**), its resistance *decreases*, so more current flows.

- As the light intensity increases, the current through the LDR *increases*.

- An LDR is made from the semiconductor compound **cadmium sulfide**. Light transfers energy that frees more electrons to carry current.

- LDRs are used in smoke detectors, automatic light controls and burglar alarms.

Symbol for a thermistor and a bead thermistor.

An LDR and its symbol.

Components in series

Effect on current of adding lamps to a series circuit

- Components in a **series circuit** are connected in one *continuous* loop.

- The diagram shows lamps 1, 2 and 3 added to a circuit one at a time.

- The brightness of the lamps gets less with each lamp added.

- At any time, the current is the *same* everywhere in the circuit.

Three lamps connected in series.

Series circuits, resistance and batteries

- In the series circuit, voltmeters V_1, V_2 and V_3 measure potential difference (p.d.) across each lamp. V_4 measures p.d. across all three lamps.
  - p.d. measured by V_4 = sum of p.d.s across lamps = p.d. of battery
  - i.e. total p.d. is shared between all components

Measuring p.d. across three lamps connected in series.

- Ohm's Law, resistance = p.d./current, gives the resistance of each lamp from p.d. and current measurements.
  - In a series circuit, total resistance = sum of resistances of components.

- A **battery** is several **cells** connected in series.

- With all cells connected plus to minus, p.d. of battery = sum of p.d.s of cells.

Top Tip!

Remember: In a series circuit, total p.d. is the *sum* of the p.d.s of each cell. This p.d. is *shared* between the components.

Questions

(Grades G-E)

1 What happens to the resistance of an LDR as light intensity increases?

(Grades D-C)

2 Give **one** use of an LDR.

(Grades G-E)

3 What happens to the brightness of a lamp in a series circuit, as more lamps are added?

(Grades D-C)

4 A cell has a p.d. of 1.5 V. What is the p.d. across four of these cells connected in series?

Components in parallel

Current and potential difference in parallel circuits

- At **junctions** (**splits**) in a **parallel circuit** (see diagram), the current divides to take different routes, then joins up again.

- Ammeters A_1 and A_5 measure the same **total current**.

- As each lamp is connected, the **brightness** remains the *same*.

- Ammeters A_2, A_3 and A_4 measuring current through each lamp show that:

$$A_1 = A_5 = A_2 + A_3 + A_4$$

- For parallel circuits, the *total current* through the whole circuit is the *sum* of the currents through the separate components.

Three lamps connected in parallel.

- Voltmeters (see diagram), show that there is the same **potential difference** across the battery and each lamp:

$$V_1 = V_2 = V_3 = V_4$$

- For parallel circuits, the *potential difference* across the battery and each component is the *same*.

Measuring potential difference.

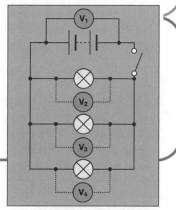

The three-pin plug

Structure and wiring of a three-pin plug

- The photograph shows the structure and wiring of a **three-pin plug**.

- The plastic covering the wires is colour coded:

  - **yellow** and **green** striped wire is **earth**, connected to top pin

  - **blue** wire is **neutral**, connected to left pin

  - **brown** wire is **live**, connected through fuse to right pin

  - **fuse** (see page 126) joins brown wire and right pin.

Top Tip!

Which goes where?

b**R**own = right

b**L**ue = left

Wiring up a three-pin plug.

Cables and plugs

- A **cable** and a three-pin plug connect most electrical appliances to the mains. **Insulating** material encloses all but the pins.

- The cable carries the three wires, live, neutral and earth, or '**twin** and **earth**'. **Insulating** plastic coats each wire. A plastic sheath encloses all three.

- Strands of each wire should be secured under their connector.

- **Twin-core cable**, with live and neutral wires only, is used in **double insulated** appliances such as a hairdryer and an electric drill.

- When the **fuse** blows, it isolates the live wire and no current flows.

- The **cable grip** prevents wires from being pulled from the plug by accident.

How Science Works

You should be able to: recognise errors in the wiring of a three-pin plug.

Questions

(Grades G-E)

1 In the first circuit diagram, the current flowing through ammeter A_1 is 12 A. What is the current measured by ammeter A_5?

(Grades D-C)

2 In the second circuit diagram, the cells provide a potential difference of 4.5 V. The three lamps are identical. What is the p.d. across each lamp?

(Grades G-E)

3 What is the colour of the earth wire in a three-pin plug?

(Grades D-C)

4 Why does a plug contain a fuse?

Domestic electricity

Direct current and alternating current

Oscilloscope trace showing the voltage of alternating current (left) and direct current (right) against time.

- **Cells** and **batteries** supply **electricity** as **direct current**, **d.c.**, in one direction. Potential difference is low, typically 1.5 V or 9 V, to power small appliances (e.g. torches, cameras).

- Domestic electricity has **alternating current**, **a.c.** It constantly changes direction. Potential difference averages 230 V, enough to power hairdryers and kettles.

Top Tip!

Mains voltage is often written as: (220–250) V, which means that it can range between this lower and upper limit; 230 V is accepted as an average value.

- The potential difference of a.c. changes (oscillates) from plus to minus (see diagram above) at 50 times per second, 50 Hz – its **frequency**.

- The **cathode ray oscilloscope** (**CRO**) displays **potential difference** measured in volts (see diagram on the right). The x-axis represents time.

- Direct current gives a straight-line, positive trace.

- As the diagram on the right shows, alternating current gives an S-shaped trace. The x-axis is at zero potential difference, values above it are positive and those below are negative. The maximum height of the wave gives the **peak** potential difference.

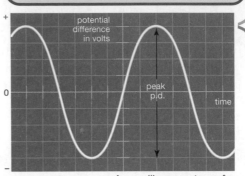

An oscilloscope trace for alternating current.

How Science Works

You should be able to: compare potential differences of d.c. supplies and the peak potential differences of a.c. supplies from diagrams of oscilloscope traces.

Safety at home

Fuses and earth wires

- A **fuse** is connected in series in a circuit. If a current is too high, the fuse wire heats and melts, breaking the circuit.

Top Tip!

The fuse and earth wire together protect the appliance and the user.

- Fuses have different ratings, appropriate to the current drawn by appliances.

- Appliances with metal casings have an **earth wire**. If the live wire touches the metal casing, a large current from the live wire to the earth wire melts the fuse, breaking the circuit.

- When electrical faults occur, fuses and earth wires protect appliances from damage and people from injury or death.

Circuit breakers

- Modern fuse boxes use **miniature circuit breakers** (**MCB**). An electromagnet or electronic mechanism immediately breaks the circuit when there is a tiny increase in current.

- A **residual current device** (**RCD**) breaks a circuit in less than 0.05 s after detecting a difference in the current in live and neutral wires. It makes lawn mowers and hedge trimmers safe to use.

A residual current device is essential wherever a cable might be cut by accident.

Questions

Grades G-E

1 What is a.c. short for?

Grades D-C

2 What does the trace of direct current look like on an oscilloscope screen?

Grades G-E

3 A fuse has a rating of 0.5 A. What size of current will melt the fuse?

Grades D-C

4 Suggest why having an RCD in a circuit is even safer than a fuse.

Which fuse?

Calculating the correct fuse rating

You should be able to: calculate the current through an appliance from its power and the potential difference of the supply, and from this determine the size of the fuse needed.

G–E

- Every appliance has a **power rating**. Power (see page 55) is measured in **watts** (**W**). Its correct **fuse** depends on the current it takes from the mains (p.d. = 230 V).

  power = p.d. × current

  so **current = power/p.d.**

- An appliance has a power rating of 1000 W.

  So current = 1000 W ÷ 230 V = 4.3 A

  – A 3 A fuse would melt when the appliance is switched on.

  – A 13 A fuse would not melt if a fault occurred.

  – The correct fuse is 5 A.

D–C

- In a resistor (e.g. electric fire), electrical energy is transformed into heat.

- The **rate** at which energy is transformed is called **power**:

  $$\text{power (watts)} = \frac{\text{energy transformed (joules)}}{\text{time taken (seconds)}}$$

  – A 100 W light bulb transforms electrical energy to light and heat at a rate of 100 J/s.

The power triangle.

The energy/power triangle.

Radioactivity

Atoms and atomic particles

particle	relative mass	charge
proton	1	+1
neutron	1	0
electron	0	−1

G–E

- An atom has a nucleus made up of **protons** and **neutrons**, surrounded by **electrons**.

- The table shows the characteristics of these atomic particles.

- Atoms have no net charge: **number of protons = number of electrons**.

- An atom that loses an electron (−1) becomes an **ion** with a charge of +1.

- An atom that gains an electron becomes an ion with a charge of −1.

- All atoms of a particular element have the *same* number of protons. Atoms of different elements have *different* numbers of protons.

D–C

- The number of:

  – protons in an atom is its **atomic number**

  – protons + neutrons is its **mass number**.

- The symbol for lithium can be written as shown:

mass number = protons + neutrons

atomic number = protons

$^{7}_{3}\text{Li}$

Isotopes

D–C

- Atoms of a particular element may have different numbers of neutrons. Such atoms have the same atomic number but different mass numbers.

- These different forms of an element are **isotopes**.

- The nuclei of some isotopes are unstable. They emit energy as particles or rays called **radiation**.

Questions

Grades G-E

1 What is the unit in which power is measured?

Grades D-C

2 A hairdryer has a power rating of 1500 W. If the mains supply is 230 V, what current flows through the hairdryer?

Grades G-E

3 Which kind of particle in an atom has a charge of +1?

Grades D-C

4 This is the full symbol for boron: $^{11}_{5}\text{B}$. How many protons does it have in its nucleus? How many neutrons does it have in its nucleus? What is its mass number?

Alpha, beta and gamma rays 2

Types of radiation

- In some **isotopes**, the *balance* of protons and neutrons in the nucleus is **unstable**.
  - To reach a stable state the nucleus emits **radiation**, which carries energy.

- Radiation can knock electrons from atoms. They become **ionised**.

- **Alpha (α) radiation** is a particle containing 2 protons and 2 neutrons – a helium nucleus. It is large and slow.

- Being large, heavy and carrying a +2 charge, it readily ionises surface atoms.

- **Beta (β) radiation** is a particle. It is a very small, fast, high energy electron.

- It is negatively charged, e⁻. Being small, it tends to miss atoms.

- **Gamma (γ) radiation** is an electromagnetic wave with no mass or charge, moving at the speed of light and carrying high energy.

- With no mass or charge, it passes between the atoms of most materials.

The ability of different radiations to penetrate different materials depends on their energy and mass (size).

Background radiation 2

Sources of background radiation

- **Background radiation** comes from the environment all the time.

- Generally, radiation is at a low level that is harmless to living things.

- Its sources include radioactive materials in rocks and soil, and cosmic rays from space. Some background radiation is due to the use or manufacture of radioactive materials by humans.

- Background radiation is detected by a **Geiger counter**, heard as intermittent clicks.

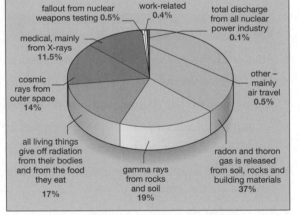

Sources of background radiation in the UK.

Sources that add to background radiation

- **Granite** rock contains radioactive materials that decay to give **radon** gas, itself radioactive and emitting harmful radiation.

- The Earth's magnetic field and the atmosphere shield the ground from most harmful **cosmic rays** from the Sun. This shielding is weaker above the Earth, so pilots and air crew are more exposed.

- Medical staff using X-rays, and others working with radioactive materials, risk higher radiation levels. They wear a radiation badge to monitor their exposure.

Questions

Grades G-E

1 What is meant by a 'radioactive isotope'?

Grades D-C

2 State the charges on: **a** an alpha particle; **b** a beta particle; **c** gamma radiation.

Grades G-E

3 How can we detect background radiation?

Grades D-C

4 What is the major source of background radiation in the UK? How can this explain the high levels of background radiation in places such as Cornwall, where there is a lot of granite?

Inside the atom

Experiments into the structure of the atom

- Alpha particles have a +2 charge. To investigate atomic structure, Rutherford and Marsden fired alpha particles at a sheet of very fine gold foil. They found that:
  - most went through, either straight or at angles
  - a few bounced back at different angles. Something repelled them
  - very few came straight back.

- They concluded that something in gold atoms was repelling the alpha particles.

- Their results suggested that:
  - most of an atom is empty space
  - it has a very small, dense, positive nucleus
  - electrons are not part of the nucleus, but orbit round it.

The scientists' model to explain what happened in their investigation.

Rutherford's model of the atom.

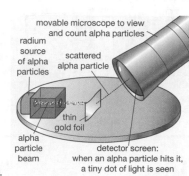

movable microscope to view and count alpha particles

radium source of alpha particles

scattered alpha particle

thin gold foil

alpha particle beam

detector screen: when an alpha particle hits it, a tiny dot of light is seen

Model of scattering apparatus.

G–E

D–C

Nuclear fission

Splitting the atom

- When an atomic nucleus is split in **nuclear fission**, some of its mass is converted to energy.

- Neutrons are fired at very large **unstable isotopes**. The nucleus of an unstable atom absorbs a neutron, splits in two and releases **nuclear energy**. It also produces more neutrons, setting up a **chain reaction**.

- Uranium-235 and plutonium-239 have unstable nuclei and so are **fissionable** substances. They are used to produce nuclear energy.

- An unstable isotope such as uranium-235 does not normally produce energy. A sample has to be pure uranium, and at least tennis ball-sized, so that neutrons do not leave the mass without setting up a chain reaction. This minimum amount is the **critical mass**.

- The nuclear bomb is an example of an uncontrolled chain reaction.

- Nuclear power stations control chain reactions very carefully.

A single atom splitting.

A chain reaction.

How Science Works

You should be able to: sketch a labelled diagram to illustrate how a chain reaction may occur.

Questions

(Grades G-E)

1 Name the **two** scientists whose experiments with gold foil and alpha particles helped to find out the structure of the atom.

(Grades D-C)

2 How did their results suggest that atoms had a very small nucleus surrounded by empty space?

(Grades G-E)

3 What does 'nuclear fission' mean?

(Grades D-C)

4 What must a uranium-235 atom absorb, to make it undergo fission?

Nuclear power station

The nuclear reactor

- Nuclear power produces a huge amount of energy as heat from a small amount of **nuclear fuel**, such as uranium-235.

- The energy heats water under high pressure. The pressurised hot water passes through a **heat exchanger** where a separate water supply is boiled to super-heated steam. This turns a turbine connected to a generator that produces **electricity**.

- To maintain a safe, controlled nuclear reaction:
  - rods of graphite, a **moderator** material, slow down neutrons so that they will be absorbed by fuel nuclei; water also acts as a coolant
  - movable **control rods** of boron or cadmium absorb excess neutrons.

G–E

In a nuclear power station, energy from nuclear fission generates steam.

Nuclear fusion

The birth of a star

- A star is formed over millions of years when dust and smaller particles in a **nebula** clump together, attracted by gravity. (See also page 66.)

- The mass of material heats up. At 15 million Kelvin, there is enough energy to fuse hydrogen nuclei together and form helium nuclei. This is **nuclear fusion**:

nuclear fusion
2 hydrogen nuclei ⟶ **1 helium nucleus + ENERGY**

A nebula in which stars are born.

G–E

Proton-proton reactions

- A hydrogen nucleus is a **proton**. Protons carry positive charge.

- It takes an immensely high temperature to *start* a **fusion reaction**. Protons need sufficient **kinetic energy** to move at very high speeds and overcome the force of repulsion between their like charges. Once started, the reaction itself produces great energy.

- As a result, the **reaction cycle** continues, generating even more energy.

- Similar reactions occur in stars and supernovae (see page 67).

In nuclear fission, the nucleus of a very large atom splits. In nuclear fusion, the nuclei of two small atoms join together.
Both processes produce huge amounts of energy.

D–C

Questions

(Grades G-E)
1 Give **one** example of a nuclear fuel.

(Grades G-E)
2 Explain the difference between 'nuclear fission' and 'nuclear fusion'.

(Grades G-E)
3 How is energy released inside a star?

(Grades G-E)
4 Name the new element that is formed when the nuclei of two hydrogen atoms fuse together.

P2b summary

Electrical current is the flow of **electrons** carrying negative **charge**. Electrical energy is transformed to heat energy when electrical charge flows through a **resistor**.

Electrical resistance in wires is due to **free electrons** colliding with atoms.

Resistance is *increased* in:
– **long** wires
– **thin** wires
– **heated** wires.

Ohm's Law states that:

$$\text{resistance} = \frac{\text{potential difference}}{\text{current}}$$

Resistance

The resistance of a **light-dependent resistor** (LDR) decreases as light intensity increases.

The resistance of a **thermistor** decreases as the temperature increases.

The resistance of a **filament lamp** increases as the temperature of the filament increases.

Insulators do not have free electrons and so are **poor conductors** with very **high resistance**.

Components can be connected in **series** and in **parallel**.

In the **UK**, domestic electricity has an **alternating current**, 230 V, a **frequency** of 50 Hz and is extremely dangerous.

An **oscilloscope** shows the **potential difference** (voltage) and **frequency** of electricity.

The **rate** at which energy is **transformed** is called **power**:

$$\text{power (W)} = \frac{\text{energy transformed (J)}}{\text{time taken (s)}}$$

Domestic electricity

Fuses and **earth wires** protect appliances from damage and people from injury or death. **Three-pin plugs** must be correctly wired and hold the correct fuse for the appliance.

The **current** through an appliance and the **size of the fuse** required can be calculated by:

power = current × potential difference

Isotopes are the different forms of an **element** with the same number of **protons** but a different number of **neutrons** in their atoms.

Isotopes of an element have the same **atomic number** but different **mass numbers**.

Isotopes with **unstable nuclei** emit **energy** as particles or rays called **radiation**.

An **alpha particle** is a helium ion. A **beta particle** is a high energy electron.
A **gamma ray** is an electromagnetic wave with no charge or mass.

Radioactive substances

Background radiation comes mainly from rocks, soil, cosmic rays, living things and medical X-rays.

Rutherford and Marsden revealed the **structure** of the **atom** using alpha particle scattering experiments.

An **atom** consists of:
– a small dense **nucleus** of **protons** and **neutrons**
– fast-moving **electrons** orbiting round the nucleus
– lots of empty **space**.

Nuclear fission is the **splitting** of an atomic **nucleus**.

Neutrons are fired at unstable isotopes and **energy** is released. This produces more neutrons and sets up a **chain reaction**.

Nuclear fusion and nuclear fission

Uranium-235 and **plutonium-239** are fissionable elements and are used to produce **nuclear energy** in nuclear power stations.

Nuclear fusion is the **joining** of two smaller **nuclei** to form a larger one.

Stars release energy by nuclear fusion:

2 hydrogen nuclei → 1 helium nucleus + ENERGY

Checklists – Science Unit 1

B1a Human biology

I know:

how human bodies respond to changes inside them and to their environment

- [ ] nerves and hormones coordinate body activities and help control water, ion and blood sugar levels and temperature
- [ ] receptors detect stimuli; information passes to the brain; and effectors (muscles and glands) respond
- [ ] reflex actions are automatic and very fast; they involve sensory, relay and motor neurones, with synapses between them, in a reflex arc
- [ ] the menstrual cycle is controlled by FSH, LH and oestrogen

what we can do to keep our bodies healthy

- [ ] a balanced diet contains the right balance of all the nutrients and energy humans need, and regular exercise helps you to stay fit
- [ ] different people have different energy needs
- [ ] obesity increases the risk of arthritis, diabetes, high blood pressure and heart disease
- [ ] too much cholesterol in the form of LDL in the blood increases the risk of heart disease

how we use/abuse medical and recreational drugs

- [ ] drugs affect people's behaviour, can damage the brain and some hard drugs are addictive
- [ ] some drugs are legal (alcohol) and some are illegal (cocaine, heroin)
- [ ] the substances in tobacco smoke cause many diseases, e.g. cancer and bronchitis
- [ ] drugs can be beneficial, but must be thoroughly tested before use

what causes infectious diseases and how our bodies defend themselves against them

- [ ] pathogenic microorganisms cause infectious diseases
- [ ] some white blood cells (phagocytes) ingest pathogens and kill them; and others (lymphocytes) make antibodies
- [ ] antibiotics kill bacteria inside the body but do not kill viruses; however, some bacteria can develop resistance to antibiotics
- [ ] immunisations and vaccinations offer protection from various diseases

B1b Evolution and environment

I know:

what determines where particular species live

- [ ] animals and plants are adapted to live in different habitats, to compete for resources and to survive attack from predators
- [ ] caribou and camels are adapted to extreme environments

why individuals of the same species are different from each other; and what new methods there are for producing plants and animals with desirable characteristics

- [ ] DNA is the genetic material that controls inherited characteristics
- [ ] reproduction can be sexual (variation in offspring) or asexual (identical offspring)
- [ ] cloning and genetic engineering can be used to produce plants and animals with desirable characteristics

why some species of plants and animals have died out and how new species of plants and animals develop

- [ ] theories of evolution have changed over time; Darwin proposed the theory of evolution by natural selection
- [ ] fossils tell us how present-day species have evolved and how they compare to prehistoric species
- [ ] species can become extinct due to environmental change and human impact

how humans affect the environment

- [ ] increases in human population use up more resources and produce more waste and pollution
- [ ] human action contributes to acid rain, air pollution, water pollution, and over-use of land
- [ ] living organisms, such as lichens, can be used as indicators of pollution
- [ ] increasing the greenhouse effect can lead to global warming
- [ ] sustainable development, e.g. using renewable energy resources and recycling, can help to safeguard the environment for future generations

C1a Products from rocks

I know:

how rocks provide building materials

- [ ] limestone, calcium carbonate, can be used to make cement, concrete and glass which are used as building materials
- [ ] an element consists of one type of atom; two atoms of the same element can join together to form a molecule; a compound consists of two or more elements joined together
- [ ] atoms are held together in molecules and lattices by chemical bonds, which involve giving, taking or sharing electrons

how rocks provide metals and how metals are used

- [ ] metals are extracted from their ores, often oxides, by reduction with carbon (iron), electrolysis (aluminium and copper) or other chemical reactions (titanium)
- [ ] metals are mixed together to make alloys (e.g. iron and other metals or carbon make steel)

- [ ] aluminium is expensive to produce, is often too soft on its own, but forms strong alloys with other metals
- [ ] copper is a hard, strong, good conductor and can be used for wiring and plumbing
- [ ] aluminium and titanium are resistant to corrosion and have low density

how we get fuels from crude oil

- [ ] crude oil is a mixture of hydrocarbon compounds that can be separated by fractional distillation. Some of the fractions can be used as fuels
- [ ] most of the compounds in crude oil are saturated hydrocarbons called alkanes, which have the general formula C_nH_{2n+2}
- [ ] the burning of fossil fuels releases useful energy but also harmful substances, e.g. sulfur dioxide causes acid rain; carbon dioxide causes global warming; smoke particles cause global dimming

C1b Oils, Earth and atmosphere

I know:

how polymers and ethanol are made from oil

☐ crude oil is made from hydrocarbons that can be cracked to form alkenes, which are unsaturated hydrocarbons and have the general formula C_nH_{2n}

☐ alkenes can be made into polymers, which are long chain molecules created when lots of small molecules called monomers are joined together in polymerisation

☐ polymers can be used to make useful substances e.g. waterproof materials and plastics, but many are not biodegradable

how plant oils can be used

☐ vegetable oils can be hardened to make margarine in a process called hydrogenation

☐ biodiesel fuel can be produced from vegetable oils

☐ oils do not dissolve in water. They can be used to produce emulsions e.g. in salad dressings

what the changes are in the Earth and its atmosphere

☐ the Earth has three main layers: the crust, mantle and the inner and outer core

☐ the Earth's atmosphere has changed over millions of years. Many of the gases that make up the atmosphere came from volcanoes

☐ for 200 million years, the proportions of different gases in the atmosphere have been much the same as they are today

☐ human activities have recently produced further changes, e.g. the levels of greenhouse gases are rising

P1a Energy and electricity

I know:

how heat (thermal energy) is transferred and what factors affect the rate at which heat is transferred

☐ heat energy can be transferred by conduction, convection and thermal radiation

☐ thermal conductors (e.g. metals) transfer heat energy easily; thermal insulators (e.g. plastic, glass) do not

☐ dark, dull surfaces emit and absorb thermal radiation better than shiny, light surfaces

☐ the bigger the temperature difference between an object and its surroundings, the faster the rate at which heat is transferred

what is meant by the efficient use of energy

☐ energy is never created nor destroyed; some energy is usually wasted as heat

☐ Sankey diagrams show energy input and output of a device

why electrical devices are so useful

☐ they transform electrical energy to whatever form of energy we need at the flick of a switch

☐ the National Grid transmits energy around the country at high voltages and low current to reduce energy losses

☐ dynamos produce electricity when coils of wire rotate inside a magnetic field

☐ how to work out the power rating of an appliance (the rate at which it transforms electrical energy)

☐ to calculate the amount of energy transferred from the mains use:

   energy transferred =   power   × time
   (kilowatt-hour, kWh) (kilowatt, kW) (hour, h)

☐ to calculate the cost of energy transferred from the mains use:

   total cost = number of kilowatt-hours × cost per kilowatt-hour

how we should generate the electricity we need

☐ we need to use more renewable energy sources, including wind, hydroelectric, tidal, wave and geothermal power

☐ most types of electricity generation have some harmful effects on people or the environment; there are also limitations on where they can be used

P1b Radiation and the Universe

I know:

what the uses and hazards of the waves that form the electromagnetic spectrum are

☐ from longest to shortest wavelength: radio waves, microwaves, infrared, visible light, ultraviolet, X-rays, gamma rays

☐ radiation has many uses e.g. radio, TV, satellites, cable and mobile phone networks

☐ some forms of electromagnetic radiation can damage living cells: ionising radiation (ultraviolet, X-rays and gamma rays) can cause cancer

what the uses and dangers of emissions from radioactive substances are

☐ the uses and hazards of radioactive substances (which emit alpha particles, beta particles and gamma rays) depend on the wavelength and frequency of the radiation they emit

☐ background radiation is all around us e.g. granite rocks can emit gamma rays and form radioactive radon gas

☐ the relative ionising power, penetration through materials and range in air of alpha, beta and gamma radiations

about the origins of the Universe and how it continues to change

☐ the Universe is still expanding; in the beginning, matter and space expanded violently and rapidly from a very small initial point i.e. the Big Bang

☐ red shift indicates that galaxies are moving apart; the further away a galaxy, the faster it is moving away from us

☐ telescopes on Earth and in space give us information about the Solar System and the galaxies in the Universe

Checklists – Additional Science Unit 2

B2a Discover Martian living!

I know:

what animals and plants are built from

- [ ] animal cells and plant cells have a membrane, cytoplasm and a nucleus; plant cells may also have a vacuole, chloroplasts and a cell wall
- [ ] in multicellular organisms, different cells are specialised for different functions
- [ ] the chemical reactions inside cells are controlled by enzymes

how dissolved substances get into and out of cells

- [ ] diffusion is the net movement of particles of gas, or substances dissolved in a solution, from a region of high concentration to a region of lower concentration
- [ ] osmosis is the diffusion of water molecules through a partially permeable membrane

how plants obtain the food they need to live and grow

- [ ] green plants use chlorophyll to trap energy from the Sun to photosynthesise
- [ ] leaves are specially adapted for photosynthesis – they can be broad, flat, thin and have lots of stomata
- [ ] the rate of photosynthesis is affected by light intensity, carbon dioxide concentration and temperature
- [ ] mineral salts in the soil are used to make proteins or chlorophyll

what happens to energy and biomass at each stage in a food chain

- [ ] energy passes along food chains but some energy is lost at every stage
- [ ] the shorter the food chain, the less energy is lost
- [ ] the mass of biomass at each stage in a food chain is less than it was at the previous stage. This can be shown in a pyramid of biomass
- [ ] reducing energy loss increases efficiency of food production
- [ ] decomposers and detritus feeders feed on dead organisms and their waste

B2b Discover DNA!

I know:

what enzymes are and what their functions are

- [ ] enzymes are proteins that act as biological catalysts, speeding up chemical reactions
- [ ] each enzyme works at an optimum temperature and pH
- [ ] they are involved in respiration, photosynthesis, protein synthesis and digestion
- [ ] enzymes are used in washing powders and in industry

how our bodies keep internal conditions constant

- [ ] blood sugar levels are controlled by the pancreas, which makes insulin to bring down blood sugar levels
- [ ] sweating cools the body down and helps to maintain a steady body temperature
- [ ] waste products e.g. carbon dioxide and urea must be removed from the body

some human characteristics show a simple pattern of inheritance

- [ ] some inherited characteristics are controlled by a single gene
- [ ] different forms of a gene are called alleles; in homozygous individuals the alleles are the same, in heterozygous individuals they are different
- [ ] in mitosis each new cell has the same number of identical chromosomes as the original
- [ ] sex chromosomes determine the sex of the offspring (male XY, female XX)
- [ ] stem cells can specialise into many types of cells

C2a Discover Buckminsterfullerene!

I know:

how sub-atomic particles help us to understand the structure of substances

- [ ] atoms consist of sub-atomic particles called protons, neutrons and electrons
- [ ] an element's mass number is the number of protons plus the number of neutrons in an atom
- [ ] an element's atomic number is the number of protons in an atom
- [ ] the arrangement of electrons can be used to explain what happens when elements react and how atoms join together to form different types of substances

how structures influence the properties and uses of substances

- [ ] ionic bonding is the attraction between oppositely charged ions
- [ ] ionic compounds are giant lattice structures with high melting points that conduct electricity when molten or dissolved
- [ ] non-metal atoms can share pairs of electrons to form covalent bonds
- [ ] giant covalent structures are macromolecules that are hard, have high melting points but do not conduct electricity
- [ ] nanoparticles are very small structures with special properties because of their unique atom arrangement

how much can we make and how much we need to use

- [ ] the relative masses of atoms can be used to calculate how much to react and how much we can produce, because no atoms are gained or lost in chemical reactions
- [ ] the percentage of an element in a compound can be calculated from the relative masses of the element in the formula and the relative formula mass of the compound
- [ ] high atom economy (atom utilisation) is important for sustainable development and economic reasons
- [ ] in some chemical reactions, the products of the reaction can react to produce the original reactants; they are called reversible reactions

C2b Discover electrolysis!

I know:

how we can control the rates of chemical reactions

☐ the rate of a chemical reaction can be found by measuring the amount of a reactant used or the amount of product formed over time

☐ chemical reactions can be speeded up by increasing the: temperature; concentration of a solution; pressure of a gas; surface area of a solid; and by using a catalyst

☐ particles must collide with sufficient energy in order to react; the minimum energy required is the activation energy

whether chemical reactions always release energy

☐ chemical reactions involve energy transfers

☐ exothermic reactions give OUT energy; endothermic reactions take IN energy

☐ if a reversible reaction is exothermic in one direction, it is endothermic in the opposite direction. The same amount of energy is transferred in each case

how can we use ions in solutions

☐ when molten or dissolved in water, ions in ionic compounds are free to move around

☐ passing an electric current through an ionic compound breaks it down into its elements: this is called electrolysis

☐ at the negative electrode, positively charged ions gain electrons (reduction) and at the positive electrode, negatively charged ions lose electrons (oxidation)

☐ electrolysis of sodium chloride solution makes hydrogen, chlorine and sodium hydroxide; copper can be purified by electrolysis

☐ metal oxides and hydroxides are bases and react with acids to form salts

☐ soluble salts can be made from acids and insoluble salts can be made by mixing solutions of ions

☐ in neutralisation reactions, hydrogen ions from acids react with hydroxide ions to produce water

P2b Discover nuclear fusion!

I know:

what the current through an electrical current depends on

☐ the symbols for components shown in circuit diagrams

☐ resistance is increased in long, thin, heated wires

☐ the current through a component depends on its resistance. The greater the resistance the smaller the current for a given p.d. across the component

☐ **potential difference = current × resistance**
   (volt, V)   (ampere, A)   (ohm, Ω)

☐ in a series circuit: the total resistance is the sum of the resistance of each component; the current is the same through each component and the total p.d. of the supply is shared between the components

☐ in a parallel circuit: the total current through the whole circuit is the sum of the currents through the separate components; the p.d. across each component is the same

what mains electricity is and how it can be used safely

☐ mains electricity is an a.c. supply of 230 V and has a frequency of 50 Hz; it is very dangerous

☐ fuses and earth wires protect appliances from damage and people from harm or death

☐ three-pin plugs must be wired correctly and hold the correct fuse

P2a Discover forces!

I know:

how we can describe the way things move

☐ distance-time graphs show how an object moves

☐ velocity-time graphs show how the velocity changes; the slope represents acceleration, and the area under the graph represents distance travelled

how we make things speed up or slow down

☐ an unbalanced force acting on an object changes its speed

☐ forces can add up or cancel out to give a resultant force; when the resultant force is *not* zero, an object accelerates:
   resultant force (N) = mass (kg) × acceleration (m/s²): $F = ma$

☐ an object falling through a fluid accelerates until it reaches a terminal velocity, when the resultant force *is* zero

☐ the stopping distance of a car is the thinking distance plus the braking distance. This increases as the speed increases

what happens to the movement energy when things speed up or slow down

☐ when a force causes an object to move, energy is transferred and work is done

☐ every moving object has kinetic energy that can be transformed into other forms

☐ the kinetic energy of a body depends on its mass and its speed

what momentum is

☐ every moving object has momentum that depends on its mass and its velocity:
   momentum (kg m/s) = mass (kg) × velocity (m/s)

☐ momentum has size and direction and is conserved in collisions and explosions

what static electricity is, how it can be used and the connection between static electricity and electric currents

☐ rubbing electrical insulators together builds up static electricity because electrons are transferred

☐ if an object gains electrons it has a negative charge; if it loses electrons it has a positive charge

☐ electrostatic charges can be used in photocopiers, smoke precipitators and paint sprayers

☐ when electrical charges move we get an electrical current

why we need to know the power of electrical appliances

☐ the power of an electrical appliance is the rate at which it transforms energy: **power (W) = energy transformed (J)**
                              time taken (s)

☐ power, potential difference and current are related by the equation:
   power   =   current   ×   potential difference
   (watt, W)   (ampere, A)   (volt, V)

☐ most appliances have their power and the p.d. they need on them so we can calculate the current and fuse required

what happens to radioactive substances when they decay

☐ isotopes (elements with the same number of protons but a different number of neutrons) with unstable nuclei emit energy as particles or rays called radiation

☐ the Rutherford and Marsden scattering experiment revealed the structure of the atom

☐ background radiation comes from rocks, soil, cosmic rays, living things and medical X-rays

what nuclear fission and nuclear fusion are

☐ nuclear fission is the splitting of an atomic nucleus; it is used in nuclear reactors

☐ nuclear fusion is the joining of two smaller nuclei to form a larger one; stars release energy by nuclear fusion

Answers

Unit B1a Human biology

Page 4
1 Brain and spinal cord.
2 Motor neurone.
3 A flash of light; a smell; a loud noise (you may think of others).
4 Adrenaline: heart, breathing muscles, eyes and digestive system. Reproductive hormones/oestrogen: ovaries.

Page 5
1 Sensory, relay and motor.
2 The neurone secretes a chemical, which diffuses across the gap and produces an impulse in the next neurone.
3 To provide cells with a constant supply of energy.
4 The water evaporates and takes heat from the skin.

Page 6
1 Ovary.
2 It makes eggs mature in the ovaries; it makes the ovaries secrete oestrogen.
3 They contain oestrogen, which stops FSH being produced, so no eggs develop.
4 Advantages: fewer unwanted pregnancies; fewer abortions; slower population growth. Disadvantages: could encourage more sexual partners; could increase spread of sexually transmitted diseases.

Page 7
1 Carbohydrates, fats, proteins, vitamins, minerals, roughage, water.
2 The greater the muscle-to-fat ratio, the faster the metabolic rate.
3 Very overweight.
4 Bones rub together as the joint moves, which is very painful. The joint may become too stiff to move.

Page 8
1 Drought, floods, war.
2 Lack of protein in the diet.
3 Unsaturated fats.
4 HDLs.

Page 9
1 All of them.
2 They experience withdrawal symptoms.
3 It is tested on tissues or animals in a laboratory.
4 Neither the experimenters nor the subjects know who is getting the drug and who is getting a placebo.

Page 10
1 Heroin, cocaine.
2 If they inject the drug and share needles, they may introduce the AIDS virus into their body.
3 It slows down reactions and may cause brain cells to shrink.
4 The liver is often the first to be really badly damaged. The brain may also be harmed.

Page 11
1 Nicotine.
2 There seems to be a link between them. When one increases, so does the other. It does not, though, necessarily mean that one *causes* the other.
3 A microorganism (e.g. a bacterium or a virus) that causes disease.
4 Doctors were no longer carrying pathogens from one patient to another.

Page 12
1 A chemical produced by white blood cells that sticks to antigens (e.g. bacteria) and destroys them. Antibodies are specific to a particular kind of antigen.
2 White blood cells, in particular lymphocytes.
3 They kill bacteria in the body without harming our cells.
4 Viruses reproduce inside body cells, so it is difficult to destroy the viruses without destroying the cells as well.

Page 13
1 A bacterium. (It is resistant to many antibiotics, including methicillin.)

2 It was a new disease, so no-one had immunity to it and there were no drugs to combat it. International travel meant it could spread quickly all over the world.
3 A small quantity of dead or weakened pathogens is injected into the blood stream. Your white blood cells make antibodies against it, so if you are infected with it in the future your immune system will destroy it immediately.
4 Many parents did not let their children have the jab, so they were not immune against measles, mumps or rubella. They could have got one of these diseases or helped to spread it to others.

Unit B1b Evolution and environment

Page 15
1 Cold climate.
2 Arctic; heat exchange takes place between warm blood in arteries and cold blood in veins in the legs.
3 Food, territory, mates.
4 Warns predators not to eat them.

Page 16
1 They are genetically identical.
2 Sexual involves gametes, asexual does not. Sexual involves fertilisation, asexual does not. Sexual produces variation in the offspring, asexual does not. (NB – it is not true to say that sexual reproduction always involves two parents, as there are many plants where one flower produces both male and female gametes and they fertilise themselves.)
3 On chromosomes in the nucleus of a cell.
4 By genes inherited from your parents.

Page 17
1 A piece cut from a plant from which a whole new plant can grow.
2 The new plants are all genetically identical to the parent. They all look the same. They all grow at the same rate. It is a quick and cheap way of getting many new plants.
3 Tissue culture.
4 A nucleus from the cell to be cloned is placed inside an egg cell from which the nucleus has been removed.

Page 18
1 It has been given genes from a different kind of organism.
2 Bacterium.   3 Lamarck and Darwin.
4 They have become adapted for eating different kinds of food.

Page 19
1 Mutation.
2 The black individuals were less likely to be eaten by birds, because they were better camouflaged. So they were more likely to reproduce, passing on the gene for black coloration to their offspring.
3 A dead organism is buried and compressed in sediment.
4 They have adaptations that suggest they lived in swampy ground.

Page 20
1 Its eggs were eaten by animals introduced to Mauritius by humans.
2 The species that is the best competitor may prevent the original species from getting enough resources to survive.
3 We produce more waste as our population increases and living standards improve.
4 The rate of increase is getting less (the graph is getting flatter).

Page 21
1 Quarrying for building materials, building houses on green sites, farming, dumping waste.
2 Getting as much production from crops or animals as possible in a small area, using lots of fertilisers and pesticides.
3 Untreated sewage, fertilisers, toxic chemicals.
4 It does not break down. It became concentrated up the food chain, so predators ended up with such large amounts in their bodies that it killed them.

Page 22
1 Sulfur dioxide.
2 It can trigger asthma attacks or cause bronchitis.
3 Sulfuric acid, nitric acid, carbonic acid.
4 It reacts with calcium carbonate, causing the stone to break down.

Page 23
1 Lichens.   2 Heavily polluted.
3 There is less photosynthesis, so less carbon dioxide is taken from the air and stored in wood. So carbon dioxide levels in the atmosphere can increase.
4 The microorganisms that are breaking them down respire, releasing extra carbon dioxide into the air.

Page 24
1 Carbon dioxide, methane.
2 They have both increased.
3 Improving the quality of life without compromising future generations.
4 Nuclear power; wind power; wave power; hydroelectric power.

Unit C1a Products from rocks

Page 26
1 One.   2 Zn
3 Neutrons and protons.
4 12. It has 12 protons in its nucleus, and 12 electrons orbiting it.

Page 27
1 An element.   2 Two more.
3 Calcium carbonate.
4 Advantages: jobs; roads; local income. Disadvantages: habitat destruction; noise; traffic (there may be other good suggestions).

Page 28
1 calcium carbonate → calcium oxide (quicklime) + carbon dioxide
2 $CaCO_3 \rightarrow CaO + CO_2$
3 By heating sand, limestone and sodium carbonate together.
4 Concrete is a mixture of cement, sand and rock chippings; it is stronger than cement.

Page 29
1 Iron oxide.
2 Oxygen is removed from it; the oxygen is taken by carbon, because carbon is more reactive than iron.
3 Impurities are removed from it.
4 They are made of a lattice of atoms all of the same size, which form layers and can easily slide over each other.

Page 30
1 Always carbon, and sometimes other metals.
2 It is very hard (and can be sharpened to a fine edge).
3 Only cobalt is a transition metal.
4 It has delocalised electrons which can move freely.

Page 31
1 Bauxite.
2 Aluminium is more reactive than carbon, so carbon is not able to take oxygen away from the aluminium.
3 Saves energy used in extracting aluminium from bauxite; reduces need for bauxite mines; reduces need to dispose of waste in landfill sites (there are other possible suggestions).
4 Destroys habitats; pollutes the air with carbon dioxide; may spread diseases between migrant workers (there are other possible suggestions).

Page 32
1 It is light (has a low density), strong and resists corrosion.
2 Titanium dioxide does not conduct electricity (because it is a covalent compound).
3 It is a good conductor of electricity and can easily be drawn out into wires.
4 There is only a limited supply of high-grade ores, so if we go on using copper we will need to get it from low-grade ores.

Page 33
1 Gold is very unreactive.
2 They can go back to their original shape after being deformed. Example: copper-zinc-aluminium alloy; nickel-titanium alloy.
3 The supplies are finite. We are using it up faster than any new oil is formed.
4 It does not produce carbon dioxide or sulfur dioxide when it burns.

Page 34
1 Hydrogen and carbon.
2 Those with small molecules.
3 C_2H_6.　　　　4 C_3H_8.

Page 35
1 More sulfur dioxide will be formed, which causes acid rain.
2 It contains carbon dioxide, which is found in the air and dissolved in rain drops to form carbonic acid.
3 It prevents acid rain from being produced; sulfur is valuable and can be made use of.
4 The gases are sprayed with a slurry of limestone, which reacts with the sulfur dioxide to form solid calcium sulfate.

Unit C1b Oils, Earth and atmosphere
Page 37
1 They do not contain any C–C double bonds.
2 They are heated to form a vapour and passed over a catalyst.
3 They contain at least one C–C double bond.
4 The double bonds can break and become single bonds, leaving a 'spare' bond to react with something else.

Page 38
1 A mixture of ethene and steam is passed over a catalyst.
2 It has an –OH group.
3 A substance made of many smaller molecules, called monomers, joined together in a long chain.
4 They have C–C double bonds, which can 'open' and react with another alkene molecule.

Page 39
1 Slippery; non-toxic; inert (unreactive); withstands high temperatures.
2 Viscosity.
3 They are not biodegradable.
4 Mix them with cellulose or starch, which microorganisms can break down.

Page 40
1 Fruits; nuts; seeds. (From named plants.)
2 Dissolve the crushed plant material in a hydrocarbon solvent. Distillation may also be used.
3 Rapeseed oil; sunflower oil; wood (there are others, e.g. elephant grass).
4 Non-renewable fossil fuels are running out.

Page 41
1 Mayonnaise; milk; emulsion paint (there are others).
2 Light is reflected from the tiny droplets of the dispersed phase.
3 A saturated fat contains no C–C double bonds, but an unsaturated fat does.
4 Add it to bromine water. If the bromine water changes from orange to colourless, the liquid contains unsaturated molecules.

Page 42
1 Vegetable oils.
2 Extra hydrogen atoms are added to the C–C double bonds, so the molecules become straight instead of bent. They can pack together more tightly, forming a solid.
3 A substance added to food to improve its taste, appearance or keeping qualities.
4 The E-number identifies the additive, which will have been tested to make sure it is safe.

Page 43
1 Chromatography.
2 Spots of the colour are placed on a pencil line on paper, which stands in a solvent. The solvent moves up the paper, taking the colours with it. Some colours travel faster than others. The height of the spot of an unknown colour can be matched with heights of spots of reference colours.

3 Probably about 4600 million years (4.6 billion years).
4 Crust, mantle and core.

Page 44
1 They thought that the surface of the Earth cracked as it cooled.
2 Heat generated by radioactive processes deep inside the Earth. (There is also heat left over from when the Earth was first formed.)
3 Tension builds up where two plates are moving in relation to each other. Then suddenly the two plates slip.
4 At a subduction zone, an oceanic plate is gradually moving underneath a continental plate. Friction stops the plates moving smoothly. The plates move in sudden jerks, causing earthquakes.

Page 45
1 Nitrogen, oxygen and water vapour.
2 0.04% (or 0.035%).
3 Argon, neon, helium, krypton and xenon.
4 Living things produced oxygen when they photosynthesised.

Page 46
1 It has increased by 1 °C.
2 It is a greenhouse gas, preventing heat from being lost into space.
3 A place where carbon is locked up. Sedimentary rocks (especially limestone), plants (especially trees) and fossil fuels.
4 Carbon dioxide dissolves in sea water to form carbonic acid, which reacts with calcium to form insoluble calcium carbonate. This collects on the sea floor, and gradually becomes compressed to form solid rocks.

Unit P1a Energy and electricity
Page 48
1 From the coffee (hotter) to your hands (colder).
2 The person is hotter than the trees, and so emits more thermal radiation which is detected by the camera.
3 More.
4 Silver surfaces emit *and* absorb thermal radiation more slowly than black surfaces.

Page 49
1 It is less dense than cold air.
2 Hotter particles vibrate faster. They are in close contact with their neighbours, so they make them vibrate faster as well.
3 Air in between the layers is a poor conductor. So less heat is conducted away from your warm body to the cold air outside your clothes.
4 Their particles are very far apart and only rarely come into contact with each other.

Page 50
1 Potential (gravitational potential).
2 10 million J.
3 Energy transfer: kinetic energy in arm to kinetic energy in gong.
Energy transformation: kinetic energy in gong to sound energy in air.
4 Heat energy in the bike, the surrounding air and the ground.

Page 51
1 400 J.
2 Exhaust gases: 215 J; moving parts: 510 J.
3 Conduction, convection and radiation.
4 The heat energy in the bath water is spread out through a much larger volume than the heat energy in the kettle water.

Page 52
1 Each bag weighs $2 \times 10 = 20$ N.
So work done = $20 \times 20 \times 3 = 1200$ J
2 $1200 \div 60 = 20$ W　　3 Machine A.
4 $(500 \div 1000) \times 100 = 50\%$

Page 53
1 It reduces the quantity of hot air leaving the room, so less heat energy needs to be used to keep the air in the room warm.
2 The fluorescent bulb, because it only loses 80% of its energy as heat.
3 Chemical (*not* electrical – you cannot store electrical energy.)
4 Usually because there is no mains electricity supply available. (It may also be cheaper.)

Page 54
1 The fuse gets hot and melts.
2 A thin wire has more resistance than a thick one. The more resistance, the more it heats up.
3 50 J.
4 number of kWh = power (kW) $\times$ time (hours)
number of kWh = $(50 \div 1000) \times 0.5 = 0.025$ kWh
cost = 8p $\times$ 0.025 = 0.2p

Page 55
1 A step-down transformer.
2 High voltage means low current, and this means less energy is lost as heat and the wires can be thinner.
3 An electric motor uses electricity to produce movement. A dynamo uses movement to produce electricity.
4 If either the magnet or coil of wire are moved relative to each other, a current will be produced in the wire.

Page 56
1 A turbine comes before a generator in a power station. A turbine is made to turn, using an energy source such as steam. The turbine then turns the generator, which generates electricity.
2 35%.
3 The water that is used is renewed when it rains, and as rivers flow into the reservoir.
4 The steam can turn a turbine connected to a generator; this is geothermal energy.

Page 57
1 They may think they look ugly; some people think they are noisy; birds may fly into them and be killed, which may upset people.
2 So that the fuel can easily be brought to them by road in large vehicles.
3 They are running out; they generate carbon dioxide which may be causing global warming.
4 They rely on water flowing rapidly downhill; they need to be in areas where there is high rainfall.

Unit P1b Radiation and the Universe
Page 59
1 Energy.
2 Radar (speed detectors); telecommunications (satellites, mobile phones).
3 Radio waves.
4 Short wavelength.

Page 60
1 (Visible) light.
2 Gamma rays, X-rays, ultraviolet, visible light, infrared, microwaves, radio.
3 It absorbs the energy in the microwaves.
4 Most go straight through, but some are absorbed.

Page 61
1 Ultraviolet.
2 It can ionise atoms and molecules, damaging DNA; this can cause cancer.
3 A signal that is sent as pulses (a series of 'ons' and 'offs').
4 A satellite that always stays above the same spot on the Earth's surface.

Page 62
1 Glass.
2 They undergo total internal reflection when they hit the edge of the fibre.
3 Yes, because they both contain the same number of protons.
4 The radiation comes from the nucleus of an atom.

Page 63
1 Alpha, beta and gamma.
2 Alpha radiation cannot penetrate skin, but it has a very high ionising effect which could damage atoms inside you if swallowed.
3 Radon gas; rocks and soil.
4 The badge shows how much radiation they have been exposed to, so they can check that they stay within safe limits.

Page 64
1 12 years.
2 The count rate will halve.
3 A radioisotope is added to the water in the pipe, and a Geiger counter can detect where most radiation is present.
4 Alpha would not go through even the thinnest paper. Gamma would go through even the thickest paper.

Page 65
1 It ionises atoms, and may damage DNA; a cell in which this has happened may divide uncontrollably.
2 The waste is radioactive, so it could harm people, e.g. by damaging their DNA.
3 The light that reaches it does not have to pass through the atmosphere, which can cause distortion.
4 More light can enter a reflecting telescope.

Page 66
1 A pull force between objects.
2 They used the gravity of Jupiter and Saturn to speed them up as they flew past.
3 The joining together of the nuclei of two atoms.
4 They become helium atoms and release energy.

Page 67
1 Jupiter, Saturn, Uranus and Neptune.
2 The gases were pushed further away than rocks when the early star exploded.
3 It will become a black dwarf.
4 There is no nuclear fusion, so no radiation pressure. So there is no outward force to act against gravity, which pulls the particles together.

Page 68
1 A huge amount of energy at a single point exploded, forming protons, neutrons and electrons.
2 Hydrogen.
3 If something is moving away from us, the frequency of the light coming to us appears to be less.
4 It shows that the galaxies are moving away from each other, and that the further away they are the faster they are moving; this could be explained if everything was spreading outwards from a single point after the Big Bang.

Unit B2a Discover Martian living!
Page 70
1 Chloroplast; cell wall; permanent vacuole.
2 Ribosome.
3 Tail for swimming; lots of mitochondria for energy; nucleus with only one set of chromosomes; enzymes in head to digest into egg.
4 In the linings of the alimentary canal and gas exchange system; they make mucus.

Page 71
1 Solution, gas.
2 It decreases the rate, because particles move more slowly.
3 Oxygen.
4 The greater the concentration gradient, the faster the rate of diffusion.

Page 72
1 Water.   2 Dilute solution.
3 Cell membrane.   4 Cell wall.

Page 73
1 carbon dioxide + water (+ light energy) → glucose + oxygen
2 Respiration.
3 Carbon dioxide can diffuse through it more quickly. Light can reach all the cells in the leaf.
4 They are near the top of the leaf, where they get sunlight. They are next to the spongy mesophyll layer, where there is carbon dioxide in the air spaces.

Page 74
1 Higher light intensity, higher carbon dioxide concentration, higher temperature (so long as it is not too high).

2 The plant is using as much light as it can; it may be limited by the amount of chlorophyll it contains, or by not having enough carbon dioxide to photosynthesise any faster.
3 Chlorophyll.
4 The nitrogen in the ions is combined with glucose to make amino acids, which are strung together to make proteins.

Page 75
1 Energy flow.
2 Some of the sunlight does not fall onto their leaves, some is reflected from the leaves, some passes straight through the leaves.
3 The mass of living organisms.
4 Mammals use a lot of food to produce heat energy to keep their body temperature constant. Reptiles just take on the heat of the surroundings.

Page 76
1 The food chain is shorter, so less energy is lost.
2 The animals need less food to produce heat energy to keep themselves warm.
3 They use central distribution points and large container lorries.
4 The farmer gets fewer eggs per hen or per square metre of land.

Page 77
1 They produce enzymes that break down waste plant and animal material.
2 Reactions happen faster at higher temperatures.
3 Earthworm.
4 Every organism, when it dies.

Page 78
1 Carbohydrates; fats; proteins; chlorophyll.
2 To reduce carbon dioxide levels in the atmosphere; to produce oxygen for us to breathe; to provide food.
3 By respiration.
4 They absorb it during photosynthesis, and store carbon in their growing bodies as carbon compounds.

Unit B2b Discover DNA!
Page 80
1 A place in an enzyme molecule where its substrate fits.
2 The enzyme molecules are denatured – they lose their shape, so the substrate does not fit in the active site.
3 It digests starch molecules to sugar molecules (maltose).
4 It is an alkali, and neutralises the acidic contents arriving from the stomach. It contains bile salts, which emulsify fats so that lipase can digest them more easily.

Page 81
1 They digest the protein (haemoglobin) in blood and make it soluble so it can be washed away.
2 They break up the stain and lift it from the cloth, so the enzymes can easily get at the fat molecules.
3 It tastes sweeter than other sugars, so less needs to be added.
4 They break down proteins into amino acids, which babies can absorb.

Page 82
1 glucose + oxygen → carbon dioxide + water
2 They keep their body temperature constant, usually above the temperature of their environment; this requires heat energy, which they release from food by respiration.
3 Intercostal muscles and diaphragm.
4 Large surface area, thin walls, rich blood supply.

Page 83
1 Liver.
2 A tiny tubule in a kidney.
3 Temperature; glucose concentration; water content; ion concentration.
4 To conserve heat; because less water may have been lost in sweat.

Page 84
1 In the brain.
2 Their thermoregulatory centre stops working when it is very cold.
3 Pancreas.

4 Banting and Best injected fluid containing it into dogs with diabetes, which became healthy.

Page 85
1 Two.   2 Mitosis.
3 So that when they fuse together the new cell will have the normal two sets.
4 They contain mixtures of alleles (different varieties of a gene) from both their parents.

Page 86
1 A cell that is not differentiated, and that can form different types of specialised cells.
2 White blood cells; red blood cells; bone cells; cartilage cells; tendon cells; fat cells.
3 DNA.
4 So that each daughter cell receives a complete set of chromosomes/DNA.

Page 87
1 Always producing offspring like themselves.
2 An allele that does not show unless the dominant allele is absent.
3 The sperm.
4 There equal numbers of sperm with a X chromosome and sperm with a Y chromosome, so there is an equal chance that the zygote will get an X or a Y from the sperm.

Page 88
1 Cystic fibrosis and Huntington's disease.
2 If they are both heterozygous, they could each pass on the recessive gene to a child.
3 Everyone's DNA is unique (except for identical twins).
4 Linking suspects to the scene of a crime; checking who is the father of a child.

Unit C2a Discover Buckminsterfullerene!
Page 90
1 Proton +1, neutron zero, electron –1.
2 39. It has a total of 39 neutrons and protons in its nucleus. (You can also work out how many of each it has. Its atomic number is 19, so it has 19 electrons and 19 protons. So it must have 20 neutrons.)
3 One.
4 By losing this one electron.

Page 91
1 Three. The atom has no overall charge.
2 They all have the same number of electrons in their outer shell, which gives them similar chemical properties.
3 The potassium atom can become stable by losing one electron, and the chlorine atom can become stable by gaining one.
4 The total charge on the positive ions is exactly balanced by the total charge on the negative ions.

Page 92
1 Molten (liquid) or in solution.
2 A compound made up of many ions all arranged in a regular lattice, held together by strong forces of attraction between the oppositely charged ions.
3 Covalent.
4 They each have a complete outer electron shell.

Page 93
1 Hydrogen; nitrogen; oxygen; chlorine.
2 They do not contain any charged particles.
3 Carbon.
4 The carbon atoms are arranged in layers so the layers can slide over one another.

Page 94
1 Their atoms lose electrons, forming positive ions.
2 Their layers of atoms can easily slide over each other.
3 One.
4 Rubidium, because metals become more reactive as you go *down* the Group.

Page 95
1 The elements in Group 7.
2 They gain one electron to form negative ions.
3 They are tiny particles made up of just a few hundred atoms.
4 They could be used as new types of catalysts.

Page 96

1 It is able to revert to its original state, after being changed by a change in its environment.
2 It darkens; it will then go back to its normal colour when the light is no longer present.
3 There is one carbon atom and four hydrogen atoms, chemically bonded together.
4 In a compound, the elements are chemically combined and they are in fixed proportions. In a mixture, they are not chemically combined and they can be in any proportion.

Page 97

1 $12 + (4 \times 1) = 16$ 2 $\dfrac{12}{16} \times 100 = 75\%$
3 16 g   4 One.

Page 98

1 They are the same. No atoms are lost or created.
2 The reaction may be reversible. Some of the product may be lost when it is separated from other products. Some of the reactants may react in unexpected ways.
3 $\rightleftharpoons$
4 nitrogen + hydrogen $\rightleftharpoons$ ammonia

Page 99

1 The forward reaction and backwards reaction happen at the same time – so there are always some reactants present.
2 It has no effect (but it does speed up the rate of the reaction).
3 Ammonia.   4 Iron.

Unit C2b Discover electrolysis!
Page 101

1 Use less concentrated acid, use cooler acid, use a thicker piece of magnesium (with less surface area).
2 Measuring the volume of a gas produced during a reaction.
3 The mass decreased as the hydrogen produced was lost to the air.
4 Time.

Page 102

1 There are more particles in the same space, so they collide more often.
2 It would double it (because particles would collide twice as frequently).
3 Sulfur is formed, which makes the mixture cloudy.
4 It is twice as fast. (At 20 °C it took 500 s for the cross to disappear, and at 30 °C it took 250 s.)

Page 103

1 Tiny chips.
2 By that time, the reaction would be complete, and there is no difference in the *total* mass of gas lost at the end of the reaction.
3 Hydrogen.
4 There are twice as many acid particles in the same volume, so the collisions between the acid particles and the magnesium particles will be twice as frequent.

Page 104

1 Manganese(IV) oxide (*or* the enzyme catalase).
2 Transition metals.
3 Exothermic (heat is given out).
4 The amount of energy taken in when the reaction goes one way is the same as the amount of energy given out when the reaction goes the other way.

Page 105

1 Equilibrium.
2 Cool things down.
3 Making nitrate fertilisers.
4 High pressure (200 atmospheres), moderate temperature (about 450 °C), iron catalyst.

Page 106

1 As a liquid (molten) or in solution.
2 Lead ions are positively charged, and opposite charges attract.
3 'Sodium' is not positive – it should say 'Sodium ions'.
4 They lose their electron and become chlorine atoms. This is oxidation – the loss of electrons.

Page 107

1 Hydrogen.
2 As a fuel; making ammonia (for fertilisers); making margarine; in weather balloons.
3 Hydrochloric acid.
4 To get rid of any unreacted pieces of zinc.

Page 108

1 Magnesium chloride and water.
2 Zinc oxide and sulfuric acid.
3 How acidic or alkaline it is – its pH.
4 Using the indicator tells you exactly how much base and acid to react together. You can then use these quantities in another reaction, to get a sample of the salt that is not contaminated with indicator.

Page 109

1 H^+
2 $H^+(aq) + OH^-(aq) \rightarrow H_2O(l)$
3 A solid that drops to the bottom of a liquid; a precipitate forms when ions in solution react to form an insoluble compound.
4 The calcium ions react with the phosphate ions to form insoluble calcium phosphate, which precipitates out.

Unit P2a Discover forces!
Page 111

1 C; the line is steepest.   2   14 km/h
3 Speed is the distance travelled in a certain time. Velocity is the same, but it includes the direction of travel as well.
4 1 m/s^2

Page 112

1 The line is horizontal (*not* 'straight').
2 The line starts off part way up the *x*-axis, then horizontal, then sloping downwards right to the bottom of the graph, then sloping upwards, then horizontal again at the same level at which it began.
3 The resultant force is 0, so the parachute continues to move at a constant velocity.
4 The box does not move, because the resultant force is 0.

Page 113

1 The small mass.   2 … directly …
3 … opposite …   4 … accelerate.

Page 114

1 The fastest, steady speed at which it falls.
2 It increases air resistance, so the resultant downward force is smaller.
3 … thinking …; … braking …
4 Having a slower reaction time (e.g. because of drinking alcohol; taking drugs; being tired); driving faster.

Page 115

1 Air resistance is the frictional force caused by movement in air; drag is the same, but in water.
2 The resistance against movement in water (drag) is greater than in air (air resistance) because water is denser than air.
3 It is transformed to sound energy, and to heat energy in the ground (and the air).
4 a At the top of the swing; b at the bottom.

Page 116

1 Joules.
2 30 J (15×1 for each parcel).
3 The lorry, because its mass is larger.
4 The elastic potential energy stored in the elastic band.

Page 117

1 … velocity.
2 Momentum has a direction (it is a vector quantity) whereas kinetic energy does not (it is a scalar quantity).
3 Zero (because velocity is zero).
4 The momentum of the balloon and the momentum of the air add up to zero (they are moving in different directions).

Page 118

1 Yes, because the direction has changed.
2 The seat belt slows down the rate of change of momentum when you suddenly slow down to a stop. The longer the time taken, the less force is used. Less force means less injury.
3 Static (electricity).
4 Electrons have moved from the shirt to the tent fabric.

Page 119

1 The charge can flow away through the conductor.
2 The strip gives electrons to the cap, and these move down the rod and the leaf. Now the rod and leaf have the same negative charge, so they repel each other.
3 They all have a positive charge, so they repel each other.
4 The charge flows away to earth – the charged object is discharged.

Page 120

1 A spark jumping from one plane to the other could cause the fuel to explode.
2 Any charge that builds up can be conducted safely away along the metal hose.
3 Opposite charges on the particles of paint and the car attract the paint to the car's surface, rather than spreading out into the air.
4 The drum had positive charge, but it only keeps its charge in the places corresponding to the dark parts of the object being photocopied. (The light parts make the drum able to conduct, so the charge flows away.)

Unit P2b Discover nuclear fusion!
Page 122

1 Check your answer against the symbols on page 122.
2 Your circuit should look just like the first circuit diagram on page 122, except that it contains another ammeter somewhere in the circuit, and only one lamp with the voltmeter going across it.
3 Electrons.
4 They have free electrons that can move easily.

Page 123

1 Circuit A.
2 The resistance increases.
3 $V = IR$ **or** $I = V \div R$ **or** $R = V \div I$
4 Because its resistance is not constant – it increases as more current flows through it making it hotter.

Page 124

1 It decreases.
2 Smoke detectors; automatic light controls; burglar alarms.
3 It decreases.
4 6 V (you *add* the p.d.s).

Page 125

1 12 A   2 1.5 V
3 Green and yellow striped.
4 The fuse melts if too much current flows through it, breaking the circuit and preventing harm.

Page 126

1 Alternating current.
2 A straight, horizontal line.
3 Anything above 0.5 A.
4 It breaks the circuit more quickly than a fuse.

Page 127

1 watts (W)   2 6.5 A
3 Proton.
4 5 protons; 6 neutrons; mass number 11.

Page 128

1 A form of an element that emits radiation.
2 a Positive; b negative; c no charge.
3 Use a Geiger counter.
4 Radon and thoron gas. Radon gas comes from granite; radon is radioactive.

Page 129

1 Rutherford and Marsden
2 Quite a lot of the alpha particles went straight through the gold foil, meaning that they must have gone through empty space rather than hitting anything.
3 The splitting of the nucleus of an atom.
4 A neutron.

Page 130

1 Uranium-235; plutonium-239.
2 In nuclear fission, the nucleus of an atom splits apart. In nuclear fusion, two atomic nuclei join together.
3 By nuclear fusion.
4 Helium.

Glossary/index

acceleration The change of velocity in a given time; it is usually measured in metres per second per second (m/s^2). 111, 112, 113, 114

acid A chemical with a pH below 7. 103, 107, 108, 109

acid rain Rain with a very low pH resulting from the release of oxides of nitrogen and sulfur into the air. 22, 35, 104

activation energy The energy a molecule must have before it can take part in a particular chemical reaction. 102

active site The part of an enzyme where the reaction occurs. 80

adaptation Characteristic that allows an animal or plant to survive in its environment. 15, 73

addiction When a person is unable to manage without a particular drug. They are dependent on it. 9, 10, 11

additives Chemicals put into food to keep it fresh or make it appetising. 42, 43

aerobic respiration The breakdown of glucose using oxygen to make energy available for chemical reactions in cells. 77, 82

agar Jelly-like substance used for growing bacteria. 71

air resistance The force air exerts on an object moving through it. 114, 115, 116

alcohol A family of compounds containing the –OH group. A general term for alcoholic drinks. 10

alkali A substance with a pH above 7. 108, 109

alkali metal A metal which burns to form a strongly alkaline oxide. 94

alkane Chemicals containing only hydrogen and carbon with the general formula C_nH_{2n+2}. 34, 37

alkene Chemicals containing only hydrogen and carbon with the general formula C_nH_{2n}. 38

allele A variety of a gene which determines different variations of a particular characteristic. 85, 87, 88

alloy A mixture of a metal element with one or more other elements. 30, 31, 33

alpha particle A helium nucleus made of two protons and two neutrons. 62, 63, 129

alpha radiation Alpha particles released from the nucleus of a radioactive isotope. 64, 65, 128

alternating current Current that rapidly reverses in direction. 126

alveolus The sac-like end of an airway in the lungs. 71, 82

amino acid The sub-unit making up protein molecules. 80, 81, 82, 83, 86

ammeter A meter that measures the current in a wire, in amps (A). 55

ammonia An alkaline gas with the formula NH_3. 45, 93, 98, 99, 104, 105, 108

ampere Or amp, is the unit used to measure electrical current (A). 127

anode The positive electrode in a circuit or battery. 31

antibiotic A substance produced by a microbe that kills other microbes. Some antibiotics can be purified and used to treat infections. 12, 13, 71

antibody Chemicals produced by white blood cells that destroy invading microorganisms. 12, 13

antigen Chemicals found in the cell membranes and cell walls of pathogens that trigger a reaction by the immune system. 12, 13

antitoxin Chemicals produced by white blood cells that neutralise bacterial toxins. 12

antiviral drug A drug that acts against viruses. 12

arthritis Inflammation and restriction of the movement of a joint. 7

artificial fertiliser A fertiliser made from inorganic chemicals. 21

asexual reproduction Reproduction involving just one parent. The offspring are genetically identical to the parent (clones). 16, 85

atmosphere The layer of gases surrounding the Earth. 24, 65

atom The smallest part of an element, atoms consist of negatively charged electrons orbiting a positively charged nucleus. 26, 27, 29, 30, 31, 62, 63, 64, 65, 90, 91, 92, 94, 96, 98, 118, 122, 123, 127, 129

atom economy The conversion efficiency of a chemical process in terms of all the atoms involved. In an ideal chemical process the amount of starting materials or reactants equals the amount of all products generated and no atom is wasted. 98

atomic number The number of protons in the nucleus of an atom. 26, 62, 90, 91, 97, 127

background radiation Radiation from space and rocks that it is around us all the time. 63, 128

bacteria Microscopic single-celled living things that have no nucleus. 11, 12, 13, 71, 77

base A substance that reacts with an acid to form a salt, or dissolves in water to form hydroxide ions. 108

bauxite Aluminium ore, aluminium oxide. 31

beta particle A high-energy electron produced when a neutron changes to a proton. 62, 62

beta radiation Beta particles released from the nucleus of a radioactive isotope. 64, 65, 128

Big Bang The event believed by many scientists to have been the start of the Universe. 68

bile An alkaline fluid produced by the liver that neutralises acid and breaks down fats. 80

biodegradable A substance which can be broken down by biological action in the environment. 39

biodiversity All aspects of variety in the living world. 23

biofuel A fuel that is produced from a living system. 38, 40

biological detergent A detergent containing enzymes. 81

biomass The mass of living material. 38, 40

biosensors A sensor that uses biological materials, usually enzymes, to detect a particular substance. 95

bladder The organ where urine is stored before it is excreted. 83

blood pressure The pressure of blood in your arteries and veins. 7

braking distance The distance it takes for the brakes to stop a moving vehicle. 114

bronchitis Inflammation and infection of the bronchi. 10, 11

camouflage Natural colouring or characteristics that enable an animal to remain unseen by predators. 19

cancer A dangerous illness in which cells grow uncontrollably, usually to form a tumour. It can be caused by radiation, smoking and some types of chemicals. 10, 11, 61, 64, 65

carbohydrase An enzyme that breaks down carbohydrates. 81

carbohydrate Chemicals found in all living things. They contain the elements carbon, hydrogen and oxygen. They are a source of energy. 7, 73, 74

carbon Carbon is present in all living things and forms a huge range of compounds with other elements. 29, 30, 34, 78, 93

carbon cycle The constant cycling of carbon as carbohydrates, fat and proteins in living things and as carbon dioxide in the air. 23, 78

carbon dioxide A gas containing only carbon and oxygen (CO_2). 22, 23, 24, 33, 35, 45, 46, 57, 73, 74, 78, 82

carbon monoxide A poisonous gas containing only carbon and oxygen (CO). It stops red blood cells carrying oxygen. 11, 33, 104

carbon sink A place where carbon is locked up, e.g. sedimentary rocks. 46

carcinogen A substance that causes cancer. 11

carnivore An animal that eats other animals. 75

carrier A person who is healthy but has a recessive allele for an inherited disorder. 88

catalyst A chemical that speeds up a reaction but is not changed or used up by the reaction. 38, 42, 80, 95, 101, 102, 104

catalytic converter A device fitted to vehicle exhausts that reduces the level of nitrogen oxides and unburnt hydrocarbons in the exhaust fumes. 33, 99, 104

cathode The negative electrode in a circuit or a battery. 31

cell differentiation The change of an unspecialised cell into a particular type of cell. 86

cell membrane The thin layer around the outside of a cell, which controls what enters and leaves it. 70, 71, 72, 88

central nervous system The brain and spinal cord. 4

chain reaction A reaction where the products cause the reaction to go further or faster. 66, 129

charge A property of matter, charge exists in two forms (negative and positive) that attract each other. 94

chemical energy Energy stored in chemicals. 50, 51, 53, 73

chlorophyll A green pigment found in plants that traps energy from sunlight which allows the plant to make glucose from carbon dioxide and water. 73, 74, 75

chloroplast An organelle found in some cells in green plants that contains the chlorophyll needed for photosynthesis. 73

cholesterol A chemical found particularly in animal fats, and foods made from them. 8

chromatography The separation of soluble substances in a mixture, producing chromatograms. The soluble substances move along paper or jelly-like films. 43

chromosome A thread-like body made of DNA and protein found in the nucleus of a cell. It only becomes visible during cell division. 16, 85, 86, 87, 88

cilia Tiny hairs found on the surface of cells and some microorganisms. 70

circuit A complete path that electricity can flow along. 54

circuit breaker A device which breaks a circuit to switch off the power when a difference in the current is detected, e.g. residual current device. 126

clone Genetically identical organisms. 16, 17

collision frequency The number of collisions in a particular area in a particular time. 102

community A group of animals or plants living or growing together in the same area. 77

competition Interaction between organisms needing a particular resource that is in short supply, e.g. plants competing for nutrients in the soil. 15, 19, 20

compound Two or more atoms of different elements bound together, in fixed proportions, by chemical bonds. 27, 33, 34, 93, 94, 96, 97

concentration The amount of a substance in a given volume. 51, 71, 74, 82, 101, 102, 103

conduction Movement of energy as heat (or electricity) through a substance. 30, 49, 92

conductor A substance that will let heat or electricity pass through it, e.g. copper. 30, 32, 49, 119

consumer Organisms in an ecosystem that use organic matter produced by other organisms. 75

convection Movement of a heated substance which carries energy with it as heat. 49

convection current Upwards movement of heated gases or liquids to float on top of the cooler, denser layers. 43, 44, 49

core The centre of the Earth. 43

cosmic ray Radiation from space which hits the atmosphere. Some passes through, while some is blocked. 63, 128

count rate The number of nuclear events in a given time, often measured by a Geiger counter. 64

covalent bond A bond between atoms where electrons are shared. 27, 32, 92, 93, 95, 96

cracking The breakdown of long-chain hydrocarbon molecules into shorter-chain ones using heat, pressure and sometimes catalysts. 37, 38

crude oil A mixture of hundreds of different compounds, mainly hydrocarbons. 20, 34, 38, 39

crust The Earth's outermost layer of solid rock. 43, 44

current The flow of electrical charge through an electrical circuit. 54, 55, 119, 122, 123, 124, 125, 126

cutting, plant Part taken from a plant that can be used to grow a new, genetically identical plant. 17

cystic fibrosis An inherited disorder that affects cell membranes. Caused by a recessive allele. 88

cytoplasm The material in the cell that is inside the cell membrane but outside the nucleus. 70, 72

decomposer An organism that breaks down dead organic matter. 78

deforestation The permanent clearing of forests. 23

denature To alter the shape of a protein molecule, e.g. by heating, so that it can no longer do its job. 80

density The mass of an object divided by its volume. 31, 94, 115

deoxyribonucleic acid See DNA.

detergent A chemical that makes grease soluble in water. 81

detritus feeder An animal that eats the dead and semi-decayed remains of living things, e.g. an earthworm. 77, 78

diabetes A disease in which blood sugar levels are not controlled because the pancreas cannot secrete insulin, or because body cells do not respond to insulin. 7, 18, 84

diaphragm A sheet of muscle below the lungs that aids breathing. 82

differentiation A process in which new cells develop specialised features in order to carry out a job. 86

diffusion The spreading of gases or liquids caused by the random movement of their molecules. 5, 71, 72, 73, 82

digestion The breakdown of large complex chemicals into smaller simpler ones, e.g. starch into glucose. 5, 77, 80, 81, 83

digestive system The gut and the other body parts linked to it that help with digestion. 80

digital signal A pulsed (on/off) signal. 61, 62

distance-time graph A graph showing time along the *x*-axis and distance covered up the *y*-axis. 111

distillation A process used to boil off a liquid from a mixture and then cool the vapours to produce a pure liquid. 34, 40

DNA Deoxyribonucleic acid. The molecule that carries the genetic information in all living organisms. 16, 18, 65, 86, 88

DNA fingerprint The sequence of bases in a DNA molecule unique to each individual. 88

DNA profile DNA displayed on a sheet of film; each individual's DNA profile (fingerprint) is unique. 88

dominant A dominant allele is one which produces a feature even when another different allele of the same gene is present. 87, 88

double insulated An electrical device in which there are at least two layers of insulation between the user and the electrical wires. 125

drag A force which slows down something moving through a liquid or a gas. 115

dynamo A device that transforms kinetic energy into electrical energy. 55, 115

earthing To be connected to the earth – a safety feature of many electrical appliances. Prevents the build up of charge. 120, 125

earthquake Sudden movements of tectonic plates against each other, causing vibrations that travel through the Earth. 44

effector An organ that responds to a stimulus. 4, 5

efficiency The ratio of useful energy output to the total energy input. 52, 56, 57, 76

egg A special cell (gamete) made by a female for sexual reproduction. 6, 16, 17, 85, 87

electrical charge Electrical energy stored on the surface of a material, also known as a static charge. 50, 53, 54, 55, 56, 57

electrode Bar of metal or carbon that carries electric current into a liquid. 106

electroluminescent A material that emits light of different colours when an alternating current passes through it. 96

electrolysis Using an electric current to split a compound, either in solution or in its molten state. 31, 106, 107

electromagnetic radiation Energy carried by a wave, which can pass through a vacuum. 48, 59, 63

electromagnetic spectrum The range of all the different types of electromagnetic waves in order of their energies. 59, 60, 65

electron A small negatively-charged particle that orbits around the nucleus of an atom. 26, 27, 30, 63, 65, 68, 90, 91, 92, 94, 118, 119, 122, 123, 124, 127, 128

electron notation Representation of the arrangement of electrons around an atom of an element as a series of numbers, e.g. magnesium, 2, 8, 2. 90

electroscope A device that detects if an object has an electrical charge (gold leaf electroscope). 119

element A substance that cannot be split into anything simpler by chemical means. 26, 33, 62, 90, 91, 93, 96, 97, 106

embryo An organism in the earliest stages of development, for example an unborn human baby or a young plant inside a seed. 6, 17, 18, 86, 88

embryo transplantation To move an embryo from one mother, or laboratory vessel, to the uterus of another. 6, 17

emphysema Breakdown of the air sacs in the lungs causing a shortage of oxygen in the blood and breathlessness. 11

emulsifier A chemical which can help to break fats down into small globules. 41, 80

emulsion A mixture with one liquid dispersed in another. 41

endothermic reactions Reactions in which energy is taken in. 104

energy The ability of a system to do something (work). We detect energy by the effect it has on the things around us, e.g. heating them up. 7, 48, 49, 50, 51, 52, 53, 55, 59, 60, 62, 63, 65, 66, 67, 68, 70, 76, 78, 82, 90, 104

energy transfer When the same form of energy moves from one place to another. 48, 49, 50, 51, 54, 61, 75, 115, 116

energy transformation When one form of energy is changed to another. 50, 51, 56, 115

environment The conditions in which a plant or animal has to survive; everything that affects a living organism. 15, 20, 39, 96

enzyme A biological catalyst that speeds up the rate of chemical reactions. 5, 18, 77, 80, 81, 86

epidemic An infectious disease affecting many people in a particular area. 12

equilibrium The point at which a forward reaction in a reversible reaction balances the reverse reaction. 105

evolution The gradual change in living organisms over millions of years caused by random mutations and natural selection. 18, 19

exothermic reaction Reactions in which energy is given out. 104, 105

extinction When all members of a species have died. 20, 23

fat Chemical made up of glycerol and fatty acids. It is used as a source of energy and to make cell membranes. 7, 8, 41, 73, 74

fertilisation When male and female sex cells (gametes) fuse together. 16, 85

fertiliser A substance added to the soil to help plants to grow. 21, 105, 109

fertility drug A drug used to make eggs mature in the ovary. 6

food chain A diagram using arrows to show how energy passes from one organism to another. 75, 76, 77

food web A diagram showing how all of the food chains in an area link together. 75

force A push or pull which is able to change the velocity or shape of a body. 52, 112, 113, 116, 118, 119, 122

fossil Preserved evidence of a dead animal or plant. 19

fossil fuel A fuel such as coal, oil and natural gas formed by the decay of dead organisms over millions of years. 22, 24, 35, 40, 46, 56, 57

fractional distillation The separation of a mixture into its many components using their different boiling points. 34

free electron An electron that is free to move between atoms in a substance. 49, 122

frequency The number of waves passing a point in one second, measured in hertz (Hz). 59, 126

friction The force that acts between two surfaces in contact with each other, resisting movement. 113, 114, 115, 116, 120

fuel A substance that gives out energy, usually as light and heat, when it burns. 38, 40

fuse A component in an electrical circuit containing a thin wire which heats up and melts if too much current flows through it, breaking the circuit. 54, 122, 125, 126, 127

gall bladder An organ associated with the liver that stores bile. 80

gamete Sex cell (egg or sperm) that joins to form a new individual during sexual reproduction. 16, 85, 88

gamma radiation Gamma rays emitted from a radioactive isotope. 64, 65, 128

gamma rays High-energy, high-frequency radiation travelling at the speed of light. 60, 61, 62, 63, 65

Geiger counter A device used to detect some types of radiation. 64, 128

gene A length of DNA on a chromosome in a cell that carries information about how to make particular proteins. 16, 18, 19, 85, 87

generator A device for converting kinetic energy into electrical energy (current flow). 55, 56

genetic engineering A range of technologies that allow scientists to manipulate individual genes. 18

genetic modification Altering the genetic make-up of an organism by transferring genes from one organism to another using DNA technology. 18

geothermal power The energy obtained by transforming the energy in hot rocks to electrical energy. 56, 57

giant covalent structure Macromolecule with strong covalent bonds arranged in a three-dimensional lattice. 93

giant ionic structure Macromolecule with strong ionic bonds arranged in a three-dimensional lattice. 92, 96

global dimming Reduction of the amount of sunlight reaching the Earth's surface because of particles (of smoke) in the atmosphere. 35

global warming An increase in the greenhouse effect caused by an increase in the atmosphere's heat-absorbing greenhouse gases. 24, 31, 35, 46, 48, 57

glucose A type of sugar. Used in respiration to release energy. 5, 7, 73, 74, 78, 81, 82, 83, 84, 96

glycogen An insoluble carbohydrate made from glucose and used for storage in cells. 83, 84

gravity The force of attraction between two bodies caused by their mass. 52, 66, 67, 116

greenhouse effect The warming of the Earth's atmosphere caused by heat becoming trapped in the atmosphere by its greenhouse gases. 24, 35

greenhouse gas Gases such as carbon dioxide and methane that cause the warming of the Earth's atmosphere. 35, 46

Haber process The industrial process developed by Fritz Haber to make ammonia from nitrogen and hydrogen. 99, 104, 105

habitat A place where plants and animals live. 15, 20, 23

haematite Iron ore, mainly iron(III) oxide. 29

half-life The average time taken for half of the atoms of a radioisotope to decay. 64

halogen A reactive non-metal with only one electron missing from its outer electron shell, e.g. chlorine and iodine. In Group 7 of the periodic table. 95

heat Energy that flows from a hot material to a colder one. 30, 34, 48, 49, 51, 53, 54, 94, 115, 123

herbicide A chemical that can kill plants, usually used to mean a chemical that kills weeds. 18, 21

herbivore An animal that eats plants. 75

hertz A unit of frequency equal to one wave per second (Hz). 59

heterozygous When the two alleles for a particular characteristic are different. 87

homeostasis Maintaining the internal conditions of the body within certain constant limits. 83, 84

homeotherm An animal that maintains a steady body temperature. 75

homozygous When the two alleles for a particular characteristic are the same. 87

hormone A chemical produced by an endocrine gland which changes the way other parts of the body work. Hormones pass around the body in the blood. 4, 6, 84

Hubble space telescope A reflecting telescope put into space to exclude atmospheric interference of the images. 65

Huntington's disease An inherited disorder of the nervous system caused by a dominant allele. 88

hydrocarbon A substance whose molecules contain only carbon and hydrogen atoms. 33, 34, 35, 37

hydroelectric power The energy obtained by transforming the kinetic energy of falling water into electrical energy. 24, 56, 57

hydrogen A colourless, odourless gas that burns easily in oxygen to form water; hydrogen is the lightest element. 26, 33, 34, 42, 68, 74, 92, 93, 96, 97, 99, 105, 107, 109, 130

hydrogenation The reaction of hydrogen with carbon-carbon double bonds. Used to make solid fats from liquid vegetable oils. 42

hydroxide A chemical containing an –OH group. 108, 109

hypothermia A condition caused by the body getting too cold. Hypothermia can lead to death if untreated. 84

immune system The parts of the body that protect against illnesses, e.g. white blood cells. 12

immunise To make immune to a disease by vaccination with a harmless pathogen. 13

indicator A substance used to show how acid or alkali a chemical is. 108, 109

infrared Radiation beyond the red end of the visible spectrum that is efficient at transferring heat. 48, 60, 61

inherit To receive something from your parents, usually used to describe characteristics that can be passed down through sperm and eggs. 16

inherited disorder A disorder caused by a defective gene that is passed on from the parents. 88

insoluble A substance that will not dissolve. 81, 109

insulation A substance that slows down the movement of energy. *Heat* insulation in the loft of a house slows down the movement of warmth to the cooler outside. *Sound* insulation cuts down the movement of sound in recording studios. *Electrical* insulation prevents the flow of electricity. 15, 24, 49, 53, 120, 125

insulator A material that does transfer heat or electrical charge. 118, 119, 122

insulin A hormone released by the pancreas that reduces blood sugar levels. 4, 5, 18, 83, 84

intensive farming Farming that uses methods (e.g. artificial fertilisers, pesticides, herbicides, battery farming) which maximise food output and minimise prices. 21

intercostal muscles Muscles between the ribs that raise and lower the ribs in breathing. 82

invertebrate An animal without a backbone. 23

ion Charged particle made when an atom, or group of atoms, gains or loses electrons. 5, 27, 30, 31, 74, 83, 91, 94, 106, 107, 109, 127

ionic bond When oppositely charged ions of different atoms are attracted to one another and held together. 27, 91, 95

ionic compound A compound in which the reacting atoms have formed oppositely charged ions that attract each other. 91, 92, 94, 95, 106

ionisation The formation of ions (charged particles). 61, 65, 120, 128

isotope A form of an element with the same atomic number but different mass numbers. 62, 63, 64, 127, 128, 129

IVF In vitro fertilisation. Eggs are fertilised outside the body. 6

joule A unit of energy (J). 50, 52, 54

kidney Organ that filters blood to make urine. 5, 83

kilowatt 1000 watts (W) = 1 kW. 54

kinetic energy Movement energy. 50, 51, 53, 55, 56, 94, 102, 114, 115, 116, 117, 118, 123, 130

kwashiorkor An illness caused by protein deficiency due to lack of food. 8

laser light Infrared and visible rays that are used to transmit information. 61, 62

lattice The ordered arrangements of atoms or ions to form a stable structure. 29, 92, 93, 106

lava Molten rock thrown up by a volcano. 44

limestone Rock made from the remains of shells and skeletons; calcium carbonate. 22, 27, 28, 29, 46

limiting factor A factor that limits the rate of photosynthesis because it is in short supply. 74

lipase An enzyme that digests fat. 81

lipoprotein A mixture of cholesterol and protein in the blood. 8

lithosphere The outer part of the Earth, consisting of the crust and upper mantle, approximately 100 km thick. 44

liver The largest organ in the body. It produces bile, processes glucose and proteins, and helps get rid of toxic substances. 5, 9, 10, 80, 83

lymphocyte A white blood cell that produces antibodies. 12

magma Molten rock inside the Earth. 44

magnesium A metal in Group 2 of the periodic table. 27, 74, 90, 97, 103

magnet An object that is magnetic is attracted by a magnet. 55

magnet field A region that attracts or repels a magnet. 55, 63, 128

malnourished When a person has an unbalanced diet that does not meet all their needs. 7, 8

mantle The layer below the outer crust of the Earth, mainly 'plastic' rock. 43, 44

mass Mass describes the amount of something; it is measured in kilograms (kg). 26, 66, 67, 101, 103, 113, 114, 115, 116, 117

mass number The mass of an atom compared with hydrogen. 26, 62, 90, 91, 127

menstrual cycle The sequence of events involving menstruation and the release of an egg from the ovary. 6

menstruation The breakdown of the lining of the uterus and its loss from the body. 6

metabolic rate The rate of chemical reactions in the cells. 7

methane A colourless, odourless gas that burns easily to give water and carbon dioxide. 24, 34, 45, 93

microorganism An organism that is only visible under a microscope. 77, 78, 81

mineral Natural solid materials with a fixed chemical composition and structure, rocks are made of collections of minerals; small amounts of minerals are needed by plants and animals to keep them healthy. 74, 83, 93

mitochondrion Cell structure that carries out respiration. 70, 82

mitosis The process of cell division that ensures that new cells have a complete copy of inherited information. 85, 86

mole The relative formula mass of a substance in grams of that substance. 97

molecule A group of atoms joined together by chemical bonds in which electrons are shared. 27, 34, 80, 82, 92, 93, 97

molten The state of a substance that has been heated to change it from a solid to a liquid. 29, 31, 44, 92, 106

momentum The product of an object's mass and velocity. It is a vector quantity (it has size and direction). 117, 118

motor A device that transforms electrical energy into kinetic energy. 55

mucus A thick liquid secreted by cells, e.g. goblet cells in the alimentary canal. 11

mutation A random change in the genotype of an organism. Mutations are almost always disadvantageous. 19, 65

nanoparticle A tiny, manufactured structure of a few hundred atoms. 95

National Grid The nationwide network of power stations and electricity cables that carries electricity to homes and businesses. 55, 56

natural selection Factors in the environment affect animals and plants so that some survive to reproduce successfully and pass on their good combinations of genes. Others survive less well and do not pass on their poor combinations of genes as often. 13, 18, 19

nebula A huge gas cloud in space. 65, 66, 130

negative ion An ion with a negative charge. 91, 92, 106

nephrons Tiny tubes in the kidneys that collect waste substances, including urea and water, from the blood. 83

nervous system The central nervous system and nerves that control the actions and functions of the body. 4

neurone A nerve cell, specialised for conducting electrical impulses. 4, 5

neutral A solution with a pH of 7 that is neither acidic nor alkaline. 108

neutralisation A reaction between an acid and an alkali to produce a neutral solution of a salt and water. 22, 104, 108, 109

neutron A particle found in the nucleus of an atom; it has no electrical charge and a mass of 1 atomic mass unit. 26, 62, 68, 90, 91, 127, 128, 129, 130

newton The unit of force (N). 113

nicotine An addictive drug found in tobacco. 11

nitrate A salt of nitric acid. 74, 82, 105

nitrogen A non-reactive gas that makes up most of the atmosphere. 27, 33, 45, 74, 99, 104, 105

non-renewable A resource that is being used up faster than it can be replaced, e.g. fossil fuels. 20, 33, 38, 39, 40, 56

nuclear fission The splitting of an atom's nucleus to release nuclear energy. 56, 129

nuclear fusion When nuclei of atoms fuse together giving out huge amounts of energy, e.g. when a star is born. 66, 67, 68, 130

nuclear power The energy obtained by transforming the energy from nuclear fission to electrical energy. 56, 57, 129, 130

nuclear radiation The energy released when the nucleus of an atom undergoes radioactive decay.

nucleus: biology The control centre of the cell, the nucleus is surrounded by a membrane that separates it from the rest of the cell. 16, 17, 86

nucleus: chemistry The central part of an atom containing the protons and neutrons. 26, 62, 63, 64, 90, 127

oestrogen A hormone produced in females by the ovary. 4, 6

Ohm's Law Current is directly proportional to potential difference across the resistor. 123, 124

optical fibre A very fine strand of silica glass, through which light passes by successive internal reflections. 62

optical telescope An instrument that uses light to form images of objects far away, e.g. in space. 65

optimum temperature The temperature range that produces the best reaction rate. 80

orbit A path, usually circular, of a smaller object around a larger object, e.g. a planet orbits the Sun; electrons orbit the nucleus in an atom. 66

organelle An internal part of a cell, e.g. ribosome, mitochondrion. 70, 82

oscilloscope A device that displays a line on a screen showing regular changes (oscillations) in something. An oscilloscope is often used to look at sound waves collected by a microphone. 126

osmosis The diffusion of water molecules from a dilute solution to a more concentrated solution through a partially permeable membrane. 72

ovary Female reproductive organ. 4, 6, 85

oxidation Loss of electrons from an atom. 104, 106

ozone A substance in the atmosphere formed from oxygen (O_3), which acts as a shield against ultraviolet radiation. 45

pancreas An organ in the abdomen that produces enzymes to break down food and the hormone insulin to reduce blood sugar level. 4, 5, 80, 84

pandemic A world-wide epidemic. 12, 13

Pangaea The supercontinent thought to be the first landmass to form. 44

parallel circuit An electrical circuit in which the current divides to take different routes then joins up again. 122, 125

partially permeable membrane A membrane through which only some molecules can pass. 72

pathogen An organism that causes a disease. 11, 12, 13

payback time How long it takes for savings to equal costs. 53

periodic table A chart grouping the elements according to their similarities, first devised by Dimitri Mendeleev. 26, 90, 91

pesticide A chemical used to kill a pest. 21

pH The level of acidity or alkalinity of a substance measured as a scale from 1 to 14: neutral, 7; acidic, below 7; alkaline, above 7. 80, 108

phagocyte A white blood cell that ingests bacteria in a process called phagocytosis. 12

photochromic A material that darkens on exposure to sunlight. 96

photosynthesis The production, in green plants, of glucose and oxygen from carbon dioxide and water using light as an external energy source. 21, 22, 23, 45, 46, 73, 74, 78, 80

pollutant A chemical that causes pollution. 31

pollution Human activity that causes damage to the environment, e.g. dumping raw sewage at sea. 20, 21, 22, 23, 33, 39

pollution indicator An organism that is sensitive to a particular level of environmental pollution. The presence or absence of a particular pollution indicator can be used to assess the degree of pollution in an environment. 23

polymer A molecule made of many repeating subunits, e.g. polythene or starch. 37, 38, 39

polymerisation The process of forming large polymers from smaller monomer molecules. 38

polyunsaturates Fats with more than one carbon-carbon double bond. 41

population All the organisms of the same species that live in a particular place. 20, 21

positive ion An ion with a positive charge. 91, 92, 94, 106

potential difference The difference between two points which will cause current to flow in a closed circuit, measured in volts (V). 119, 122, 123, 124, 125, 126

potential energy Stored energy in chemicals or objects. 50, 53, 115, 116

power The rate that a system transfers energy, i.e. does work, usually measured in watts (W). 52, 55

power rating The rate at which an appliance transforms electrical energy. 54, 127

power station A place that generates electricity. 22, 35, 56, 57

precipitate To fall out of solution. A insoluble solid. 109

predator An animal that hunts and kills other animals. 15, 20, 75

pressure The force acting on a surface divided by the area of the surface, measured in newtons per square metre (N/m^2). 102

prey Animals that are hunted by other animals. 75

producer An organism that makes organic material from inorganic substances. Green plants are primary producers because they use energy in sunlight to make sugar. 75

product Something made by a chemical reaction. 98, 99, 101, 107

protease An enzyme that digests proteins. 81

protein A complex molecule that contains carbon, hydrogen, oxygen, nitrogen, and sometimes sulfur. It is made of one or more chains of amino acids. It is used for growth, repair and energy. 7, 73, 74, 80, 81, 83, 86

protein synthesis To make protein in a cell from amino acids. 70, 80

proton A particle found in the nucleus of an atom with a charge of +1 and a mass of 1 atomic mass unit. 26, 62, 63, 68, 90, 91, 127, 128, 130

pyramid of biomass A diagram to show the masses of living organisms present at each step in a food chain. 75

radiation Energy that travels as light or other forms of electromagnetic waves. 59, 60, 61, 62, 63, 64, 65, 127, 128

radio wave A form of electromagnetic radiation used to carry radio signals. 50, 60, 65

radioactive Material which gives out radiation. 56

radioactive decay The breakdown of an unstable nucleus of an atom, releasing energy. 43, 62, 64, 128

radioactive waste Waste produced by radioactive materials used at power stations, research centres and some hospitals. 65

radioisotope A radioactive isotope. 62, 63, 64

radiotherapy Using radiation (gamma rays) to treat certain types of disease, e.g. cancer. 64

radon A naturally occurring radioactive gas, found near granite rocks. 63, 128

reactant A chemical taking part in a chemical reaction. 98

reaction time: biology/physics The time taken to respond to a stimulus. 114

reaction time: chemistry The time taken for a reaction to finish. 101, 103, 104, 105

reactivity A measure of how easily a chemical will react. 32, 37, 38, 94, 95, 107

reactivity series A series of metals in order of their ability to react with water and acids. 107

recessive A recessive allele is one which produces a feature only when the dominant allele of the same gene is not present. 87, 88

receptor A special cell that detects a stimulus. 4, 5

reduction Gain of electrons by an atom. 106

reflex action A rapid automatic response to a stimulus. 5

reflex arc The route of the electrical impulse in a reflex action along a sensory, a relay and a motor neurone. 5

relative atomic mass The mass of an atom (A_r) as the sum of its protons and neutrons (same as its mass number). 91, 97

relative formula mass The mass of a molecule (M_r) in comparison to the mass of hydrogen, which is taken as 1. 97

relative mass The mass of a particle (proton, neutron or electron) in an atom relative to one another. 90

renewable A resource that can be replaced, e.g. wind power, wave power and solar power are all renewable sources of energy. 33, 38, 40, 56, 57

resistance: biology When an organism's genetic make-up changes so that it is unaffected by an agent used to destroy it. 12, 13, 18

resistance: physics The amount by which a conductor prevents the flow of electric current. 54, 122, 123, 124

resistor A conductor that reduces the flow of electric current. 122, 123, 127

respiration The chemical process that releases energy from food in all cells. All living things must respire. 23, 70, 71, 73, 75, 76, 77, 78, 80, 82

resultant force The sum of all the different forces acting on an object. 112, 114

reversible reaction A reaction whose direction can be changed by a change in conditions. 96, 98, 99, 104, 105

ribosome An organelle in the cell in which protein synthesis takes place. 70, 98

salt A compound made when an acid reacts with an alkali. Common salt is sodium chloride. 8, 74, 94, 95, 107, 108, 109

Sankey diagram A diagram that shows the input and output of energy in a device. 51, 52

satellite A body orbiting around a larger body. Communications satellites orbit the Earth to relay messages from mobile phones. 61, 65

saturated Fats or oils containing only single bonds between carbon atoms. 8, 34, 37, 41

secretion The production of useful substances by organs or cells, e.g. hormones by glands. 4, 5, 81

sedimentation Particles settling out of suspension in water. 19

series circuit An electrical circuit where components are connected in one continuous loop. 122, 124

sewage Waste matter and rubbish carried in sewers. 21, 23

sex chromosomes Chromosomes that carry the genes that determine sex. 87

sexual reproduction Reproduction involving gametes and fertilisation. There are usually two parents each contributing half the genes of the offspring. 16

shape memory When a smart alloy that can be deformed is able to return to its original shape when heated. 33, 96

shell A grouping of electrons in an energy level around an atom. 26, 27, 30, 90, 92, 94, 95

silicon dioxide Sand. Covalently bonded silicon and oxygen atoms. 93

skin graft Surgical replacement of damaged skin by undamaged skin. 17

smart alloy An alloy that has a shape memory. 33

smart material A material that responds to a change in the environment by changing in some way that is reversible. 96

smog Smoke particles plus water droplets. 22

smoke Tiny particles (e.g. carbon particles) suspended in air, produced by burning materials. 22, 64

solar cell A device that converts light energy into electrical energy. 57

solar energy Energy from the Sun. 53, 57, 75

Solar System The collection of planets and other objects orbiting around the Sun. 43, 66, 67

species A group of living things of the same type. Humans belong to the species *Homo sapiens*. 18, 19, 20

spectroscope A device that splits light into bands of different colours. 68

spectral pattern The pattern formed when an element is heated and emits light, as viewed by a spectroscope. 68

speed How fast an object is moving. 101, 111, 112, 113, 114, 115, 116

sperm A special cell (gamete) made by a male for sexual reproduction. 16, 70, 85, 87

star A large gas body in space that emits light from internal nuclear reactions. 66, 67, 130

static electricity Electric charge on the surface of a non-conductor, i.e. it is charge that cannot flow easily. 102, 118, 119

stem cell An undifferentiated cell found in embryos and certain parts of adults that can develop into many other types of cell. 86

stimulus A change inside or outside of an organism that brings about a response, e.g. altered blood sugar levels; a flash of light. 4, 5

stoma Tiny hole in the surface of a leaf which allows gases to diffuse in and out. 73

stopping distance The distance required to bring a moving vehicle to a halt. The sum of the thinking distance and the braking distance. 114

subduction zone A region where one tectonic plate moves under another. 44

substrate Something that an enzyme causes to react to make a product. 88

surface area The area of a surface. Surface area has a significant effect on the rate of many chemical reactions. 15, 82, 95, 102, 103

sustainable development Development that does not make life more difficult for future generations. 24, 98, 105

synapse The gap between the ends of two neurones. 5

target organ An organ affected by the presence of a particular hormone, e.g. adrenaline from the adrenal gland affects the heart. 4

tectonic plate Cracked sections of the lithosphere that ride on top of the fluid mantle. 44

telecommunications Communication over a distance by cable, telephone or broadcasting. 61

temperature A measure of how hot something is. 46, 48, 71, 74, 77, 80, 83, 84, 101, 102, 123, 124

temperature receptor Cell in the brain that monitors blood temperature. 84

testis Male reproductive organ. 4, 85

testosterone A male hormone made by the testes. 4

thermal decomposition Breaking down a chemical using heat. 28, 104

thermal energy *See* heat.

thermal radiation The transfer of heat energy by rays. Also called infrared radiation. 48

thermochromic Material that changes colour when it gets hotter or colder. 96

thermopile An instrument that measures temperature. 48

thermoregulatory centre An area of the brain containing temperature receptors that monitors blood temperature. 84

thinking distance The distance travelled by a moving vehicle before the brakes are applied. The product of speed and reaction time. 114

tidal power The energy obtained by transforming the energy in the rise and fall of the tides into electrical energy. 56, 57

tissue A group of cells of the same type, e.g. nervous tissue contains nerve cells. 70, 73

total internal reflection When light hits a glass/air interface at an angle of more than 42°, it is reflected back at the boundary. 62

toxic chemical A chemical that harms living things, a poison. 9, 21, 42

toxin A poison. Usually used to mean a poison produced by a living organism. 11

transformer A device that alters the voltage of an alternating current. 55, 56

transition element A metal belonging to the transition group in the periodic table. 30

turbine A device that transfers kinetic energy into circular movement, usually to drive a generator. 56, 57, 130

ultraviolet radiation Radiation just beyond the blue end of the spectrum of visible light. UV light causes tanning and some sorts of skin cancer. 60, 61

Universe Everything, everywhere. 68

upthrust An upward force, e.g. of water on a floating object. 113

uranium A radioactive metal used in nuclear power stations and bombs. 129, 130

urea A substance made in the breakdown of proteins in the liver. 5, 83

urine A solution of urea and other wastes in water, made in the kidneys for excretion by the bladder. 5, 83

uterus The organ in the female where the baby grows during pregnancy; also known as the womb. 6

vaccine A chemical developed to immunise people against a particular disease. 13

vaccination Injection of a vaccine into the bloodstream. 13

vacuum A space with no particles in it. 48, 59

Van de Graaff generator A device used to generate static electrical charges. 119

variation The existence of a range of individuals of the same group with different characteristics. 16, 19, 23, 85

vegetable oils Oils obtained from the fruits, seeds and nuts of some plants, e.g. olive, rape, coconut. 40

velocity The speed of an object in a given direction (v). 111, 113, 114, 115, 117, 118

virus A microorganism (a thousand times smaller than a bacterium) that invades cells, reproduces and bursts cells, making people ill. 11, 12, 13

vitamin A substance found in foods that is essential to maintain health. 40

volcano The upthrust of the crust of the Earth at a point of weakness, allowing magma from the mantle to be forced out. 44

volt The international unit of electrical potential (V). 126

voltage The potential difference across a component or circuit (V). 55, 56, 94, 120

voltmeter A meter that measures the voltage between two points. 122, 124, 125

watt A unit of power (W), 1 watt equals 1 joule of energy being transferred per second. 52, 54, 127

wave power The energy obtained from transforming the energy of waves into electrical energy. 56, 57

wavelength The distance between two identical points on a wave. 59, 60, 61, 68, 75

weight The force of gravity acting on a body measured in newtons (N). 52, 114

wind farm A group of wind turbines used to generate electricity from the kinetic energy in moving air. 56

withdrawal symptoms The combination of physical and psychological symptoms produced when an addictive drug is withheld for a period of time. 9, 10

work Work is done when a force moves. The greater the force or the larger the distance, the more work is done. 52, 115, 116

X-ray Electromagnetic radiation used for imaging or to destroy some types of cancer cells. 60, 61, 88

XX chromosomes The sex chromosomes present in a human female. 87

XY chromosomes The sex chromosomes present in a human male. 87

yeast A unicellular fungus used extensively in the brewing and baking industries. 38

yield The ratio of product to starting materials; a high yield means that most of the starting material is converted to useful products. 98, 105

zygote A fertilised egg, produced when a male and female gamete join. 17

Collins Revision

GCSE Foundation Science

Exam Practice Workbook

FOR AQA A + B

Coordination

1 a Name the **two** parts of the body that make up the central nervous system.

_____ and _____   [1 mark]

b The diagram shows a neurone from the human body.

i What type of neurone is this? _____   [1 mark]

ii Draw an arrow on the diagram to show the direction in which the nerve impulse travels along the neurone.   [2 marks]

iii Complete the sentences by choosing the **correct** words from the box.

central	chemical	effectors	electrical	receptors	spinal

Neurones carry information in the form of _____ impulses.

Sensory neurones carry impulses from _____ to the

_____ nervous system.   [3 marks]

Receptors

2 The drawing shows some parts of the head where there are receptors. Match statements, **A**, **B**, **C** and **D**, with labels **1 – 4** of the head.

A contains receptors that detect chemicals
B contains touch receptors
C contains light receptors
D contains sound receptors

1	2	3	4

[4 marks]

3 a The diagram shows some glands in the human body.

Match labels, **A**, **B**, **C** and **D**, with the statements **1 – 4** in the table.

1	secretes insulin
2	secretes oestrogen
3	secretes adrenaline
4	secretes testosterone

1	2	3	4

[4 marks]

b Adrenaline is secreted by the adrenal glands when you get a sudden fright. It makes the heart beat faster.

i Name the target organ of adrenaline.

_____   [1 mark]

ii Suggest how adrenaline could help you to escape from danger.

_____   [1 mark]

Reflex actions

1 a The list contains the names of some parts of the body involved in a reflex action.

| effector | motor neurone | receptor | sensory neurone |

In what sequence does a nerve impulse travel?

A effector, motor neurone, receptor, sensory neurone

B sensory neurone, receptor, effector, motor neurone

C receptor, sensory neurone, motor neurone, effector

D motor neurone, effector, sensory neurone, receptor [1 mark]

b The diagram shows a synapse.

i What is contained in the structures labelled X? _____ [1 mark]

ii Describe how the nerve impulse travels across the synapse.

_____

_____ [2 marks]

In control

2 a A label on a bottle of a sports drink lists these ingredients:

| glucose | water | sodium ions | chloride ions |

Which ingredient gives you energy?

A glucose

B water

C sodium ions

D chloride ions [1 mark]

b List **three** ways in which water is lost from the body.

1_____ 2_____ 3_____ [3 marks]

c Explain why an athlete might need to drink liquids containing sodium and chloride ions.

_____

_____ [2 marks]

d If you have too much glucose in your blood, your pancreas secretes a hormone called insulin. Describe how insulin helps to make the blood glucose level go back to normal.

_____

_____ [2 marks]

Reproductive hormones

1 The diagram shows the menstrual cycle.

a At which stage could an egg be fertilised? _____ [1 mark]

b Which hormones help to control the menstrual cycle?

  A adrenaline, oestrogen, LH, testosterone

  B FSH, LH, oestrogen, progesterone

  C FSH, oestrogen, insulin, adrenaline

  D insulin, progesterone, testosterone, adrenaline [1 mark]

Controlling fertility

2 a In IVF treatment, where is an egg fertilised?

  A in a Petri dish   **B** in the oviduct   **C** in the uterus   **D** in the vagina   [1 mark]

b This table shows the success rate for one IVF treatment in women of different ages.

Age of woman in years	Chance of having a baby after IVF
23–35	More than 20%
36–38	15%
39	10%
40 and over	Less than 6%

  i Describe how a woman's age affects the success rate for IVF treatment.

  _____ [1 mark]

  ii Using the data in the table, and your own ideas, suggest arguments supporting the idea that women over 55 years of age should not be given IVF treatment on the National Health Service.

  _____

  _____ [2 marks]

3 a Oral contraceptives (the pill) prevent the production of a hormone. Which hormone is this?

  A FSH   **B** oestrogen   **C** progesterone   **D** testosterone   [1 mark]

b Explain how the use of oral contraceptives could increase the spread of sexually transmitted diseases such as HIV/AIDS.

  _____

  _____ [2 marks]

Diet and energy

1 a Complete the sentences by choosing the **correct** words from the box.

balanced	fats	malnourished	obese	slimming	vitamins

We get our energy from the food that we eat, especially _____ ,

carbohydrates and proteins. A _____ diet contains the right amount of

energy. People who eat a poor diet are said to be _____. [3 marks]

b Katie has a high metabolic rate. Olivia's metabolic rate is lower. Olivia is jealous of Katie because Katie can eat more food than Olivia without putting on weight.

    i What is **metabolic rate**?

    _____

    [1 mark]

    ii Explain why Katie's high metabolic rate means that she can eat more food than Olivia without putting on weight.

    _____

    _____

    [2 marks]

    iii Suggest what Olivia could do to increase her metabolic rate.

    _____

    [1 mark]

Obesity

2 The graph shows the percentage of obese men and women in five different countries.

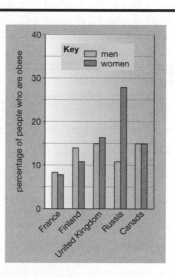

a Which country has the lowest percentage of obese people?

_____ [1 mark]

b What is the percentage of obese women in Russia?

_____ [1 mark]

c In what way does the pattern of obesity in Russia differ from the pattern in Canada?

_____ [1 mark]

d Explain why obese people are more likely to suffer from arthritis than people of normal weight.

_____

_____ [2 marks]

Not enough food

1 a In some parts of the world people do not have enough to eat. Explain how this can lead to increased rates of tuberculosis (TB) and other infectious diseases.

_____ [1 mark]

G–E

D–C

b A slimming programme advertises that you can lose 18 lbs (8 kg) in four days. It says:

> The 18-in-4 program identifies common, but specific FRUITS, VEGETABLES, CHICKEN AND BEEF that need to be eaten in specific combinations. These combinations of specific foods work with the body's chemistry and result in accelerated weight loss.

i Explain why the **combination** of food that a person eats, rather than the **quantity** of food that they eat, is unlikely to help them lose weight.

_____ [1 mark]

ii Suggest **two** reasons why it is not good for a person to lose this much weight so quickly if they are trying to improve their health.

1 _____ [1 mark]

2 _____ [1 mark]

Cholesterol and salt

2 This table shows information about the cholesterol concentration in the blood of men and women in two cities in the United Kingdom. A blood cholesterol concentration above 6.6 units is dangerous to health.

G–E

City	Men		Women	
	Average cholesterol concentration	Percentage with cholesterol above 6.6 units	Average cholesterol concentration	Percentage with cholesterol above 6.6 units
A	5.9	27	5.9	31
B	6.1	35	6.1	36

a What conclusion can be drawn from this data?

A People in City B eat more fast food than people in City A.

B People in City B have a higher risk of heart disease than people in City A.

C Men have higher cholesterol levels than women.

D Women have a lower risk of heart disease than men. [1 mark]

b Complete the sentences by choosing the **correct** words from the box.

| cholesterol   less   monounsaturated   more   salt   saturated |

The blood carries _____ in two forms, HDL and LDL. It is better

to have _____ HDL than LDL. Eating _____ fats

can help to reduce blood cholesterol. [3 marks]

c Explain how a high blood cholesterol level can lead to heart disease.

_____

_____ [3 marks]

D–C

Drugs

1 a The table is about the uses of drugs. Match words, **A**, **B**, **C** and **D**, with the statements **1 – 4** in the table.

A recreational **B** legal **C** illegal **D** dependency

1	A person uses the drug because they like the way it makes them feel.
2	A person uses the drug because they cannot manage without it.
3	It is against the law to use this kind of drug.
4	It is not against the law to use the drug, even though it may be harmful.

1	2	3	4

[4 marks]

b Explain what is meant by being **addicted** to a drug.

_____

_____ [2 marks]

Trialling drugs

2 a Describe how a drug is tested **before** it is tried out on humans.

_____ [1 mark]

b This table shows the results of a trial to test a new drug to help people to recover from flu.

		Given zanamivir	Given a placebo
A	Mean age of subjects in years	19	19
B	Number of days until their temperature went down to normal	2.00	2.33
C	Number of days until they lost all their symptoms and felt better	3.00	3.83
D	Number of days until they felt just as well as before they had flu	4.5	6.3
E	Average score the volunteers gave to their experience of the five major symptoms of flu	23.4	25.3

Answer these questions using information from the table.

i Give the letter of the row in the table that shows how researchers keep one variable constant in their trial. _____

[1 mark]

ii Give the letters of **all** the rows in the table that show that the drug helps people to get over flu more quickly.

_____ [1 mark]

iii Explain why some of the people in the trial were given a placebo.

_____

_____ [2 marks]

iv The trial was a **double-blind** trial, where neither the researchers or the subjects knew whether they were being given the drug or a placebo. How can this make the results more valid?

_____

_____ [2 marks]

Illegal drugs

1 a Which drug is illegal?

    A alcohol

    B aspirin

    C cannabis

    D nicotine   [1 mark]

  b **i** Explain how injecting heroin into the blood can increase the risk of getting HIV/AIDS.

       _____

       _____ [2 marks]

    ii Explain why people who have begun taking cocaine may find it almost impossible to stop taking the drug.

       _____

       _____ [2 marks]

Grades: G–E, D–C

Alcohol

2 The table shows the number of people who were admitted to hospital in one area of the USA in one year because they had been taking drugs.

Drug they had taken	Number of men admitted to hospital	Number of women admitted to hospital
Alcohol	380	140
Heroin and similar drugs	90	50
Cocaine	40	30
Cannabis	100	20

  a Which drug caused the most hospital admissions?

    _____ [1 mark]

  b Which drug in the table is a legal drug in the United Kingdom?

    _____ [1 mark]

  c Amy says that the table shows that more people use heroin than cocaine. Is she right? Explain your answer.

    _____

    _____ [2 marks]

  d Zachary says that the table shows that more men than women take these drugs. Is he right? Explain your answer.

    _____

    _____ [2 marks]

  e Excessive drinking of alcohol damages the liver. Why does this happen?

    _____ [1 mark]

Grades: G–E, D–C

Tobacco

1 a The table is about smoking tobacco. Match words, **A**, **B**, **C** and **D**, with the statements **1 – 4** in the table.

A bronchitis   **B** emphysema   **C** nicotine   **D** tar

1	contains carcinogens
2	inflammation and infection in the bronchi
3	breakdown and loss of air sacs in the lungs
4	the addictive substance in cigarette smoke

1	2	3	4

[4 marks]

b This graph shows information about the numbers of men who smoked and the numbers of men who died of lung cancer between 1911 and 2001.

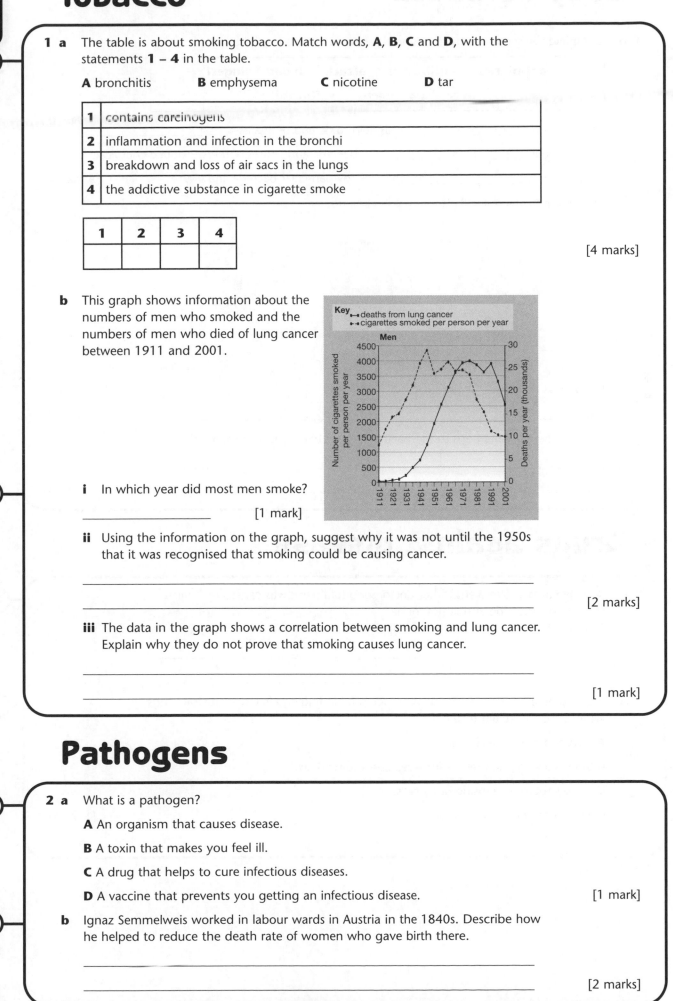

i In which year did most men smoke?

_____   [1 mark]

ii Using the information on the graph, suggest why it was not until the 1950s that it was recognised that smoking could be causing cancer.

_____

_____

[2 marks]

iii The data in the graph shows a correlation between smoking and lung cancer. Explain why they do not prove that smoking causes lung cancer.

_____

_____

[1 mark]

Pathogens

2 a What is a pathogen?

A An organism that causes disease.

B A toxin that makes you feel ill.

C A drug that helps to cure infectious diseases.

D A vaccine that prevents you getting an infectious disease.

[1 mark]

b Ignaz Semmelweis worked in labour wards in Austria in the 1840s. Describe how he helped to reduce the death rate of women who gave birth there.

_____

_____

[2 marks]

Body defences

1 a Complete the sentences by choosing the **correct** words from the box.

antibiotics	antibodies	attack	blood	ingest
liver	pathogens	suffocate		

White _____ cells help to defend us against _____ .

Some of them _____ bacteria and viruses. Others produce chemicals

called _____ that destroy particular bacteria and viruses. [4 marks]

b The diagrams show a cell attacking a bacterium.

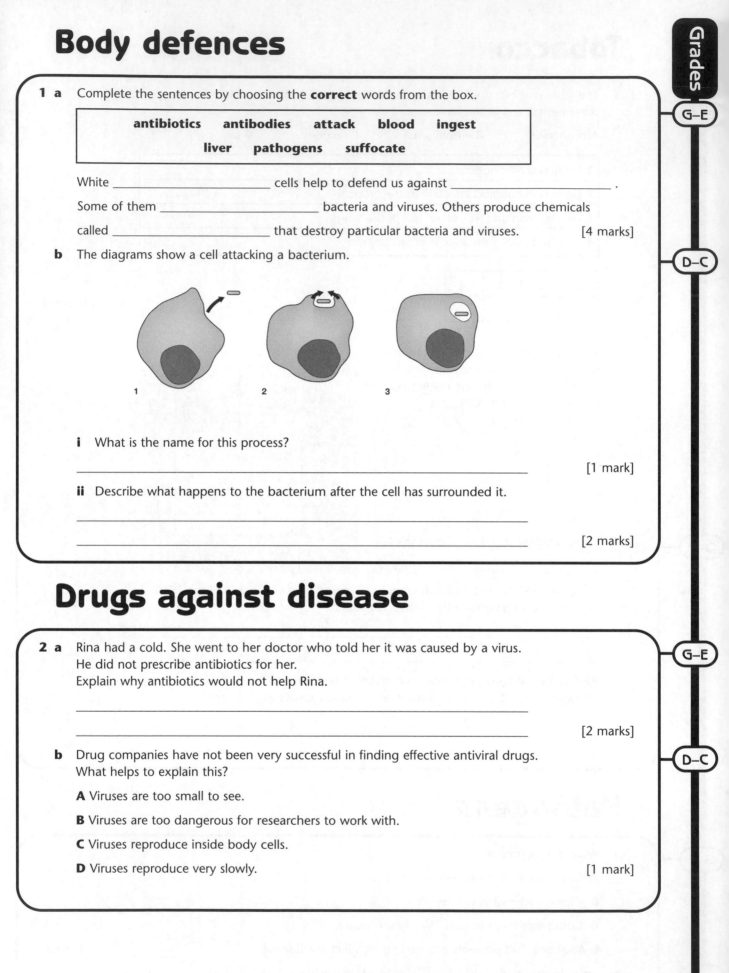

i What is the name for this process?

_____ [1 mark]

ii Describe what happens to the bacterium after the cell has surrounded it.

_____

_____ [2 marks]

Drugs against disease

2 a Rina had a cold. She went to her doctor who told her it was caused by a virus.
He did not prescribe antibiotics for her.
Explain why antibiotics would not help Rina.

_____

_____ [2 marks]

b Drug companies have not been very successful in finding effective antiviral drugs.
What helps to explain this?

A Viruses are too small to see.

B Viruses are too dangerous for researchers to work with.

C Viruses reproduce inside body cells.

D Viruses reproduce very slowly. [1 mark]

Arms race

1 Complete the sentences by choosing the **correct** words from the box.

| fewer | immune | more | MMR | MRSA | natural | random | resistant |

Bacteria can become _____ to antibiotics by the process of

_____ selection. One example is _____ . We can help

to prevent this happening by using _____ antibiotics.   [4 marks]

2 Most deaths from MRSA happen in hospitals. Which **cannot** help to explain this?

A A lot of antibiotics are used in hospitals, increasing the chance of antibiotic resistance developing.

B It is easy for bacteria such as MRSA to spread from one person to another in a hospital ward.

C Many people in hospital are already ill and weak.

D MRSA is only found in hospitals.   [1 mark]

Vaccination

3 Which diseases does the MMR vaccine protect against?

A measles, mumps and rubella

B meningitis, mumps and rickets

C meningitis, MRSA and rubella

D measles, MRSA and rickets   [1 mark]

4 TB is an infectious disease caused by a bacterium. A vaccine called BCG was introduced in 1951. It is given to most school pupils between the ages of 10 to 14. The table shows the number of new cases of TB in the United Kingdom in some of the years between 1950 and 1990.

Year	Number of new cases of TB
1950	50 000
1960	25 000
1970	16 000
1980	9000
1990	6000

a How many more people caught TB in 1960 than in 1970?

_____   [1 mark]

b Do the figures in the table prove that the BCG vaccination has been effective? Explain your answer.

_____

_____   [2 marks]

c Suggest why the number of new cases of TB took many years to drop as low as 6000 cases per year.

_____

_____   [2 marks]

d The TB vaccine contains an extract made from the bacteria that cause TB. Explain how the vaccine can make a person immune to TB.

_____

_____

_____

_____   [4 marks]

B1a revision checklist

I know:

how human bodies respond to changes inside them and to their environment

☐ nerves and hormones coordinate body activities and help control water, ion and blood sugar levels and temperature

☐ receptors detect stimuli; information passes to the brain; and effectors (muscles and glands) respond

☐ reflex actions are automatic and very fast; they involve sensory, relay and motor neurones, with synapses between them, in a reflex arc

☐ the menstrual cycle is controlled by FSH, LH and oestrogen

what we can do to keep our bodies healthy

☐ a balanced diet contains the right balance of all the nutrients and energy humans need, and regular exercise helps you to stay fit

☐ different people have different energy needs

☐ obesity increases the risk of arthritis, diabetes, high blood pressure and heart disease

☐ too much cholesterol in the form of LDL in the blood increases the risk of heart disease

how we use/abuse medical and recreational drugs

☐ drugs affect people's behaviour, can damage the brain and some hard drugs are addictive

☐ some drugs are legal (alcohol) and some are illegal (cannabis, cocaine, heroin)

☐ the substances in tobacco smoke cause many diseases, e.g. cancer and bronchitis

☐ drugs can be beneficial, but must be thoroughly tested before use

what causes infectious diseases and how our bodies defend themselves against them

☐ pathogenic microorganisms cause infectious diseases

☐ some white blood cells (phagocytes) ingest pathogens and kill them; and others (lymphocytes) make antibodies

☐ antibiotics kill bacteria inside the body but do not kill viruses; however, some bacteria can develop resistance to antibiotics

☐ immunisations and vaccinations offer protection from various diseases

Hot and cold

1 Arctic foxes live where it is very cold. They have thick white fur and small ears. Elephants live where it is hot. They have almost no hair and large ears.

Match words, **A, B, C** and **D**, with the spaces **1 – 4** in the sentences.

A increases   **B** decreases

C cools   **D** camouflages

The thick fur of an Arctic fox traps air and ____ **1** ____ the rate of heat loss from its body to the air.

The white colour of the fur ____ **2** ____ the fox in the snow.

Elephants have large ears. This ____ **3** ____ their surface area in relation to their

volume and ____ **4** ____ their bodies.

1	2	3	4

[4 marks]

2 The diagram shows the blood flow in the leg of a caribou. Caribou live in very cold places.

a What is the temperature of the blood in the artery as it flows into the top of the caribou's leg?

_____ °C

b Explain why the blood in the caribou's foot has a lower temperature than this.

_____

_____

[1 mark]

c How does the arrangement of blood vessels shown in the diagram help to stop the temperature deep inside the caribou's body from dropping too low?

_____

[1 mark]

Adapt or die

3 Complete the sentences by using the **correct** words from the box.

adapted   camouflaged   competitors   poisons
predators   thorns

Plants and animals are _____ for survival in their habitat. Some plants

have _____ in their leaves, which harm insects that eat the leaves.

Some plants have warning colours to deter _____ .

[3 marks]

4 For which resources do plants compete with one another?

A light, water and nutrients

B nutrients, warmth and oxygen

C water, warmth and carbon dioxide

D soil, oxygen and carbon dioxide

[1 mark]

Two ways to reproduce

1 a Complete the sentences by using the **correct** words from the box.

| adults | cells | clones | genes | genetically | visually |

In asexual reproduction, the offspring are all _____ identical.

They are _____. In sexual reproduction, the offspring get a mixture

of _____ from both their parents.

[3 marks]

b The diagram shows what happens during sexual reproduction.

Match statements, **A**, **B**, **C** and **D**, with the labels **1 – 4** in the diagram.

A This is a female gamete.

B The new cell contains genes from both parents.

C Two gametes fuse together.

D This is a male gamete.

1	2	3	4

[4 marks]

G–E

D–C

Genes and what they do

2 What is DNA?

A A cell containing chromosomes.

B A type of protein.

C The chemical from which genes are made.

D The nucleus of a reproductive cell.

[1 mark]

G–E

3 Kim and Leah are non-identical twins. Kim has a straight nose but Leah has a turned-up nose. Kim has natural brown hair and Leah has natural blonde hair. Kim is heavier than Leah. Kim is better at maths than Leah.

a Which of Kim's and Leah's characteristics could be caused **entirely** by their genes?

_____ [1 mark]

b Suggest **one** way in which the twins' environment could have produced one of the differences between them.

_____ [1 mark]

c The shape of the earlobes is caused by just one gene. Kim has attached earlobes. Leah has unattached earlobes.
Explain how this could have been caused.

_____

_____ [2 marks]

G–E

D–C

Cuttings

1 Which statement describes an advantage of growing new plants from cuttings rather than seeds?

A The new plants will be stronger.

B The new plants will be genetically different from each other.

C The new plants will have a bigger range of flower colours.

D The new plants can be produced more quickly and cheaply. [1 mark]

2 Oak trees reproduce sexually. They produce seeds called acorns. Two acorns from an oak tree grow into new trees. One of the trees is much taller than the other.

a Suggest **two** reasons why one tree is taller than the other.

1_____

2_____ [2 marks]

b Gardeners can also grow new trees from cuttings. They take a small piece of stem from a tree and place it in soil. The piece of stem grows roots and becomes a new plant.

Mia wants to grow an oak tree exactly like one she already has in her garden. Explain why using cuttings would be better than planting acorns.

_____

_____ [2 marks]

Clones

3 This diagram shows how new cattle can be produced using a technique called **embryo transplants**.

a Complete the sentences using words from the diagram.

Fertilisation happens when a _____

and an egg fuse together. They form a new cell

called a _____ . This divides to form

a ball of cells. The cells in the ball can be split apart.

Each cell grows into an embryo, which can be put

into a different _____ .

[3 marks]

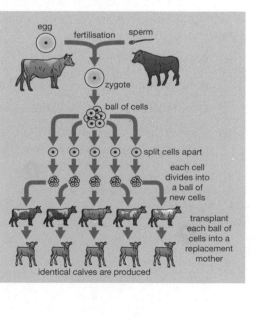

b Explain why the embryos that are transplanted into the replacement mothers are clones.

_____

_____

_____

_____ [2 marks]

Genetic engineering

1 Read this article about genetically engineered maize.

> Maize seeds are a protein- and carbohydrate-rich food. Maize is grown in many parts of the world. Insect pests reduce yields of maize. Many farmers spray the maize plants with pesticides to kill the pests.
>
> Now some new varieties of maize plants have been produced using genetic engineering. A gene for making a toxin has been transferred into the maize cells. The plants make a toxin (poison) in their leaves, which kills any insects that eat them.

How could growing the genetically engineered maize plants be an advantage for farmers?

A The farmers do not have to spend time and money on spraying pesticides.

B The plants will produce better tasting seeds.

C The plants will produce maize seeds containing more protein.

D The toxin will cause the insects to become resistant to pesticides. [1 mark]

2 The diagram shows how genetic engineering has been used to produce insulin from bacteria.

Explain why the insulin that the bacteria make is exactly the same as the insulin made by human cells.

human cell containing normal insulin gene

insulin gene is cut out by special enzyme

bacterium with chromosome

chromosome is taken out of bacterium and split open by an enzyme

the gene is attached to the bacterial chromosome (plasmid)

gene is transferred into a bacterial cell

the gene makes the bacteria cell produce insulin

insulin

_____

_____

[2 marks]

Theories of evolution

3 The drawings show how Lamarck thought that giraffes evolved their long necks. He thought that their necks stretched as they reached for high-up leaves. The giraffes passed on their long necks to their offspring.

a What is the best term to describe Lamarck's idea?

  A conclusion   **B** experiment

  C mutation   **D** theory   [1 mark]

b Charles Darwin thought that there was a different explanation. He suggested that there was natural variation in neck length amongst giraffes. The ones with the longest necks could get more food, so they were most likely to survive and reproduce.

What is the name for this process?

  A acquired characteristics   **B** extinction

  C natural selection   **D** speciation   [1 mark]

c Explain how the process suggested by Darwin could eventually produce a population of giraffes all with long necks.

_____

_____

[2 marks]

Natural selection

1 Complete the sentences by choosing the **correct** words from the box.

| adapted   compete   die   fight   generations   genes   looking   reproduce   strength |

Organisms _____ for survival. Only the best _____ ones live long enough

to _____ . They pass on their _____ to their offspring. Over many

_____ , most of the organisms will contain these advantageous genes.   [5 marks]

2 Peppered moths have pale wings with dark speckles. This camouflages them against tree bark so they are less likely to be eaten by birds. In the industrial revolution, air pollution made tree bark darker. A variety of peppered moths with dark wings became more common than the pale moths.

   a What is the term for the relationship between the birds and the moths?

    A camouflage   **B** competition   **C** evolution   **D** predation   [1 mark]

   b What can explain the dark moths becoming more common?

    A Any moths that by chance had dark wings were more likely to survive.

    B The moths changed their colour so that they were better camouflaged from birds.

    C The pale peppered moths flew away to less polluted areas.

    D The polluted air made the moth's wings get darker.   [1 mark]

Fossils and evolution

3 What is a fossil?

  A A rock formed by a volcanic eruption.

  B An ancient species of plant or animal that is now extinct.

  C An old bone.

  D The remains of an organism, turned to stone.   [1 mark]

4 The diagram shows some horses that lived long ago.

   a How do we know about the structure of the legs of horses that lived millions of years ago?

    _____

    _____

    _____ [1 mark]

Years ago	What it looked like and idea of size	Bones of its front leg	How it lived
10 million	height: 1.0 m		lived in very dry places; it was a fast runner
40 million	height: 0.6 m		lived in dryer conditions; needed to be able to run away from predators
60 million	height: 0.4 m		lived on soft ground near water; its feet could support its weight without sinking into the mud

   b Use the theory of natural selection, and the information in the diagram, to suggest why the feet of horses that lived 40 million years ago had three toes but the feet of horses that lived 10 million years ago had only one toe.

   _____

   _____

                                                  [2 marks]

Extinction

1 Dodos were flightless birds that lived on Mauritius, in the Indian Ocean. When people arrived on the island, they killed and ate dodos. The people introduced rats that ate dodo eggs. Now dodos are extinct.

Which of these statements explains why dodos became extinct?

A A new competitor was introduced.

B A new predator was introduced.

C New diseases were introduced.

D There were changes to the dodo's habitat.

[1 mark]

2 Crayfish live in streams and rivers in England. A larger species of crayfish has been introduced from America. The introduced species eats the same food as the English crayfish. It carries a disease that does not affect it but that kills English crayfish. Explain why the English crayfish may become extinct.

_____

_____

_____

[2 marks]

More people, more problems

3 Complete the sentences by using the **correct** words from the box.

| biomass | oil | population | metals | plastics | pollution | reproduction | species |

As the human _____ increases, we use more and more non-renewable fuels

such as _____ . We also use more _____ , which we

extract from ores. We cause more _____ . [4 marks]

4 The graph shows how the human population has changed since 1750.

a Which part of the graph is a prediction?

_____
[1 mark]

b How many people lived on Earth in 1920?

_____
[1 mark]

c By how many times did the world population increase between 1920 and the year 2000?

_____
[1 mark]

Land use

1 The expanding human population causes many changes to natural landscapes.

Match words, **A**, **B**, **C** and **D**, with the spaces **1 – 4** in the sentences.

A fields   **B** forests   **C** landfill   **D** quarries

We take more building stone from _____ **1** _____ .

We make more _____ **2** _____ to grow crops and keep animals for food.

We use more _____ **3** _____ sites to dispose of rubbish.

We remove more _____ **4** _____ to make space for farmland.

1	2	3	4

[4 marks]

Pollution

2 Untreated sewage should not be allowed to flow into rivers or the sea.
Explain why it would be dangerous to swim in water polluted by sewage.

_____

_____ [2 marks]

3 A farmer sprayed some fertiliser, containing nitrate ions, onto a field of wheat growing next to a pond. Shortly afterwards, it rained and some of the nitrate ions washed into the river.

a After a few days, there were many more bacteria living in the pond. Explain why this happened.

_____

_____ [2 marks]

b After another day, there was so little dissolved oxygen in the water that many of the fish in the pond died.
Explain why the oxygen levels in the water decreased.

_____

_____ [2 marks]

4 DDT is a persistent insecticide. It does not break down inside the bodies of animals and plants. The diagram shows what happens as DDT passes along a food chain.

a How does the information in the diagram explain why animals at the top of food chains are more likely to be poisoned by DDT than other animals?

_____

_____

_____ [2 marks]

grebes (1600 p.p.m. of DDT)
fish (250 p.p.m. of DDT)
plankton (5 p.p.m. of DDT)
water (0.02 p.p.m. of DDT)

b DDT is now banned in Europe and America but DDT is still found in small amounts in animals' bodies. Suggest why this is so.

_____

_____ [2 marks]

Air pollution

1 a Sulfur dioxide is a pollutant.
Where does most sulfur dioxide come from?

A car exhausts

B coal-burning power stations

C landfill sites

D untreated sewage [1 mark]

b Describe **one** way in which sulfur dioxide in the air can damage a
person's health.

_____ [1 mark]

What causes acid rain?

2 The table shows the pollutants emitted in car exhausts in the United Kingdom in one year.

Pollutant	Emissions in thousands of tonnes
Carbon monoxide	3300
Carbon dioxide	31 200
Smoke particles	130
Nitrogen oxides	714
Sulfur dioxide	12

a Which pollutant is produced in the largest amount?

_____ [1 mark]

b Carbon monoxide combines irreversibly with haemoglobin in our blood.
Why is this dangerous?

_____ [1 mark]

c Explain how breathing in smoke particles can damage health.

_____

_____ [2 marks]

d Nitrogen oxides and sulfur dioxides both contribute to the formation of acid rain.
Describe how acid rain is formed.

_____

_____ [2 marks]

e Describe **two** harmful effects of acid rain on living organisms.

1_____

2_____ [2 marks]

f Explain how adding a base, such as limestone, to acidified lakes can help to
reduce the harmful effects on wildlife.

_____

_____ [2 marks]

Pollution indicators

1 The table shows the minimum pH level in which five different aquatic organisms can live.

a Which organism can survive in the most acidic water?

_____

[1 mark]

Organism	Minimum pH in which it can live
Freshwater shrimp	6.0
Mayfly larva	5.4
Caddis fly larva	5.1
Black fly larva	4.5
Chironomid larva	4.3

b A researcher found caddis fly larvae in a lake but no mayfly larvae or freshwater shrimps. What does this tell her about the pH of the lake?

_____

[1 mark]

2 The table shows some freshwater animals and the level of oxygen that they need to live.

Animal	Oxygen level needed
Mayfly larva	High
Cased caddis fly larva	Fairly high
Snail	Fairly low
Blood worm	Low
Rat-tailed maggot	Very low

a Explain how a scientist could use this information to quickly find out how much oxygen there is in a river.

_____

_____

[2 marks]

b A stream has raw sewage flowing into it. Which animal would you be most likely to find in the water in the stream? Explain your answer.

_____

_____

[2 marks]

Deforestation

3 What is deforestation?

A Cutting down the largest trees in a forest.

B Removing the largest branches from the trees.

C The permanent clearing of forests.

D Replanting trees after cutting them down.

[1 mark]

4 How can deforestation increase the greenhouse effect?

A Fewer trees means that less carbon dioxide is used in photosynthesis.

B Fewer trees means that less oxygen is added to the air by photosynthesis.

C Fewer trees means that less water vapour is added to the air by transpiration.

D Fewer trees means that there are fewer habitats for animals.

[1 mark]

The greenhouse effect – good or bad?

1 Which **two** of these gases contribute to the greenhouse effect?

A carbon dioxide and methane

B methane and oxygen

C oxygen and nitrogen

D nitrogen and carbon dioxide

[1 mark]

G–E

2 Complete the sentences by using the **correct** words from the box.

| colder | conducted | radiated | space | warmer |

Carbon dioxide traps energy that is _____ from the Earth. This stops the

energy from escaping into _____ . It keeps the Earth _____

than it would otherwise be.

[3 marks]

D–C

3 The graphs show the changes in average world temperature and carbon dioxide concentration in the air between 1700 and 2000.

What can we definitely conclude from these graphs?

A Both carbon dioxide and temperature have increased since 1700.

B Increasing carbon dioxide causes an increase in temperature.

C Increasing temperature causes an increase in carbon dioxide.

D By 2050, the average temperature will be about 1.5 °C higher than in 1700.

[1 mark]

D–C

Sustainability – the way forward?

4 a State **one** way in which insulating homes helps to reduce our use of natural resources.

_____

[1 mark]

G–E

b Jack likes to cycle to school, rather than being taken in the car by his father. He says it keeps him fit and is a small contribution towards sustainable development.

i Explain what is meant by **sustainable development**.

_____

_____

[2 marks]

D–C

ii Suggest how Jack's decision about his journey to school makes a small contribution to sustainable development.

_____

_____

[2 marks]

B1b revision checklist

I know:

what determines where particular species live

☐ animals and plants are adapted to live in different habitats, to compete for resources and to survive attack from predators

☐ caribou and camels are adapted to extreme environments

why individuals of the same species are different from each other; and what new methods there are for producing plants and animals with desirable characteristics

☐ DNA is the genetic material that controls inherited characteristics

☐ reproduction can be sexual (variation in offspring) or asexual (identical offspring)

☐ cloning and genetic engineering can be used to produce plants and animals with desirable characteristics

why some species of plants and animals have died out and how new species of plants and animals develop

☐ theories of evolution have changed over time; Darwin proposed the theory of evolution by natural selection

☐ fossils tell us how present-day species have evolved and how they compare to prehistoric species

☐ species can become extinct due to environmental change and human impact

how humans affect the environment

☐ increases in human population use up more resources and produce more waste and pollution

☐ human action contributes to acid rain, air pollution, water pollution, and over-use of land

☐ living organisms such as lichens can be used as indicators of pollution

☐ increasing the greenhouse effect can lead to global warming

☐ sustainable development, e.g. using renewable energy resources and recycling can help to safeguard the environment for future generations

Elements and the periodic table

1 Use the periodic table to help you answer these questions.

Complete each sentence by crossing out the **incorrect** symbol in the box.

The symbol for **hydrogen** is

H
He

The symbol for **sodium** is

So
Na

The symbol for **beryllium** is

Be
B

[2 marks]

Atomic structure 1

2 This is a diagram of an atom.

Which **one** of the following statements about the atom is **correct**?

A The atom has 4 electrons.

B The atom has 4 protons.

C The nucleus has a charge of +3.

D The atomic mass of the atom is 3.

[1 mark]

3 Sulfur is an element in the periodic table.

a **i** Complete the sentence to show the numbers of the different types of particle in an atom of sulfur.

An atom of sulfur contains _____ protons,

_____ electrons and _____ neutrons. [2 marks]

32	**S**
16	sulfur

ii When sulfur reacts with other elements, it gains electrons. Gaining electrons does not affect the mass of the sulfur atoms. Explain why.

_____ [1 mark]

b **i** Give the symbol of the element that contains atoms with one more proton than sulfur.

_____ [1 mark]

ii Give the symbol of an element that is in the same group as sulfur but has fewer protons.

_____ [1 mark]

Bonding

1 Match formulae, **A**, **B**, **C** and **D**, with the sentences **1 – 4** in the table.

A NaCl

B N_2

C H_2O

D KNO_3

1	**2**	**3**	**4**

1	This is a compound that contains three different types of atom.
2	This is a molecule with two identical atoms joined together.
3	This compound contains a covalent bond between hydrogen and oxygen atoms.
4	This compound contains chloride ions.

[4 marks]

G–E

D–C

Extraction of limestone

2 Complete the sentences by crossing out the **incorrect** words in each box.

Limestone contains the chemical compound

| calcium chloride |
| calcium carbonate |
| lead carbonate |

The same chemical compound is found in other rocks, for example

| marble |
| granite |
| slate |

[2 marks]

D–C

3 A company wants to start to extract limestone from a new quarry.
Local people protest against the quarry. This is one of the protesters.

Noise and lorries all night! Say NO to the quarry!

a What could the company do to solve the problems
of 'noise and lorries all night'?

_____

_____

_____

_____   [2 marks]

b Give **one advantage** of living near a limestone quarry.

_____

_____   [1 mark]

Thermal decomposition of limestone

1 Limestone contains calcium carbonate. This is the formula of calcium carbonate: $\boxed{CaCO_3}$

Complete the table to show the three types of atoms in calcium carbonate.

Name of atom	Symbol	Number of atoms in formula
Calcium		1
	C	

[3 marks]

2 a When calcium carbonate is heated it reacts to make a solid compound and a gas.

Complete the word and symbol equation for the reaction by filling in the boxes.

calcium carbonate → [　　　　　　] + [　　　　　　]

$CaCO_3$　　→　　CaO　　+　[　　　　　] 　[3 marks]

b Eve did an experiment. She weighs a lump of limestone before and after heating it. She finds that the lump is lighter after heating. Which sentence is the best explanation for this?

A The atoms get smaller when they are heated.

B A gas is given off.

C The limestone has been burned.

D The lump got wet when it was heated. [1 mark]

Uses of limestone

3 The diagram shows a double glazed window in a new house.

mortar holds window in place

window is made of **plastic**

glass panes are held in place by **aluminium** metal

glass

Which **two** materials named on the diagram are made using limestone?

_____ and _____ [2 marks]

4 Concrete is a mixture of cement, sand and rock chippings. Give **one advantage** of using concrete as a building material instead of cement.

_____ [1 mark]

The blast furnace

D–C

1 Iron is extracted from iron ore in the blast furnace.

Match words, **A**, **B**, **C** and **D**, with the numbers **1 – 4** in the labels.

A oxygen

B calcium carbonate

C iron oxide

D carbon

haematite contains mainly 1

coke contains mainly 2

limestone contains mainly 3

hot air ← air contains 4

1	2	3	4

[4 marks]

D–C

2 Complete the sentences by choosing the **correct** words from the box.

carbon monoxide	silicon dioxide	oxidation	oxygen	reduction

In the blast furnace, iron oxide reacts to form iron by reacting with _____.

During the reaction, the iron oxide loses _____.

This type of reaction is called _____. [3 marks]

Using iron

G–E

3 Cast iron contains 96% iron, and the rest is other impurities. The table shows the difference between cast iron and wrought iron.

	Cast iron	**Wrought iron**
Percentage of iron	96%	100%
Properties	Harder	Softer
	Stronger	Weaker
	More brittle	Easily shaped

a Calculate the percentage of impurities in cast iron.

_____ [1 mark]

b An artist uses wrought iron to make a statue. He twists lengths of iron together. Why is it better for him to use wrought iron instead of cast iron?

_____

_____ [2 marks]

Using steel

1 Steel is an **alloy**. Which of the following statements about steel alloys is true?

A An alloy contains all the same type of atoms.

B Steel is softer and weaker than iron.

C Alloys are mixtures of more than one element.

D Alloys are non-metal compounds.

[1 mark]

2 The diagrams show how the atoms are arranged in wrought iron and stainless steel.

wrought iron   **stainless steel**

a Describe the difference in the **structure** of wrought iron and stainless steel.

_____

_____

_____ [3 marks]

b Explain why stainless steel is more suitable than wrought iron for making cutlery.

_____

_____ [2 marks]

Transition metals

3 Draw a ring round **two** metals that are transition metals. Use the periodic table to help you.

aluminium   copper   lithium   manganese   magnesium [2 marks]

4 The diagram shows the structure of a metal.

a Name the **two** types of particles shown in the diagram.

_____ and _____ [2 marks]

b Use the diagram to explain why metals conduct electricity.

_____

_____ [2 marks]

Aluminium

G–E

1 Match the properties of aluminium, **A**, **B**, **C** and **D**, with its uses **1 – 4** in the table.

A low density   **B** good conductor   **C** non-toxic   **D** does not corrode

1	electricity cables
2	greenhouse frames
3	drinks cans
4	aircraft

1	2	3	4

[4 marks]

D–C

2 Aluminium is extracted from aluminium oxide using electrolysis.
Complete the sentences about electrolysis by crossing out the **incorrect** word in each box.

Aluminium cannot be extracted by reacting with carbon because it is too
reactive
unreactive
.

During electrolysis the positive aluminium ions are attracted to the
positive
negative
electrode.

The ions form aluminium atoms when they
gain
lose
electrons.   [2 marks]

Aluminium recycling

D–C

3 Here is some information about aluminium.

42% of aluminium cans in the UK are recycled

Most aluminium cans are made from mined bauxite

a Bauxite is the main ore that is used to make aluminium. Give **two disadvantages** of mining bauxite.

_____

_____   [2 marks]

b Suggest reasons why **less than half** the aluminium cans we use are recycled.

_____

_____   [2 marks]

Titanium

1 The table shows some information about titanium, aluminium and iron.

	Titanium	Aluminium	Iron
Density in g/cm³	4.5	2.7	7.9
Melting point °C	1675	660	1535
Corrosion resistance	Does not corrode	Does not corrode	Corrodes
Main oxide in ore	TiO_2	Al_2O_3	Fe_2O_3
Bonding in oxide	Covalent	Ionic	Ionic
Cost of extraction	Very high	High	Low

a Titanium and aluminium are used to make bicycle frames. Use information from the table to explain why they are better metals to use than iron.

_____

_____

_____ [3 marks]

b i Aluminium is extracted by electrolysis of molten aluminium oxide. Use information from the table to explain why molten aluminium oxide conducts electricity, but molten titanium oxide does not.

_____

_____ [2 marks]

ii Aluminium is electrolysed at 900 °C. At this temperature, the aluminium leaves the electrolysis tank as a liquid. Explain why titanium is not a liquid at this temperature.

_____ [1 mark]

Copper

2 Chalcocite is a compound found in some copper ores. The chemical name for chalcocite is copper sulfide. This is the formula of copper sulfide: CuS_2

a Name the **two** elements in copper sulfide.

_____ and _____ [2 marks]

b The first step to extract copper from copper sulfide is by heating in air. This reaction happens:

copper sulfide + oxygen → copper oxide + sulfur dioxide

What type of reaction is this? Draw a ring round the **correct** answer.

oxidation   **neutralisation**   **reduction**   **titration**   [1 mark]

3 Another way of extracting copper is by leaching.

Put the stages of the leaching process, **A**, **B**, **C** and **D**, into the correct order **1 – 4**.

A An electric current is passed through the solution.

B Copper deposits are formed at the negative electrode.

C A solution of chemicals runs through the copper ore.

D Low grade ores are found on old spoil heaps.

1	2	3	4

[4 marks]

Smart alloys

1 Match words, **A**, **B**, **C** and **D**, with the correct descriptions **1 – 4** in the table.

A aluminium

B gold

C steel

D iron

1	A very unreactive metal that is found naturally as a pure metal.
2	A pure metal that is extracted from its ore by heating with carbon.
3	An alloy made by mixing different elements together.
4	A pure metal that is extracted by electrolysis.

1	2	3	4

[4 marks]

2 One use of smart alloys is for making braces to straighten teeth.
The diagram shows some information about smart alloys.

This smart alloy is made from nickel and titanium

Smart alloys are mixtures of different elements

Smart alloys have a 'shape' memory

Smart alloys can be bent and stretched into different shapes

Smart alloys contain mainly metal atoms

a Complete the sentences about smart alloys.
Smart alloys are **similar** to normal alloys because _____

_____

Smart alloys are **different** to normal alloys because _____

_____ [4 marks]

b What do you have to do to a smart alloy to return it to its original shape?

_____

_____ [2 marks]

Fuels of the future

3 a Complete the sentences by choosing the **correct** words from the box.

non-renewable   renewable   incomplete   carbon dioxide
water   hydrogen   oxygen

Petrol is made from crude oil. Crude oil will run out because it is _____ .

When petrol burns it makes _____ and _____ .

Carbon monoxide comes from a car exhaust due to _____ combustion.

[4 marks]

b Suggest **one** reason why pollution from cars is a bigger problem in cities than in the countryside.

_____ [1 mark]

Crude oil

1 Crude oil is separated by fractional distillation.

Match descriptions, **A**, **B**, **C** and **D**, to the numbers **1 – 4** on the diagram.

A The oil is headed to vapour point here.

B Condensation happens here.

C This is the coolest part of the column.

D Molecules with the highest boiling point leave the column here.

1	2	3	4

[4 marks]

Alkanes

2 Alkanes are hydrocarbons that contain single bonds.

Look at the structures of these molecules.

a i Which of the molecules is **not** an alkane?

Molecule _____ [1 mark]

ii Explain how you can tell.

_____

_____ [2 marks]

b i Which of the molecules is **ethane**?

Molecule _____ [1 mark]

ii Draw a similar diagram to show the structure of **methane**.

[1 mark]

Pollution problems

1 These equations show what happens when fossil fuels burn.

Match words, **A**, **B**, **C** and **D**, to the numbers **1 – 4** in the equations.

A water vapour

B oxygen

C sulfur dioxide

D carbon

1	2	3	4

_____ **1** _____ + oxygen → carbon dioxide

hydrocarbon + _____ **2** _____ → carbon dioxide + _____ **3** _____

sulfur + oxygen → _____ **4** _____

[4 marks]

Reducing sulfur problems

2 Sulfur dioxide is a pollutant gas that is given off when coal burns.

a Sulfur dioxide dissolves in rain water to make it **slightly** acidic (acid rain). What is the pH of acid rain? Draw a ring round the **correct** answer.

1   **5**   **7**   **9**   **14**   [1 mark]

b Tom goes on a field trip to visit a park and a lake that he thinks has been damaged by acid rain. What evidence should he look for?

_____

_____

_____

[3 marks]

3 All petrol garages in the United Kingdom sell low sulfur petrol.

Which of these statements about sulfur in petrol are true?

A Petrol companies used to mix sulfur compounds into petrol.

B Sulfur is now removed from petrol using a solvent.

C Sulfur in petrol harms the ozone layer.

D Sulfur in petrol is oxidised when the petrol burns.

_____

000.00
000.00

ULTRA LOW SULFUR PETROL

[2 marks]

4 Sulfur dioxide can be removed from the gases in power stations by reacting it with powdered limestone.

a Give **one** reason why it is important to remove sulfur dioxide from the waste gases.

_____

[1 mark]

b Give **one** reason why using large amounts of limestone can cause harm to the environment.

_____

[1 mark]

C1a revision checklist

I know:

how rocks provide building materials

☐ limestone, calcium carbonate, can be used to make cement, concrete and glass which are used as building materials

☐ an element consists of one type of atom; two atoms of the same element can join together to form a molecule; a compound consists of two or more elements joined together

☐ atoms are held together in molecules and lattices by chemical bonds, which involve giving, taking or sharing electrons

how rocks provide metals and how metals are used

☐ metals are extracted from their ores, often oxides, by reduction with carbon (iron), electrolysis (aluminium and copper) or other chemical reactions (titanium)

☐ metals are mixed together to make alloys (e.g. iron and other metals or carbon make steel)

☐ aluminium is expensive to produce, is often too soft on its own, but forms strong alloys with other metals

☐ copper is a hard, strong, good conductor and can be used for wiring and plumbing

☐ aluminium and titanium are resistant to corrosion and have low density

how we get fuels from crude oil

☐ crude oil is a mixture of hydrocarbon compounds that can be separated by fractional distillation. Some of the fractions can be used as fuels

☐ most of the compounds in crude oil are saturated hydrocarbons called alkanes, which have the general formula C_nH_{n+2}

☐ the burning of fossil fuels releases useful energy but also harmful substances, e.g. sulfur dioxide causes acid rain; carbon dioxide causes global warming; smoke particles cause global dimming

Cracking

1 Crude oil contains alkanes with both long and short chains.

long chain alkane	short chain alkane

a Which two elements are found in hydrocarbons? Draw a ring around the **two** correct answers.

calcium   carbon   helium   hydrogen   nitrogen   oxygen   [2 marks]

b The table shows some information about the alkanes in crude oil.

Type of alkane	Uses	Annual use in UK per year
Short chain	Fuels	Very high
Long chain	Making chemicals Road surfacing	Medium

Use the information in the table to answer these questions.

i Explain why we use much larger quantities of short chain alkanes compared to long chain alkanes.

_____

_____   [2 marks]

ii Cracking involves breaking up long chain molecules. Explain why this makes long chain molecules more useful.

_____

_____   [2 marks]

2 This molecule is made in a cracking reaction.

Which of the following statements is **not** true for this molecule?

A It is called propene.   **B** It is a hydrocarbon.

C It is a saturated molecule.   **D** It contains some single bonds.   [1 mark]

Alkenes

3 When alkanes in crude oil are cracked, some alkenes are formed.

a Butene is an alkene. It has the formula C_4H_8. Pentene has one more carbon atom than butene. What is the formula for pentene?

_____   [1 mark]

b Some of the following statements are true for alkanes and some are true alkenes. Put letters, **A**, **B**, **C** and **D**, into the table in the correct places.

A contains a double bond   **B** is saturated

C contains more hydrogen atoms than carbon atoms   **D** has the general formula $C_nH_{(2n+2)}$

True for alkanes	True for alkenes

[3 marks]

Making ethanol

G–E

1 Ethene is made from crude oil. Ethene is used to make ethanol.
Look at the structures of ethene and ethanol.

ethene   **ethanol**

Give **two** differences between the structures of ethene and ethanol.

_____

_____   [2 marks]

D–C

2 This information appeared in a newspaper article about ethanol.

BIOFUELS FOR THE FUTURE

Fuels such as petrol and diesel could soon be replaced by renewable biofuels that contain ethanol made from sugar beet. These new biofuels could solve our problems of relying on finite crude oil reserves.

Many farmers already use biofuels for farm vehicles.

a Explain how sugar beet can be used to produce ethanol.

_____

_____   [2 marks]

b Explain why ethanol made in this way is a 'renewable biofuel'.

_____

_____   [3 marks]

Plastics from alkenes

G–E

3 a The sentences below are about making polyethene. Match words, **A**, **B**, **C** and **D**, to the numbers **1 – 4** in the sentences.

A monomers   **B** polymers   **C** polymerisation   **D** cracking

Polyethene contains long chain molecules called _____ **1** _____ .

These are made when many _____ **2** _____ join together.

This process is called _____ **3** _____ .

Ethene is used to make polyethene. It is from molecules in crude oil by _____ **4** _____ .

1	2	3	4

[4 marks]

b i Which of these molecules can form a long chain in a similar way to ethene?
Draw a ring round the **correct** molecule.

A   **B**   **C**   [1 mark]

ii Explain why you made your choice.

_____   [1 mark]

Polymers are useful

1 Look at the structures of polymers **A**, **B**, **C** and **D**.

  a Which polymer contains **cross links**?

    _____ [1 mark]

  b Which polymer has the highest **density**?

    _____ [1 mark]

  c Polymers A and C are heated until they go soft and runny. Explain why polymer A is softer and runnier than polymer C.

    _____

    _____ [2 marks]

2 Look at the table about two polymers.

Polymer	Softening temperature	Stiffness and strength	Other information
Polyethene	100 °C	Stiff, breaks and cracks easily	Cheap to make Waterproof
PTFE	Does not soften when heated	Flexible and strong	Expensive Very smooth Waterproof

  a Explain why PTFE is used as a coating for frying pans.

    _____

    _____ [2 marks]

  b Explain why polyethene **can** be used to make cups for hot drinks, but **cannot** be used for pan handles.

    _____

    _____ [2 marks]

  c Explain why PTFE is a better material to use to coat waterproof coats than polyethene.

    _____

    _____ [2 marks]

Disposing of polymers

3 **A**, **B**, **C** and **D** are methods to dispose of used polymers. Match letters, **A**, **B**, **C** and **D**, with the sentences **1 – 4** in the table.

A landfill

B burning

C processing to make padding for sleeping bags

D designing new polymers that are biodegradable

1	This involves microorganisms.
2	This can be adapted to generate electricity.
3	This is an example of recycling.
4	This takes up land for a long time.

1	2	3	4

[4 marks]

Oil from plants

1 a Which of the following parts of plants are good sources of vegetable oils?
Put a ring around **two correct** answers.

    roots   nuts   leaves   seeds   petals   [2 marks]

b What is the first stage in extracting oil from plants?

   A crushing in a press

   B roasting in an oven

   C boiling in water

   D dissolving in acid   [1 mark]

2 Oil from plants needs to be purified before it can be used.
Complete the sentences about purifying oil by crossing out the **incorrect** word in each box.

The oil is added to a solvent so that it | dissolves / melts | .

The mixture of oil and solvent is then | filtered / distilled | .

The oil and solvent separate because they have different | boiling points / melting points | .   [2 marks]

Green energy

3 Harriet is not sure that biofuels are a good idea. This is what she says.

> Growing plants for biofuels uses up a lot of land that we need to grow food.
> Some oils that come from plants are far too thick to be used for most engines.
> They cost a lot more to produce than petrol.

a Harriet says that some oils are too thick to be used in engines. How is this problem overcome?

   _____ [1 mark]

b Harriet talks about the disadvantages of using biofuels. Give **two advantages** of using biofuels.

   _____

   _____ [2 marks]

c In the future, petrol will become more expensive than biofuels. Explain why.

   _____

   _____ [2 marks]

Emulsions

1 The diagram shows how oil and water are mixed together in a sample of milk.
Match words, **A**, **B**, **C** and **D**, to the numbers **1 – 4** in the diagram labels.

A emulsion

B emulsifier

C water

D oil

tiny droplets of **1**

the oil and water mix together because milk contains a natural **3**

large amount of **2**

the mixture of oil in water is called an **4**

1	2	3	4

[4 marks]

2 The diagram shows what some different types of emulsions contain.

Shaving foam contains air bubbles spread out in a liquid.

After shave cream has water droplets spread out in an oil.

air

liquid

oil

water

shaving foam

after shave cream

a Complete the table about these emulsions choosing the **correct** word from the box.

air	oil	liquid	water

Type of emulsion	Dispersed phase	Continuous phase
Shaving foam		
After shave cream		

[2 marks]

b The ingredients in these emulsions are immiscible. What does 'immiscible' mean?

_____

[1 mark]

Polyunsaturates

3 Olive oil is a **monounsaturated oil**. It contains one double bond.

a How is the bonding in a **saturated fat** different from olive oil?

_____

[1 mark]

b How is the bonding in a **polyunsaturated oil** different from olive oil?

_____

[1 mark]

c Complete the table to show the observations you would expect when each oil or fat reacts with bromine water.

Type	Olive oil	Saturated fat	Polyunsaturated oil
Observations	Orange to colourless		

[2 marks]

Making margarine

1 Margarine for spreading on bread is made from vegetable oils. The vegetable oils are treated in a chemical process to make them more suitable to use for spreading.

   a   Give **one** reason why vegetable oil does not make a good spread.

   _____

   _____   [1 mark]

   b   What is the chemical process called that is used to turn the vegetable oil into margarine? Draw a ring round the **correct** answer.

   condensation   hydrogenation   polymerisation   solidification   [1 mark]

2 The diagram in box 1 shows part of a molecule of vegetable oil.

   box 1   box 2

   a   Draw a similar diagram in box 2 to show the same molecule after it has been hydrogenated.   [2 marks]

   b   Give **two** ways this process is speeded up when it is carried out on an industrial scale.

   _____

   _____   [2 marks]

Food additives

3 The table shows some information about food additives with E-numbers.

Purpose	Number
Colouring	E100-181
Preservative	E200-285 and 1105
Antioxidant	E300-340
Emulsifier	E400-499

   a   Match words, **A**, **B**, **C** and **D**, with the sentences **1 – 4** in the table.

   A colouring   **B** preservative   **C** antioxidant   **D** emulsifier

1	E321 is an example of this type of additive.
2	This additive makes the food last longer when it is stored.
3	This additive improves the appearance of food.
4	This additive helps oil and water to mix in the food.

1	2	3	4

   [4 marks]

   b   Which of the following statements about additives is true?

   A Additives with E-numbers have been proved to be totally safe.

   B Additives with E-numbers are allowed to be used in foods in Europe.

   C Additives are essential for a healthy diet.

   D All additives improve the taste of food.   [1 mark]

Analysing chemicals

1 Ruby set up an experiment to separate the dyes in some ink. This is the diagram that she drew in her book. She makes some mistakes on her diagram.

stop when all the filter paper is soaked with solvent

level of solvent

mixture of ink and water

spot of ink

a Give **three** mistakes that Ruby made on her diagram.

1_____

2_____

3_____ [3 marks]

b Ruby did a second experiment. She tested inks, **A**, **B**, **C** and **D**, and an 'unknown' ink. This is her result.

start line

A   B   C   D unknown

i Which ink, **A**, **B**, **C** or **D**, contains a mixture of colours?

_____ [1 mark]

ii Which inks does the unknown ink contain?

_____ and _____ [2 marks]

iii What is the name for this technique of analysing colours in ink?

_____ [1 mark]

The Earth

2 The diagram shows the structure of the Earth.

Match words, **A**, **B**, **C** and **D**, to the sentences **1 – 4** in the table.

A mantle

B inner core

C outer core

D crust

crust

mantle

core

1	this is the furthest from the Earth's surface
2	a thin layer of solid rock
3	convection currents here reshape the Earth's surface
4	this contains a liquid mixture of iron and nickel

1	2	3	4

[4 marks]

Earth's surface

1 One old theory about how the Earth's surface was formed is the 'shrinking Earth' theory. Why did scientists start to think that this theory was wrong?

A The Earth's surface is folded into mountains and valleys.

B The Earth's interior is very hot.

C The crust of the Earth is made of rock.

D New evidence that continents have moved was discovered. [1 mark]

2 The diagram shows the continent of Pangaea 200 million years ago.

The sentences describe how the continents moved to their current positions.

Match words, **A**, **B**, **C** and **D**, to the numbers **1 – 4** in the sentences.

A lithosphere

B convection currents

C tectonic plates

D mantle

The upper mantle and crust form the Earth's _____ **1** _____ .

The continents are carried on large masses of rock called _____ **2** _____ .

Heat from the Earth's core causes the rocks to move by _____ **3** _____ .

This happens in the _____ **4** _____ . [4 marks]

1	2	3	4

Earthquakes and volcanoes

3 a Look at the diagram that shows where two plates meet. Match sentences, **A**, **B**, **C** and **D**, with the numbers **1 – 4** on the diagram.

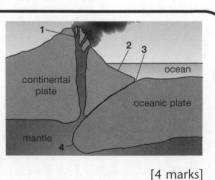

A Earthquakes are most likely to happen here.

B Rock is melting here.

C Rock is being pushed up here.

D Lava is cooling down most quickly here.

1	2	3	4

[4 marks]

b Which of the plates, the continental plate or the oceanic plate is the **least** dense? Explain your reasoning.

_____ [1 mark]

The air

1 Match gases, **A**, **B**, **C** and **D**, to the descriptions **1 – 4** in the table.

A argon   **B** carbon dioxide   **C** oxygen   **D** nitrogen

1	This gas is a noble gas.
2	The gas is found in the largest amount in air.
3	This gas is used by animals for respiration.
4	This gas is made during combustion.

1	2	3	4

[4 marks]

2 Which of the following boxes shows the gases in modern air in the correct order?

	A
most	carbon dioxide
↓	oxygen
least	nitrogen

	B
least	carbon dioxide
↓	oxygen
most	nitrogen

	C
most	oxygen
↓	nitrogen
least	carbon dioxide

_____

[1 mark]

Evolution of the air

3 This table shows gases coming from a modern active volcano on a Pacific island.
The **temperature** of the volcano is over **1000 °C**.

Name of gas	Percentage of gas
Water	37.1
Carbon dioxide	48.9
Sulfur dioxide	11.8
Hydrogen	0.49
Carbon monoxide	1.51

a **i** Why is water a gas when it comes out of the volcano?

_____

[1 mark]

ii The Earth's early atmosphere contained large amounts of water vapour.
Describe what happened to all this water vapour.

_____

_____

_____

[3 marks]

b Scientists think that our early atmosphere came from volcanoes.
i Which gas is present in the **largest quantity** in the gases from this volcano?

_____

[1 mark]

ii What **two** processes removed most of this gas from the early atmosphere?

_____

_____

[2 marks]

c Give **two** other important differences between the gases from this volcano and our modern atmosphere.

_____

_____

[2 marks]

Atmospheric change

1 Complete the sentences about climate change by crossing out the **incorrect** word in each box.

Over the last century the average temperature of our atmosphere has

decreased
increased

.

This is causing the ice caps to

dissolve
melt

and sea levels to

rise
fall

.

Another effect is that the weather is becoming more

settled
extreme

.

[3 marks]

2 Some friends are discussing the environment. This is what they say:

I'm really worried about the effect of using CFCs on the ozone layer

I think we should plant more trees in forests

I think we should all burn biofuels

I think it is ridiculous the amount of fossil fuels we burn

Put ticks (✓) in the table to show whether each activity increases, decreases or makes no difference to the amount of carbon dioxide in the air.

Activity	Decreases amount of carbon dioxide in air	Increases amount of carbon dioxide in air	Does not affect amount of carbon dioxide in air
Using CFCs			
Planting more trees			
Burning biofuels			
Burning fossil fuels			

[4 marks]

3 The Earth stores carbon dioxide in carbon sinks. Plants are one type of carbon sink.

a Choose the names of two other types of carbon sinks. Draw a ring around **two correct** answers.

crude oil   rainwater   limestone   photosynthesis   respiration   [2 marks]

b Plants act as carbon sinks. Trees are much better carbon sinks than short-lived plants, such as grass. Explain why.

_____

_____

[2 marks]

c The seas contains large amounts of dissolved carbon dioxide. More carbon dioxide can dissolve in cold water than in warm water. Scientists are worried that if the climate gets warmer, the amount of carbon dioxide in the **air** will increase very quickly. Explain how this could happen.

_____

_____

[2 marks]

C1b revision checklist

I know:

how polymers and ethanol are made from oil

☐ crude oil is made from hydrocarbons that can be cracked to form alkenes, which are unsaturated hydrocarbons and have the general formula C_nH_{2n}

☐ alkenes can be made into polymers, which are long chain molecules created when lots of small molecules called monomers are joined together in polymerisation

☐ polymers can be used to make useful substances e.g. waterproof materials and plastics, but many are not biodegradable

how plant oils can be used

☐ vegetable oils can be hardened to make margarine in a process called hydrogenation

☐ biodiesel fuel can be produced from vegetable oils

☐ oils do not dissolve in water. They can be used to produce emulsions e.g. in salad dressings

what the changes are in the Earth and its atmosphere

☐ the Earth has three main layers: the crust, mantle and the inner and outer core

☐ the Earth's atmosphere has changed over millions of years. Many of the gases that make up the atmosphere came from volcanoes

☐ for 200 million years, the proportions of different gases in the atmosphere have been much the same as they are today

☐ human activities have recently produced further changes, e.g. the levels of greenhouse gases are rising

Heat energy

1 a What is meant by **heat transfer**?

_____

_____ [2 marks]

b Explain why a hot cup of tea cools down if it is left in a cool room.

_____ [2 marks]

G–E

2 a **Thermal radiation** is a type of **electromagnetic** wave.
Write down **three** properties of thermal radiation.

_____

_____ [3 marks]

b Emily has just moved to a new house which has large south-facing windows.
During the summer the house gets very hot.

i Explain why the house stays cooler if she draws the curtains in the daytime.

_____

_____ [2 marks]

ii Her friend suggests that she could paint the outside of her house a different
colour. Explain how this would help the house feel cooler in summer.

_____

_____ [3 marks]

D–C

Thermal radiation

3 Match words, **A**, **B**, **C** and **D**, with the numbers **1 – 4** in the sentences.

A thermometer   **B** heat   **C** temperature   **D** thermopile

The ____ **1** ____ of an object indicates how hot it is, and this is measured using a ____ **2** ____.

The amount of ____ **3** ____ radiating from hot objects gets less as the

object cools down. A ____ **4** ____ measures the amount of heat

radiating from the object.

1	2	3	4

[4 marks]

G–E

4 Match each thermopile, **A**, **B**, **C** and **D**, with the numbers **1 – 4** in the sentences.

The thermopile detecting most heat is ____ **1** ____.

The thermopile detecting least heat is ____ **2** ____.

Of the two light coloured surfaces, the thermopile
giving out most heat is ____ **3** ____.

Of the two dark coloured surfaces, the thermopile
giving out least heat is ____ **4** ____.

shiny black   dull white
dull black   shiny white
B   C   A   D

1	2	3	4

[4 marks]

G–E

Conduction and convection

G–E

1 Match words, **A**, **B**, **C** and **D**, with the numbers **1 – 4** in the sentences.

A air pockets   **B** wooden   **C** metal   **D** plastic

The base of a saucepan is made of ____ **1** ____ because it conducts heat well. The handle is made of ____ **2** ____ because it is an insulator. Oven gloves prevent burns because they contain ____ **3** ____.

A hot pan can safely be put on a ____ **4** ____ mat which insulates the work surface.   [4 marks]

1	2	3	4

D–C

2 a Match words, **A**, **B**, **C** and **D**, with the numbers **1 – 4** in the sentences.

A rise   **B** sink   **C** transfer   **D** density

When a liquid is heated, its ____ **1** ____ changes. The warm liquid will ____ **2** ____ and the cool liquid will ____ **3** ____.

The convection current helps heat to ____ **4** ____ through the liquid.   [4 marks]

1	2	3	4

b One end of a metal rod is placed in the flame of a Bunsen burner. After a while, the other end of the rod becomes hot.

   i What is the name of the type of heat transfer that takes place?

     _____ [1 mark]

   ii Explain how heat is transferred from one end of the metal rod to the other.

     _____

     _____

     _____   [3 marks]

heat-proof mat   copper   iron   aluminium
rods heated in Bunsen flame
drawing pin held on with petroleum jelly
tripod
Bunsen burner

The experiment is repeated with three rods each made of different metals.

G–E

c Suggest **one** way to tell which type of metal is the best conductor.

     _____

     _____   [2 marks]

Heat transfer

G–E

3 Hanif's house has a woollen carpet.

a Explain how the carpet helps to stop heat loss through the floor.

     _____

     _____   [3 marks]

D–C

b What is meant by **free electrons**?

     _____

     _____   [2 marks]

c Use this idea to explain why wool is a better insulator than aluminium foil.

     _____

     _____

     _____   [3 marks]

Types of energy

1 Match words, **A**, **B**, **C** and **D**, with the numbers **1 – 4** in the sentences.

A chemical   **B** kinetic   **C** potential   **D** sound

The ____ **1** ____ energy in chips changes into ____ **2** ____ energy

when a person speaks. A girl sitting on the top of a wall

has ____ **3** ____ energy, and a person riding on a skateboard

has ____ **4** ____ energy.

1	2	3	4

[4 marks]

G–E

Energy changes

2 George drives his car along a level road.

a What is the main energy transformation that takes place in the engine?

_____ energy → _____ energy   [2 marks]

b Some of the energy is wasted. Write down **two** forms of wasted energy produced by the car.

_____ [2 marks]

c Explain what happens to the wasted energy.

_____
_____ [2 marks]

d Describe how the energy from the fuel is transformed as George drives his car up a hill and then stops it.

_____
_____
_____ [4 marks]

G–E

D–C

3 **a** Energy is measured in

    A joules

    B watts

    C metres

    D kilograms   [1 mark]

   b The main energy change in a radio is

    A sound → electrical

    B chemical → electrical

    C electrical → sound

    D heat → sound   [1 mark]

   c Which of these is an energy transfer?

    A A battery that changes chemical energy to electrical energy.

    B A plant that changes sunlight into chemical energy.

    C A kettle that uses electrical energy to heat water.

    D A hot radiator that warms the air in a room.   [1 mark]

   d The law of Energy Conservation says that

    A energy is created in the Sun.

    B energy cannot be created or destroyed.

    C energy cannot be destroyed.

    D energy cannot be created.   [1 mark]

G–E

D–C

Energy diagrams

1 The diagram shows a Sankey diagram for a light bulb.
Match numbers, **A**, **B**, **C** and **D**, with the numbers **1 – 4**
in the sentences.

A 6   **B** 57   **C** 60   **D** 100

The total energy output of the light bulb is ____ **1** ____ J. ____ **2** ____ %

of the input energy is transformed into output energy. The useful energy

output of the light bulb is ____ **3** ____ J, and ____ **4** ____ J is wasted energy.

1	2	3	4

[4 marks]

2 The picture shows a Sankey diagram for a petrol engine.

a Write down the total amount of wasted energy.

_____ J   [1 mark]

b In what form is most of the energy lost from the engine?

_____ [1 mark]

c An improved engine has been designed. It changes
1000 J of chemical energy into 500 J of kinetic energy. The rest is wasted heat.
Draw a Sankey diagram to show this.

[4 marks]

Energy and heat

3 a Wasted energy

A disappears

B can be stored for future use

C spreads to the surroundings

D can change back into the original form of energy   [1 mark]

b In a Sankey diagram

A the total input energy is equal to the output energy.

B the total input energy is more than the output energy.

C the total energy is less than the output energy.

D the total energy and the output energy are not linked.   [1 mark]

c Useful sources of energy are normally

A diluted

B concentrated

C spread out

D hard to find   [1 mark]

d When a solid piece of plastic heats up

A its particles spread out.

B its particles move closer together.

C its particles start to vibrate.

D its particles vibrate more.   [1 mark]

Energy, work and power

1 Match words, **A**, **B**, **C** and **D**, with the numbers **1 – 4** in the sentences.

A energy   **B** time   **C** power   **D** work

When Alex climbs a flight of steps, he does some ____ **1** ____ and

gains ____ **2** ____. The next day he climbs more quickly and

takes less ____ **3** ____, so his ____ **4** ____ increases.

1	2	3	4

[4 marks]

---

2 The weight of Molly and her skis is 500 N. A ski lift carries her to the top of the slope, which is a height of 60 m.

a How much **work** is done by the ski lift?

_____ J   [3 marks]

b If this took 75 seconds, work out the **power** of the ski lift.

_____ W   [3 marks]

c Write down **one** reason why the energy supplied to the lift is greater than this.

_____ [1 mark]

d The ski lift is replaced by a more **efficient** model. Explain why the energy used by the new ski lift will be less than the energy used by the first lift.

_____

_____ [2 marks]

Efficiency

3 a The **energy used per second** by a 60 W bulb is

A 60 W

B 60 J

C 6 J

D 6 W   [1 mark]

b The efficiency of a normal bulb is 5%. A fluorescent bulb wastes 75% of electrical energy supplied. LED lights use 50% of the energy supplied to produce light. Which list shows the bulbs in **order of efficiency** with the most efficient first?

A LED, normal, fluorescent

B fluorescent, LED, normal

C normal, LED, fluorescent

D LED, fluorescent, normal   [1 mark]

c The power of a kettle is 200 kW. What is the energy used by the kettle in one minute?

A 12 000 kJ

B 200 kJ

C 12 000 kW

D 200 kW   [1 mark]

d A kettle uses 170 kJ to heat water. 200 kJ of electrical energy is supplied to the kettle. The kettle's efficiency is

A 118%

B 60%

C 85%

D 8.5%   [1 mark]

Using energy effectively

1 Luke is worried that he is wasting energy at home.

  a Write down **two** reasons to try and save energy.

    _____

    _____ [2 marks]

  b Luke has not got any money to spend on home improvements.
    Explain why he will save energy if
    i he turns off lights in empty rooms.

    _____ [1 mark]

    ii he closes the curtains at night.

    _____ [1 mark]

  c Describe **one** other way to save energy in the home.

    _____

    _____ [2 marks]

2

Type of insulation	Installation cost (in £)	Annual saving (in £)	Payback time (in years)
Loft insulation	240	60	4
Cavity wall insulation	360	60	6
Draught-proofing doors and windows	45	15	3
Double glazing	2500	25	100

  a Explain what is meant by **payback time**.

    _____

    _____ [2 marks]

  b Draught proofing has the shortest payback time.
    Write down **two** benefits of installing draught proofing.

    _____

    _____ [2 marks]

  c The Brown family are planning to stay in their house for ten years and would like
    to install some insulation. Use the table to explain why draught proofing is not
    the best insulation method for them. Include a calculation for full marks.

    _____

    _____

    _____ [3 marks]

Why use electricity?

3 Match phrases, **A**, **B**, **C** and **D**, with the numbers **1 – 4** in the sentences.
  A batteries **B** solar energy **C** different energy forms **D** sunny places

Electrical energy has many uses because it can be changed into ____ **1** ____.

Portable supplies of electricity are ____ **2** ____.

Some batteries can be recharged using ____ **3** ____,

but solar energy can only be used in ____ **4** ____.

1	2	3	4

[4 marks]

Electricity and heat

1 Megan is preparing breakfast.

a Write down the main energy transformation for a toaster.

_____ [1 mark]

b The toaster stops working. Mum checks the fuse. Explain how the fuse works.

_____
_____ [3 marks]

c Fuse wires have different ratings. Explain why the thickness of 15 A fuse wire is not the same as 5 A fuse wire.

_____
_____ [2 marks]

The cost of electricity

2

Appliance	Power rating	Appliance	Power rating
Laptop computer	100 W	Television	50 W
Electric kettle	2.4 kW	Microwave oven	1.2 kW
Hairdryer	1.5 kW	Electric fire	2.0 kW
Radio	4 W	Lawn mower	1.3 kW

a What is meant by **power rating**?

_____
_____ [2 marks]

b Write down the power rating of the electric kettle in watts.

_____ W [1 mark]

c Calculate the total energy used in when
i the electric fire is switched on for 4 hours.

_____ kWh [2 marks]

ii the hairdryer is switched on for 15 minutes.

_____ kWh [3 marks]

3 Match numbers, **A**, **B**, **C** and **D**, with the spaces **1 – 4** in the sentences.

A 12 000   **B** 1200   **C** 10 000   **D** 1000

A microwave oven rated at 1.2 kW uses ____ **1** ____ joules per second. When it is switched on for 10 seconds it uses ____ **2** ____ joules. The reading on an electricity meter changes from 31667 to 32667 when ____ **3** ____ kWh are used. If each kWh costs 10p, the cost of this electricity is ____ **4** ____ pence.

1	2	3	4

[4 marks]

The National Grid

G–E

1 Match words, **A**, **B**, **C** and **D**, with the statements **1 – 4** in the table.

A cables   **B** National Grid   **C** power station   **D** substation

I	a network of power stations and cables carrying electricity across the country
2	generates electricity
3	reduces the voltage of electricity
4	carries electricity between pylons

1	2	3	4

[4 marks]

D–C

2 Match numbers, **A**, **B**, **C** and **D**, with spaces **1 – 4** in the table.

A 6   **B** 3   **C** 2   **D** 4

Item	Power (watts)	Current (amps)	Voltage (volts)
Electric blanket	920	**1**	230
Remote controlled car	12	2	**2**
Television	460	**3**	230
Torch	**4**	0.5	6

1	2	3	4

[4 marks]

G–E

3 Electricity can travel through the National Grid at voltages of 400 000 V.

a What is the voltage of electricity supplied to our homes?

_____ [1 mark]

b What does a step-down transformer do?

_____ [1 mark]

D–C

c Explain why energy losses are reduced if electricity is transported at a very high voltage.

_____

_____ [2 marks]

Generating electricity

G–E

4 a Anna's bicycle has a **dynamo** attached to its lights. A dynamo is used to change kinetic energy into electrical energy.
   i Why do the lights shine when the wheels are turning?

_____ [1 mark]

   iI How does the brightness of the lights change as she cycles faster?

_____ [1 mark]

Power stations

1 The diagram shows how electricity is generated in a power station.

boiler   steam   National Grid

burning fuel   condenser   turbine   generator   step-up transformer   pylon

a Explain how the steam is created.

_____

_____   [2 marks]

b Write down the energy change that takes place
i in the boiler.

_____   [2 marks]

ii in the generator.

_____   [1 mark]

c One disadvantage of burning fossil fuels is that they will run out one day. Write down **one advantage** of using fossil fuels to generate electricity.

_____   [1 mark]

d Some of the energy released from burning fossil fuels is wasted. Explain what happens to it.

_____

_____   [2 marks]

2 Match words, **A**, **B**, **C** and **D**, with the statements **1 – 4** in the table.

A biomass   **B** coal   **C** hydroelectricity   **D** uranium

1	a non-renewable fuel
2	a nuclear fuel
3	a renewable energy source
4	a renewable fuel

1	2	3	4

[4 marks]

Renewable energy

3 a Explain why wind turbines do not always produce electricity.

_____   [1 mark]

b France has a large tidal power station.
i Explain how the tidal power station generates electricity.

_____

_____

_____   [3 marks]

ii Write down **one disadvantage** of using tidal power stations.

_____   [1 mark]

Grades: G–E, D–C, G–E, D–C, G–E, D–C

Electricity and the environment

1 a Write down **two** harmful effects of burning fossil fuels.

_____

_____ [2 marks]

b Explain why wind turbines are useful in remote areas.

_____

_____ [2 marks]

Making comparisons

2 a Sienna is comparing the effects on the environment of coal-fired power stations and nuclear power stations.

   i Which of these power stations release greenhouse gases?

_____ [1 mark]

   ii Which of the fuels used in these power stations, will run out one day?

_____ [1 mark]

b Power stations need to be located on suitable sites. Write down **three factors** that a company may consider before choosing a site for a coal-fired power station.

_____

_____

_____ [3 marks]

3 Match words, **A**, **B**, **C** and **D**, with the numbers **1 – 4** in the sentences.

A cheap   **B** efficient   **C** flexible   **D** reliable

Fossil fuel power stations are designed to waste less energy, so they are more ____ **1** ____.

At the moment, fossil fuels are ____ **2** ____ but their price will rise when supplies get low.

Solar cells cannot generate electricity when it is dark or cloudy because they are not ____ **3** ____.

In some places, spare electricity is used to pump water behind dams so hydroelectric stations can

provide electricity in seconds, making it very ____ **4** ____.

1	2	3	4

[4 marks]

4 It has been decided to build a new power station. The choice is between a large hydroelectric plant and an oil-fired power station.

a What **two** things affect the total cost of electricity produced?

_____

_____ [2 marks]

b How do you expect the fuel costs to change for these power stations over the next 50 years?

_____

_____ [2 marks]

P1a revision checklist

I know:

how heat (thermal energy) is transferred and what factors affect the rate at which heat is transferred

☐ heat energy can be transferred by conduction, convection and thermal radiation

☐ thermal conductors (e.g. metals) transfer heat energy easily; thermal insulators (e.g. plastic, glass) do not

☐ dark, dull surfaces emit and absorb thermal radiation better than shiny, light surfaces

☐ the bigger the temperature difference between an object and its surroundings, the faster the rate at which heat is transferred

what is meant by the efficient use of energy

☐ energy is never created nor destroyed; some energy is usually wasted as heat

☐ Sankey diagrams show the energy input and output of a device

why electrical devices are so useful

☐ they transform electrical energy to whatever form of energy we need at the flick of a switch

☐ the National Grid transmits energy around the country at high voltages and low current to reduce energy losses

☐ dynamos produce electricity when coils of wire rotate inside a magnetic field

☐ how to work out the power rating of an appliance (the rate at which it transforms electrical energy)

☐ to calculate the amount of energy transferred from the mains use:

    energy transferred   =   power   ×   time
    (kilowatt-hour, kWh)   (kilowatt, kW)   (hour, h)

☐ to calculate the cost of energy transferred from the mains use:

    total cost = number of kilowatt-hours × cost per kilowatt-hour

how we should generate the electricity we need

☐ we need to use more renewable energy sources, including wind, hydroelectric, tidal, wave and geothermal power

☐ most types of electricity generation have some harmful effects on people or the environment; there are also limitations on where they can be used

Uses of electromagnetic radiation

G–E

1 a What is meant by a **spectrum**?

_____ [1 mark]

b Write down **three properties** that all electromagnetic waves share.

_____

_____

_____ [3 marks]

Electromagnetic spectrum 1

G–E

2 The diagram shows the members of the electromagnetic spectrum.

| Radio waves | **1** | Infrared | **2** | Ultraviolet | **3** | Gamma rays |

Increasing frequency

⟶

Increasing energy

⟶

a Fill in the names missing on the diagram.

1 _____ **2** _____ **3** _____ [3 marks]

D–C

b Write down the relationship between frequency and energy.

_____ [1 mark]

D–C

3 Match words, **A**, **B**, **C** and **D**, with the numbers **1 – 4** in the sentences.

A gamma rays

B infra red

C radio waves

1	2	3	4

D ultra violet

____ **1** ____ are used for broadcasts. ____ **2** ____ are used to detect parts of the body

that are warmer due to inflammation. Our skin develops a suntan when exposed

to ____ **3** ____. ____ **4** ____ can be used to kill cells, and sterilise surgical instruments. [4 marks]

D–C

4 a Add a label to the diagram to show one wavelength. [1 mark]

b Under the wave, add a diagram of another wave which carries more energy. [2 marks]

Electromagnetic spectrum 2

1 Match words, **A**, **B**, **C** and **D**, with the numbers **1 – 4** in the sentences.

A absorbs   **B** carry   **C** reflects   **D** transmits

All electromagnetic waves ____ **1** ____ energy. A mirror ____ **2** ____ light

so we can see an image. A window ____ **3** ____ light helping us see

during the day. Black material ____ **4** ____ light so it looks dark.

1	2	3	4

[4 marks]

G–E

2 The table shows some equipment that uses electromagnetic radiation.

Wave type	Wavelength	Sources	Detectors
Gamma rays	10^{-12} m	Radioactive nuclei	Geiger-Müller tube
X-rays	10^{-10} m	X-ray tubes	Geiger-Müller tube
Ultraviolet	10^{-7} m	Sun, very hot objects	Photographic film, fluorescent material
Visible light	0.0005 mm	Hot objects, Sun, lasers, LEDs	Eyes, photographic film
Infrared	0.1 mm	Warm or hot objects	Skin, thermometer
Microwaves	1–10 cm	Radar, microwave ovens	Aerial, mobile phone
Radio waves	10– 1000+ m	Radio transmitters	Aerial, TV, radio

a Write down **one** type of radiation that is

i detected by aerials and given out by radar.

_____ [1 mark]

ii detected by Geiger-Müller tubes and given out by radioactive nuclei.

_____ [1 mark]

iii detected by the skin and given out by hot objects.

_____ [1 mark]

b Complete these sentences using the words **absorbed**, **transmitted** or **reflected**.

i Satellites use microwave communication because these waves are

_____ by the atmosphere.

ii Ultraviolet waves are _____ by our skin, causing suntan.

iii White cars stay cool in the Sun because infrared is _____
by the light colour. [3 marks]

D–C

Waves and matter

3 a What type of electromagnetic radiation is used to take shadow pictures of bones?

_____ [1 mark]

b Explain how a picture of a person's bones can be taken using electromagnetic radiation.

_____
_____
_____
_____ [3 marks]

c Suggest **one** reason why these pictures are taken only when necessary.

_____ [1 mark]

G–E

D–C

Dangers of radiation

1 Match words, **A**, **B**, **C** and **D**, with the numbers **1 – 4** in the sentences.

A absorb   **B** burn   **C** ionise   **D** kill

Cells ____ **1** ____ many types of radiation which can damage them.

High doses of gamma rays ____ **2** ____ cancer cells.

Low doses of X-rays and gamma rays ____ **3** ____ cells, damaging them.

Infrared radiation and microwaves ____ **4** ____ cells.

1	2	3	4

[4 marks]

2 Rachel has put sun-block on to protect her skin before she goes out in the Sun.

  a   What type of electromagnetic radiation does sun-block protect her from?

_____ [1 mark]

  b   Write down **three** risks from too much exposure to this type of radiation.

_____

_____

_____ [3 marks]

  c   Explain why darker skins are less likely to be damaged by the Sun's radiation.

_____

_____

_____ [3 marks]

Telecommunications

3 a   What is the name given the type of signal shown in A?

_____

[1 mark]

  b   Explain how digital signals carrying a cable TV programme are sent.

_____

_____

_____

[2 marks]

  c   Suggest **one** reason why underground telephone cables are no longer used.

_____ [1 mark]

  d   Many communication links are now made using satellite links.
Write down **two advantages** of using satellite communications.

_____

_____ [2 marks]

Fibre optics: digital signals

1 a Explain how an **optical fibre** is used to send digital signals.

_____

_____ [2 marks]

b The diagram shows light travelling down an optical fibre.

i Continue the path of the light ray until it emerges from the optical fibre. [3 marks]

ii What is the **name** given to the process that occurs inside the optical fibre?

_____ [1 mark]

iii Describe **three advantages** of using optical fibres to send digital signals on a telephone network.

_____

_____

_____ [3 marks]

Radioactivity

2 Match words, **A**, **B**, **C** and **D**, with the arrows **1 – 4** in the diagram.

A electron

B neutron

C nucleus

D proton

1	2	3	4

[4 marks]

3 The table gives some information about different atoms.

Element	Number of protons	Number of neutrons
A	6	6
B	6	7
C	7	7
D	8	8

a Which elements are isotopes of each other?

_____ [1 mark]

b How many electrons are orbiting the nucleus of element A? _____ [1 mark]

c What is the mass number of element C? _____ [1 mark]

d Explain what is meant by **radioactive decay**.

_____

_____ [2 marks]

Alpha, beta and gamma rays 1

1 The diagram shows three different types of radiation and their penetrating properties.

a Which types of radiation can be safely stored in
   i a box lined with lead?

   _____ [1 mark]

   ii a cardboard box?

   _____ [1 mark]

   iii a metal tin $\frac{1}{2}$ cm thick?

   _____ [1 mark]

α particles

β particles

γ rays

lead block | radioactive source | sheet of paper | thin aluminium sheet | thick block of lead

2 a Explain what is meant by an **alpha particle**.

   _____

   _____ [2 marks]

b Alpha particles are very ionising. Inside the body, they damage cells. Explain why alpha particles are unlikely to cause damage to people in most cases.

   _____ [1 mark]

c Write down **two** uses for alpha radiation.

   _____

   _____ [2 marks]

d Why are gamma rays **less ionising** than alpha particles?

   _____

   _____ [2 marks]

Background radiation 1

3 a Background radiation is
   A very rare.
   B found only in a few places.
   C found all around us.
   D only caused by natural sources. [1 mark]

b Radon gas is found in some areas of the UK. It is
   A not a health risk.
   B a health risk causing raised levels of lung cancer in these regions.
   C caused by man-made sources only.
   D not possible to change your exposure to radon gas. [1 mark]

c Cosmic radiation is caused by the Sun. It is
   A mainly absorbed by the atmosphere.
   B a significant health risk on all flights.
   C not a health risk.
   D the same intensity at sea level and on high altitude flights. [1 mark]

d Radiation badges are worn
   A to protect people from exposure to radioactivity.
   B as a source of radioactivity.
   C to give an instant read-out of exposure to radioactivity.
   D to monitor exposure to radioactivity over a period of time. [1 mark]

Half-life

1 a After two half-lives, in a radioactive sample
  A all of the atoms will decay.
  B none of the atoms will decay.
  C exactly half of the atoms will decay.
  D about half of the atoms will decay. [1 mark]

c The half-life of a sample is the time taken for
  A half of its mass to disappear.
  B its original activity to fall to half.
  C half of the samples measured to decay completely.
  D its count rate to change. [1 mark]

b The length of a sample's half-life depends on
  A the type of material that is decaying.
  B the temperature the sample is kept in.
  C the pressure the sample feels.
  D the country the sample was found in. [1 mark]

d A sample of cobalt has a half-life of 5 years, and its count rate is 1200 counts per second. After 15 years, the count rate is
  A 600   **B** 300   **C** 150   **D** 0 [1 mark]

Uses of nuclear radiation

2 Gamma rays can penetrate the body and kill cells.

a What is meant by **radiotherapy**? [1 mark]

b Explain why cancer cells are killed by gamma rays without damaging healthy cells. [2 marks]

c Another medical use for radioactivity is a tracer. Explain how a tracer can be used to monitor the flow of blood in a patient's body. [2 marks]

d Write down **one** more use of radioactivity in hospitals. [1 mark]

3 The diagram shows one method of controlling the thickness of paper in a factory.

a Write down **one** reason why is it important to monitor the thickness. [1 mark]

b Why is alpha radiation not used to detect the thickness of the paper? [1 mark]

c Explain how the machine should adjust the rollers if the amount of radiation reaching the detector gets smaller. [2 marks]

thickness detector
source of Beta radiation
paper sheet

Safety first

1 a A person is working with a radioactive sample. Which precaution will make no difference to her exposure?

A using tongs

B putting the sample back in its case after use

C wearing a lead apron

D reducing the temperature of the room [1 mark]

b Alpha particles

A are only absorbed by cells inside the body.

B are absorbed by cells inside and outside the body.

C are only absorbed by cells outside the body.

D are not absorbed by any cells. [1 mark]

c When radioactivity is absorbed, which of these effects never occurs?

A infection

B cell mutation

C absorption of energy

D ionisation [1 mark]

Searching space

2 a Land-based telescopes are often placed on mountains because

A they are at risk of vandalism.

B of light pollution.

C dust pollution in the atmosphere blocks starlight.

D this avoids vibrations. [1 mark]

b Telescopes are used to examine the night sky. What is used in an optical telescope to focus light?

_____ [1 mark]

c Explain why a reflecting telescope produces better images than a simple optical telescope.

_____
_____ [2 marks]

d The Hubble telescope is a space telescope. Explain what is meant by a **space telescope**.

_____
_____ [2 marks]

e Describe **two advantages** of having a telescope in orbit.

_____
_____ [2 marks]

Gravity

1 Match words, **A**, **B**, **C** and **D**, with the numbers **1 – 4** in the sentences.

A force   **B** gravity   **C** mass   **D** separation

All objects are attracted to each other because of ____ **1** ____.

The attractive ____ **2** ____ is greater if the ____ **3** ____ of the

objects is small or if the ____ **4** ____ of the objects is big.

1	2	3	4

[4 marks]

2 The diagram shows how one spacecraft used gravity to travel across the Solar System.

a Explain what caused the spacecraft to change direction as it passed the different planets.

_____

_____

[1 mark]

Voyager 2 launch

Earth's orbit

Jupiter — Sun

Saturn

Uranus

Neptune

Voyager 2

b When the spacecraft travelled close the Sun, how did the effect of the Sun's gravity change?

_____

[1 mark]

Birth of a star

3 Match words, **A**, **B**, **C** and **D**, with the sentences **1 – 4**.

A gravity   **B** nebulae   **C** nuclear fusion   **D** star

1 a huge cloud of dust and gas molecules in space

2 a large body giving out its own heat and light

3 this forces particles closer together in a nebulae

4 when two nuclei join together

1	2	3	4

[4 marks]

4 The statements, **A**, **B**, **C** and **D**, describe the stages in the formation of a star. Put them in the correct order, matching the statements with the correct place **1 – 4** in the boxes.

A gas molecules are pulled together by gravity to form a cloud

B nuclear fusion starts

C the cloud attracts more particles and gets larger

D the core of the nebula heats up dramatically

1		2		3		4		the star begins to shine
	→		→		→		→	

[4 marks]

Formation of the Solar System

G–E

1 Match words, **A**, **B**, **C** and **D**, with the statements **1 – 4** in the table.

A Jupiter   **B** Mercury   **C** Solar System   **D** Sun

1	the name of our closest star
2	one of the gas giants
3	the central star and its orbiting planets
4	one of the rocky planets

1	2	3	4

[4 marks]

D–C

2 The diagram shows one idea of how the Solar System was formed.

a Explain what happens to the dust cloud when the Sun ignited.

_____

_____

[1 mark]

b Explain **one** reason why the rocky planets formed closest to the Sun and gassy planets further away.

_____

_____

[2 marks]

Life and death of a star

G–E

3 Stars begin as a cloud of dust and gas which are crushed by gravity until nuclear fusion reactions begin.

a Explain why the Sun is constantly losing mass.

_____ [1 mark]

b Write down the three stages that a star, like our Sun, will go through before it eventually dies.

_____ → _____ → _____ [3 marks]

D–C

c What **two** factors are balanced when a star is in the stable phase?

_____

_____

[2 marks]

D–C

4 Match words, **A**, **B**, **C** and **D**, with the numbers **1 – 4** in the sentences.

A black hole   **B** neutron star   **C** red giant   **D** supernova

A star the size of our Sun will eventually expand to form a ____ **1** ____.

Stars larger than our Sun become red supergiants, which collapse and

explode as a spectacular ____ **2** ____. The core remains an incredibly

dense ____ **3** ____. If the star was very massive, a ____ **4** ____ remains instead.

1	2	3	4

[4 marks]

In the beginning

1 a When is the Big Bang thought to have taken place?

_____ [1 mark]

b Explain what is meant by the **Big Bang theory**.

_____
_____ [2 marks]

c Explain how heavier elements were created from **quarks**.

_____
_____
_____ [3 marks]

G–E

D–C

The expanding Universe

2 Astronomers looking at distant galaxies have noticed an effect they called the red shift.

a Explain what is meant by the **red shift**.

_____
_____ [2 marks]

b What does the red shift tell astronomers about the **motion** of galaxies?

_____ [1 mark]

G–E

3 a Galaxies close to us have a smaller red shift than galaxies further away. This tells us that galaxies are moving

A at the same speed.

B away from us at different speeds.

C towards us at different speeds.

D slowly. [1 mark]

b A spectroscope is used to

A measure your heart.

B display the electromagnetic spectrum.

C split light into different colours.

D rejoin light that has been split up. [1 mark]

c An element emits a unique pattern of light when it is

A heated.

B looked at under high pressure.

C reacting with other elements.

D cooled. [1 mark]

d The change in the colour of light from galaxies occurs because of

A the great distance the light travels.

B the different temperatures of the stars in the galaxies.

C interference from our atmosphere.

D the change in wavelength of the light. [1 mark]

D–C

P1b revision checklist

I know:

what the uses and hazards of the waves that form the electromagnetic spectrum are

☐ from longest to shortest wavelength: radio waves, microwaves, infrared, visible light, ultraviolet, X-rays, gamma rays

☐ radiation has many uses e.g. radio, TV, satellites, cable and mobile phone networks

☐ some forms of electromagnetic radiation can damage living cells: ionising radiation (ultraviolet, X-rays and gamma rays) can cause cancer

what the uses and dangers of emissions from radioactive substances are

☐ the uses and hazards of radioactive substances (which emit alpha particles, beta particles and gamma rays) depend on the wavelength and frequency of the radiation they emit

☐ background radiation is all around us e.g. granite rocks can emit gamma rays and form radioactive radon gas

☐ the relative ionising power, penetration through materials and range in air of alpha, beta and gamma radiations

about the origins of the Universe and how it continues to change

☐ the Universe is still expanding; in the beginning, matter and space expanded violently and rapidly from a very small initial point i.e. the Big Bang

☐ red shift indicates that galaxies are moving apart; the further away a galaxy, the faster it is moving away from us

☐ telescopes on Earth and in space give us information about the Solar System and the galaxies in the Universe

Cells

1 The diagram shows a plant cell.

cell membrane	cell wall	cytoplasm
chloroplast	nucleus	vacuole

a Use words from the box to name parts **A**, **C**, **E** and **F**.

A_____   C_____

E_____   F_____   [3 marks]

b Write the letters of **all** the labelled parts on this cell that are also present in animal cells.

_____   [1 mark]

c This cell is found in a plant leaf. Its function is to make food by photosynthesis.

Describe **one** way in which the cell is adapted for photosynthesis.

_____   [1 mark]

d Describe the roles of each of these organelles.

Ribosomes _____

Mitochondria _____   [2 marks]

G–E

D–C

Specialised cells

2 The diagram shows an egg cell.

a Most cells have **two** sets of chromosomes.
Why does the egg cell have only **one** set?

_____

_____ [1 mark]

b Explain why the egg cell needs many mitochondria.

_____

_____   [2 marks]

cell membrane, *little vesicles containing fat and other nutrients*, *mitochondria*, *cytoplasm*, *nucleus containing one set of chromosomes*

c The diagram shows a ciliated cell.
Cells like this are found in
the lining of the trachea and
bronchi leading down to the lungs.

i Describe the function of ciliated cells.

_____   [1 mark]

ii State **two** ways in which ciliated cells are adapted for their function.

1_____

2_____   [2 marks]

G–E

D–C

Diffusion 1

1 a Complete the definition of diffusion by using the **correct** words from the box.

| gas | higher | lower | net | particles | solid | smell |

Diffusion is the spreading of the _____ of a _____ , or of any

substance in solution. This results in a _____ movement from a region

where they are in a _____ concentration. **[4 marks]**

b Explain why diffusion happens faster when the temperature is higher.

_____

_____ **[2 marks]**

2 The diagram shows a cell.

high concentration
of oxygen

low concentration
of oxygen

cell membrane,
which is permeable
to oxygen

a Is this a plant cell or an animal cell? Explain your answer.

_____ **[1 mark]**

b Explain why there is a low concentration of oxygen inside the cell.

_____

_____ **[2 marks]**

c The cell membrane is **permeable** to oxygen. Explain what this means.

_____ **[1 mark]**

d Name the part of the cell that uses oxygen. (These parts are not shown on

the diagram.) _____ **[1 mark]**

e Explain why oxygen diffuses into the cell more rapidly, when these parts are
working faster.

_____

_____ **[2 marks]**

Diffusion 2

3 The cells inside a plant leaf photosynthesise during the daytime. There are tiny holes
in the underside of the leaf called stomata. Gases are able to diffuse in and out of the
leaf through the stomata.

a Which gas will diffuse **into** the leaf during the daytime?

_____ **[1 mark]**

b Explain why this gas diffuses into the leaf.

_____

_____ **[2 marks]**

Osmosis 1

1 Complete the sentences by choosing the **correct** words from the box.

concentrated	diffusion	dilute	flow	fully	liquid	partially	water

Osmosis is the _____ of _____

from a _____ to a more concentrated solution through a

_____ permeable membrane. [4 marks]

2 The diagram shows a piece of Visking tubing containing sugar solution. The tubing is in a beaker of water.

glass tubing

dilute sugar solution

Visking tubing – a partially permeable membrane

concentrated sugar solution

a Water passes through the Visking tubing by osmosis. In which direction does osmosis take place?

From the _____ into the _____ . [1 mark]

b Describe what happens to the level of the liquid in the glass tubing.

_____ [1 mark]

c Explain why this happens. Refer to water molecules, sugar molecules and the Visking tubing membrane in your answer.

_____

_____ [3 marks]

Osmosis 2

3 The diagram shows an animal cell that has been placed in water.

distilled water

cytoplasm – a fairly concentrated solution

cell membrane – a partially permeable membrane

a Which part of the cell is a partially permeable membrane?

_____ [1 mark]

b Water enters the cell by osmosis. What will happen to the cell?

_____ [1 mark]

c Explain why this does **not** happen to a plant cell in pure water even though water enters the cell by osmosis.

_____

_____ [2 marks]

d A plant cell is placed in a concentrated sugar solution. What will happen to the cell? Explain your answer.

_____

_____ [3 marks]

Photosynthesis

1 a Complete the equation for photosynthesis by crossing out the **two incorrect** words in each box.

| oxygen nitrogen ~~carbon dioxide~~ | + | glucose water oxygen | → | nitrogen glucose carbon dioxide | + | oxygen water nitrogen |

[2 marks]

b Describe the role of chlorophyll in photosynthesis.

_____

_____

[2 marks]

Leaves

2 a Photosynthesis happens inside the leaves of plants. The diagram shows a plant leaf.

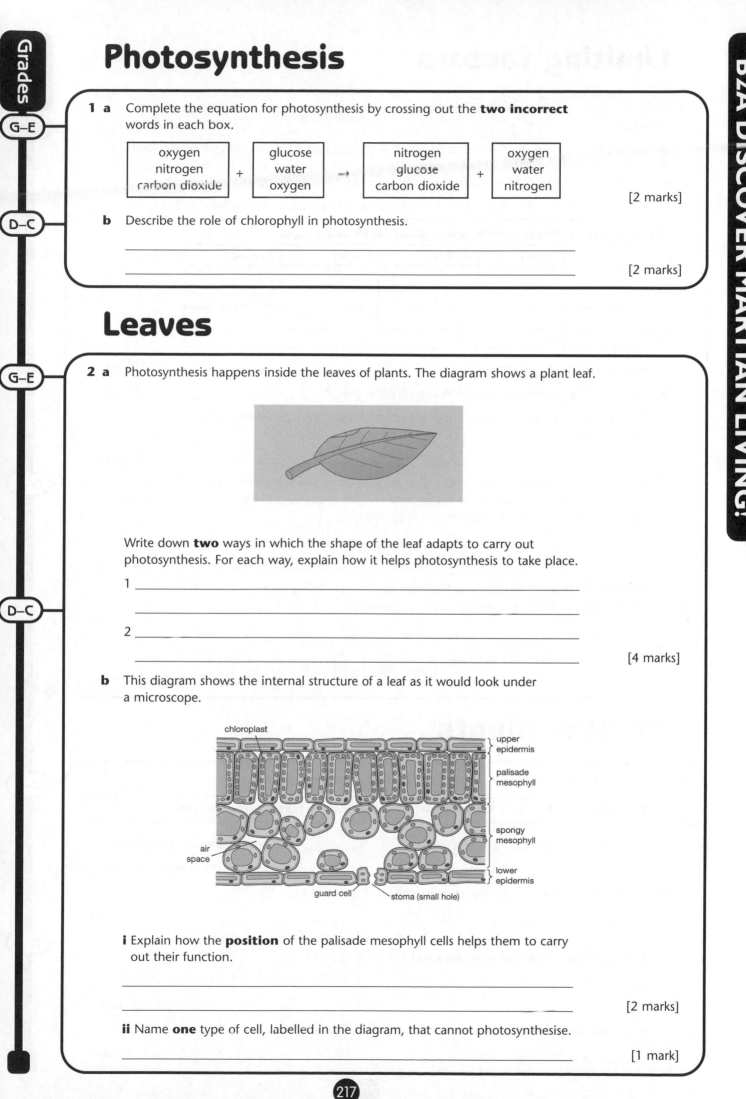

Write down **two** ways in which the shape of the leaf adapts to carry out photosynthesis. For each way, explain how it helps photosynthesis to take place.

1 _____

_____

2 _____

_____

[4 marks]

b This diagram shows the internal structure of a leaf as it would look under a microscope.

chloroplast — upper epidermis — palisade mesophyll — spongy mesophyll — lower epidermis — guard cell — stoma (small hole) — air space

i Explain how the **position** of the palisade mesophyll cells helps them to carry out their function.

_____

_____

[2 marks]

ii Name **one** type of cell, labelled in the diagram, that cannot photosynthesise.

_____

[1 mark]

Limiting factors

1 Grace set up an experiment to find out if a water plant makes oxygen more rapidly when it has more light.

She put a piece of pond weed into a test tube of water. She took it into a room where the light was dim. She counted the number of bubbles that the pond weed gave off in one minute. She did this twice more at the same light intensity. She then repeated this at three other light intensities.

The table shows her results. The higher the number for light intensity, the brighter the light.

Light intensity	Number of bubbles given off in one minute			
	1st try	2nd try	3rd try	Average
1	6	9	10	8.3
2	16	18	20	18.0
3	28	27	29	28.0
4	30	32	34	

a Fill in the empty box in the table. [1 mark]

b What was the dependent variable in Grace's experiment? _____ [1 mark]

c What do Grace's results suggest about the effect of light intensity on the rate of photosynthesis?

_____ [1 mark]

d Grace's teacher said that Grace's results suggest she should let the plant settle down at each new light intensity, before beginning to count the bubbles. Explain how Grace's results support this suggestion.

_____

_____ [2 marks]

e Grace thinks that her results show that light can be a **limiting factor** for photosynthesis. Do you agree? Explain your answer.

_____

_____

_____ [3 marks]

Healthy plants

2 a Complete the sentences by choosing the **correct** words from the box.

| amino | calcium | carbohydrates | cellulose | fatty | magnesium | proteins |

Plants need nitrate ions for producing _____ acids, which are then used to form

_____ . They need _____ ions for making chlorophyll. [3 marks]

b Describe how plants obtain the mineral ions that they need.

_____ [2 marks]

c Explain these observations.
i A plant that is deficient in nitrate ions has stunted growth.

_____

_____ [2 marks]

ii A plant that is deficient in magnesium ions has yellow leaves.

_____ [1 mark]

Food chains

G–E

1 In the Arctic Ocean, tiny organisms called phytoplankton float in the upper layers of the water. They contain chlorophyll and photosynthesise. They are eaten by many animals, including the huge whale shark and tiny shrimp-like organisms called krill. Krill are eaten by several kinds of fish and by penguins. Seals eat fish and penguins. Killer whales eat seals.

a Using the information above, draw the longest Arctic Ocean food chain you can.

_____ [1 mark]

b What do the arrows in your food chain mean?

_____ [1 mark]

c Which organisms are the producers in your food chain?

_____ [1 mark]

d Draw a ring around each of the terms that **correctly** describe the position of penguins in the Arctic Ocean food web.

  carnivore   **predator**   **prey**

  primary consumer   **secondary consumer**   [1 mark]

D–C

2 a Energy enters a food chain as solar energy, in light. In what form is the energy passed along the food chain?

_____ [1 mark]

b Explain why green plants only capture a small part of the solar energy that reaches them.

_____
_____ [2 marks]

Biomass

G–E

3 This diagram shows a pyramid of biomass for a food chain.

mass of lion
mass of antelope
mass of grass

a What is biomass?

_____ [1 mark]

b Explain why not all of the antelope biomass becomes lion biomass.

_____
_____ [2 marks]

D–C

4 Mammals and birds keep their body temperature constant, even when the temperature around them is low.
Explain why this means that energy transfer from mammals or birds to the next organism in a food chain is not very efficient.

_____
_____ [2 marks]

Food production

1 Karen and Jo are eating lunch. Karen is eating an apple. Jo is eating a cold chicken leg.

   a Draw a food chain for Karen's meal, and another for Jo's.

      Karen _____

      Jo _____   **[2 marks]**

   b Using the food chains that you have drawn, explain why growing apples is a more efficient method of food production than rearing chickens.

      _____

      _____   **[2 marks]**

   c Some farmers rear chickens indoors, in heated sheds. Explain how this can help to reduce energy losses and improve the efficiency of production of chickens.

      _____

      _____   **[2 marks]**

2 Food that is sold in supermarkets is distributed from central warehouses using fleets of large lorries. The food arrives at the central warehouses from all over the world. The distance the food travels from where it is made to where it is eaten is sometimes called 'food miles'.

Emma's mother looks carefully to see where food has been produced and tries to buy things that were grown locally. Explain how this could help to lessen damage to the environment.

_____

_____   **[2 marks]**

The cost of good food

3 Hens that are kept to produce eggs may be kept in three different ways:

Ways of keeping hens	Freedom of hens	Cost of a box of 6 eggs
Free range	Can roam outside	99p
In barns	Can move around freely	73p
In battery cages	Cannot move freely	54p

   a Explain why free range eggs are more expensive than barn eggs and battery eggs.

      _____

      _____   **[2 marks]**

   b Suggest why many people buy free range eggs even though they are more expensive than battery eggs.

      _____

      _____   **[2 marks]**

G–E

D–C

G–E

D–C

Death and decay

1 Decay of biological waste materials and bodies is caused by microorganisms.

a Name **two** kinds of microorganisms that cause decay.

1_____

2_____ [2 marks]

b Explain why decay happens most rapidly when there is plenty of oxygen available.

_____

_____ [2 marks]

2 The diagram shows a thermometer.

a Write these words in the correct boxes.
- the temperature in a fridge
- all microorganisms and spores are destroyed

b Explain why food keeps longer in a freezer than in a fridge.

_____

_____ [2 marks]

Cycles

3 a Complete the sentences by choosing the **correct** words from the box.

community detritus photosynthesis population respiration trophic

Organisms such as earthworms, which eat dead leaves and other plant remains,

are called _____ feeders. They help to recycle the materials in

the plant remains, so that they become available to other members of

the _____ of organisms in the ecosystem. For example, they release

some of the carbon in the leaves back into the air, in the form of carbon dioxide,

by the process of _____. [3 marks]

b Which of the organisms in a food web may be eaten by detritus feeders?

_____ [1 mark]

c Explain why detritus feeders are said to be **consumers** in a food chain.

_____

_____ [2 marks]

The carbon cycle 1

1 About 0.04% of the air around us is carbon dioxide.

a Name the process by which some living organisms take carbon dioxide from the air.

_____

[1 mark]

b Name the process by which all living organisms return carbon dioxide to the air.

_____

[1 mark]

c We need to eat carbon-containing substances in our diet. Draw a circle around each of the nutrients that contains carbon.

carbohydrates   fats   minerals   proteins   water

[2 marks]

The carbon cycle 2

2 The diagram shows parts of the carbon cycle. The numbers indicate how many billions of tonnes of carbon pass along each pathway in one year.

fossil fuel burning 3   land use 2.6   photosynthesis 120   plant respiration 50   decay of dead plants and animals and wastes 60   sea – surface gas exchange 115   circulation

a Explain how decay of dead plants and animals releases carbon dioxide into the air.

_____

_____

[2 marks]

b Calculate the difference between the quantity of carbon removed from the air by photosynthesis by land plants and the quantity returned to the air by plant respiration.

_____

[1 mark]

c Explain how this supports the idea that deforestation could contribute to global warming.

_____

_____

[3 marks]

d 'Land use' refers to the disturbance of soil, which contains a lot of carbon locked up in the form of carbon compounds. Using what you know about decomposers, suggest how this carbon got into the soil.

_____

_____

[2 marks]

e Suggest what is meant by 'sea surface gas exchange'.

_____

_____

[2 marks]

B2a revision checklist

I know:

what animals and plants are built from

☐ animal cells and plant cells have a membrane, cytoplasm and a nucleus; plant cells may also have a vacuole, chloroplasts and a cell wall

☐ in multicellular organisms, different cells are specialised for different functions

☐ the chemical reactions inside cells are controlled by enzymes

how dissolved substances get into and out of cells

☐ diffusion is the net movement of particles of gas, or substances dissolved in a solution, from a region of high concentration to a region of lower concentration

☐ osmosis is the diffusion of water molecules through a partially permeable membrane

how plants obtain the food they need to live and grow

☐ green plants use chlorophyll to trap energy from the Sun to photosynthesise

☐ leaves are specially adapted for photosynthesis – they can be broad, flat, thin and have lots of stomata

☐ the rate of photosynthesis is affected by light intensity, carbon dioxide concentration and temperature

☐ mineral salts in the soil are used to make proteins or chlorophyll

what happens to energy and biomass at each stage in a food chain

☐ energy passes along food chains but some energy is lost at every stage

☐ the shorter the food chain, the less energy is lost

☐ the mass of biomass at each stage in a food chain is less than it was at the previous stage. This can be shown in a pyramid of biomass

☐ reducing energy loss increases efficiency of food production

☐ decomposers and detritus feeders feed on dead organisms and their waste

Enzymes – biological catalysts

1 Complete these sentences by choosing the **correct** words from the box.

| active | catalysts | digestion | DNA | organisms | protein | substrate | surface |

Enzymes are biological _____. They are _____ molecules.

Each kind of enzyme only works on a particular kind of _____ , which fits

perfectly into a fold in the enzyme called the _____ site.   [4 marks]

2 This graph shows how temperature affects the activity of an enzyme.

a   At what temperature does the enzyme work most quickly?

_____   [1 mark]

b   Using what you know about rates of reaction, explain why the activity of the enzyme increases as the temperature is increased from 10 °C to 20 °C.

_____

_____

_____   [3 marks]

c   Lucy says that the enzyme stops working at high temperatures because it has been killed. Explain why Lucy is wrong, and give the correct explanation for the shape of the graph at temperatures above 40 °C.

_____

_____

_____   [3 marks]

Enzymes and digestion

3 The diagram shows the sites of production of some of the enzymes that help with digestion.

a   Name **two** organs, shown on the diagram, where amylase is produced.

1 _____   2 _____

[2 marks]

b   What does amylase do?

_____

_____   [1 mark]

c   Protease enzymes and hydrochloric acid are produced in the stomach. Suggest a value for the optimum pH for the protease enzymes in the stomach.

_____   [1 mark]

d   Bile flows into the small intestine. Bile contains a base called sodium hydrogencarbonate. Suggest how this affects the partly digested stomach contents as they move from the stomach into the small intestine.

_____   [1 mark]

Enzymes at home

1 If blood gets onto clothes it makes a stain that is difficult to remove. The stain is caused by haemoglobin.

    a Explain how proteases in biological washing powders can help to remove blood stains from clothes.

    _____

    _____ [2 marks]

    b Explain why most biological washing powders work best at about 40 °C.

    _____

    _____ [2 marks]

    c When biological washing powders were first introduced, some people found that they harmed their hands. Now, in most biological washing powders, the enzymes are inside tiny particles, surrounded by a harmless covering.

    Suggest how biological washing powders could harm people's hands if the enzymes were not enclosed like this.

    _____

    _____ [2 marks]

Enzymes and industry

2 a The boxes on the left contain the names of enzymes. The boxes on the right contain short descriptions of industrial processes. Draw lines linking each enzyme with the appropriate process.

proteases		convert starch into sugar syrup
carbohydrases		pre-digest protein in some baby foods
isomerase		convert glucose into fructose

    b Suggest how pre-digested foods can help a baby to grow well.

    _____

    _____ [2 marks]

    c Explain why foods sweetened with fructose may contain fewer kilojoules than foods sweetened with glucose.

    _____

    _____ [2 marks]

Respiration and energy

1 Respiration is a metabolic reaction that takes place inside all living cells. Respiration produces carbon dioxide as a waste product.

a Write the **word** equation for aerobic respiration.

_____ [2 marks]

b Explain why living cells need to respire.

_____ [1 mark]

c Name the organ through which carbon dioxide is lost from the body.

_____ [1 mark]

d Sperm cells contain a large number of mitochondria. With reference to respiration, explain why this is so.

_____

_____ [2 marks]

Removing waste: lungs

2 The diagram shows the contents of the human thorax.

a Use label lines and letters to label these two structures:

A the diaphragm

B a part of the lungs where gas exchange takes place   [2 marks]

b Describe how carbon dioxide from respiring cells is transported to the lungs.

_____

[2 marks]

3 a The table shows the approximate composition of inhaled air and exhaled air. Complete the table by writing in values for the percentage of oxygen in exhaled air and the percentage of nitrogen in inhaled air.

[2 marks]

Name of gas	In inhaled air (approximate %)	In exhaled air (approximate %)
Carbon dioxide	0.04	4
Oxygen	21	
Nitrogen		79
Water vapour	Variable	High

b Explain why there is less oxygen in exhaled air than in inhaled air.

_____

_____ [2 marks]

c The diagram shows one of the alveoli in the lungs, and a blood capillary.
Using what you know about diffusion, and the information on the diagram, explain why there is still quite a lot of oxygen in the air that we breathe out.

_____

_____

[2 marks]

inhaled air   exhaled air

thin walls of alveolus

thin layer of water lining the airspace

carbon dioxide made by respiring cells

O_2 O_2
O_2 O_2

O_2

diffusion gradient for oxygen

CO_2
CO_2
CO_2

diffusion gradient for carbon dioxide

blood capillary

Removing waste: liver and kidneys

1 Complete the sentences by choosing the **correct** words from the box.

> amino   bile   fatty   kidneys   liver   lungs   urine

Urea is a waste product that is made in the _____ from excess _____

acids. The urea dissolves in the blood and is carried to the _____ , where it is excreted

from the body dissolved in a liquid called _____ .   [4 marks]

2 The kidneys filter the blood and allow some of the substances to pass out of the body in urine. The table shows the concentrations of six substances in the blood and in urine.

Substance	Found in blood plasma (%)	Found in urine (%)
Water	92	95
Amino acids	0.05	0
Proteins	8	0
Glucose	0.1	0
Salt	0.37	0.6
Urea	0.03	2

a Explain the difference between the concentration of urea in blood plasma and in urine.

_____

                                                                                [1 mark]

b Suggest why the kidneys do not allow glucose to pass out of the body in urine.

_____

                                                                                [2 marks]

c Salt is made up of sodium ions and chloride ions. The kidneys are able to adjust the quantity of these ions that are lost in urine. Suggest how this can help to keep the ion content of the body constant.

_____

_____

                                                                                [2 marks]

Homeostasis

3 a One of the functions of the kidneys is to help to control the water content of the blood. Using what you know about osmosis, suggest how having too much water in the fluids that surround the cells might harm the cells.

_____

_____

                                                                                [2 marks]

b When there is too much water in the body, the kidneys excrete large amounts of dilute urine. When there is not enough, they excrete small amounts of concentrated urine.
State **two** ways in which the body loses water, other than in urine.

1 _____

2 _____   [2 marks]

c Explain why a person's kidneys produce less urine on hot days than on cold days.

_____

_____

                                                                                [2 marks]

Keeping warm, staying cool

1 The body's thermoregulatory centre constantly monitors blood temperature.

a Describe **one** way in which the body can automatically cool itself down.

_____

[1 mark]

b Suggest why cells cannot work properly if the core body temperature gets too high.

_____

[2 marks]

Treating diabetes

2 This graph shows a person's blood glucose concentration before and after they were given an injection of insulin.

a Using information in the graph, state the person's normal blood glucose concentration.

_____

[1 mark]

b Describe what happens to the blood glucose concentration after the insulin injection is given.

_____

[2 marks]

c Andrew has diabetes. His body does not produce insulin.
Andrew injects himself with insulin each day. What else can he do to control the concentration of glucose in his blood?

_____

[1 mark]

3 In 1922, F.G. Banting and C.H. Best published a research paper in which they described experiments that they had done on dogs that had had their pancreases removed. They were testing their hypothesis that the pancreas produced something that reduced blood glucose concentration. This is a summary of one of their experiments.

Time	Treatment	Blood glucose concentration	Other results
10.30 am	None	0.35%	
11.00 am	Injection of glucose given	0.40%	Glucose was excreted in the urine for the next four hours
3.00 pm	Injection of pancreas extract given	Rapidly fell to 0.09%	

a Why did Banting and Best use a dog without a pancreas in this experiment?

_____

[1 mark]

b What made the dog's blood glucose concentration rise to 0.40% at 11.00 am?

_____

[2 marks]

c Explain why glucose was excreted in the dog's urine between 11.00 am and 3.00 pm.

_____

[1 mark]

d Did Banting's and Best's results support their hypothesis? Explain your answer.

_____

_____

[2 marks]

Cell division – mitosis

1 A human body cell contains 46 chromosomes.

   a How many **pairs** of chromosomes are there in a human cell?

   _____ [1 mark]

   b How many copies of each gene does a body cell contain?

   _____ [1 mark]

   c When a cell divides, each new cell must receive a full set of chromosomes.
      What must happen to the chromosomes before the cell divides?

   _____ [1 mark]

2 This cell contains one pair of chromosomes.

   a The cell divides by mitosis. Complete the diagrams to show the chromosomes in the two
      daughter (new) cells produced after the cell has completed its division.

                                                  [2 marks]

   b Choose the best word to complete this sentence.

      The new cells that are produced by mitosis are_____ identical to the parent cell.

                                                  [1 mark]

Gametes and fertilisation

3 Sperm cells and egg cells are gametes.

   a Explain why gametes must have only 23 chromosomes each.

   _____

   _____ [2 marks]

   b Explain the meaning of the term **allele**.

   _____ [1 mark]

   c Explain how alleles can cause variation in the characteristics of offspring that
      are produced by sexual reproduction.

   _____

   _____

   _____ [3 marks]

Stem cells

1 Match words, **A**, **B**, **C** and **D**, with the descriptions **1 – 4** in the table.

A bone marrow   **B** differentiation   **C** nerve cell   **D** stem cell

1	A cell that can divide to form other kinds of specialised cell.
2	The process in which a cell becomes specialised for a particular function.
3	A cell that has become specialised to transmit electrical impulses.
4	A part of the body where cells become specialised as red or white blood cells.

1	2	3	4

[4 marks]

2 Stem cells can be obtained from very early human embryos. They are able to divide to form every kind of cell in the body.

a Explain how embryo stem cells differ from the stem cells found in adult bone marrow.

_____

_____ [2 marks]

b Describe how the use of embryo stem cells might be able to treat conditions such as paralysis after the spinal cord has been damaged.

_____

_____ [2 marks]

Chromosomes, genes and DNA

3 This diagram shows a cell, and some of the contents of its nucleus.

A cell   cell membrane

nucleus

gene

chromosome

a Name the chemical that chromosomes are made of.

_____ [1 mark]

b Each chromosome contains many genes. They store information, in the form of a code, about how the cell should make proteins by linking smaller molecules together.

What are these smaller molecules?

_____ [1 mark]

c Name the part of the cell in which proteins are made.

_____ [1 mark]

d Outline how the genes determine the kind of proteins that the cell makes.

_____

_____ [2 marks]

Inheritance

G–E

1 Complete the sentences by choosing the **correct** words from the box.

| alleles | amino acid | bases | dominant | gene | recessive |

Some characteristics, such as flower colour In pea plants, are controlled by a single

_____. Genes often have different forms, called _____.

If an allele produces a characteristic only when a different allele is not present,

it is said to be a _____ allele. [3 marks]

D–C

2 In pea plants, height is controlled by a single gene. The gene has two alleles, **T** and **t**. A plant with genes **T** and **t** is tall.

a Which allele is dominant? Explain your answer.

_____

_____ [2 marks]

b Write down the alleles that a short plant has.

_____ [1 mark]

How sex is inherited?

G–E

3 a Which **two** chromosomes carry the genes that determine a person's sex?

_____ [1 mark]

b Explain how these chromosomes differ in a man and a woman.

_____

_____ [2 marks]

D–C

c Which determines a child's sex: its mother's egg, or its father's sperm? Explain your answer.

_____

_____ [2 marks]

4 This diagram shows the chromosomes in a person's cells.

a How many **sets** of chromosomes are there in this cell?

_____ [1 mark]

b Is the person male or female?

_____ [1 mark]

Inherited disorders

1 Huntington's disease is a disorder of the nervous system. It is an inherited disorder.

a Explain what is meant by an **inherited disorder**.

_____

_____ [2 marks]

b A person with only one allele for Huntington's disease will have the disease. State whether the allele for Huntington's disease is dominant or recessive.

_____ [1 mark]

2 Angela and Sam have two children. Their son has cystic fibrosis. Neither Angela nor Sam has cystic fibrosis, and their daughter does not have it either.

a Explain how Angela and Sam had a son with cystic fibrosis, even though they do not have it themselves.

_____

_____

_____

_____ [4 marks]

b When their daughter grows up, could she have a child with cystic fibrosis? Explain your answer.

_____

_____

_____ [3 marks]

DNA fingerprinting

3 A DNA fingerprint is an image made from a tiny part of a person's DNA.

a Explain why a DNA fingerprint can be used to identify an individual.

_____

_____ [2 marks]

b A mother wants to know which of two possible men is her child's father. DNA fingerprints are made of the mother, child and the two men. Their DNA fingerprints are shown in the diagram.

possible father A   mother   child   possible father B

i On the diagram, draw circles around **one** band that is shared by the mother and her child. [1 mark]

ii On the diagram, draw a square around **one** band that the child has, but that is not present in its mother's DNA. [1 mark]

iii Which man is the child's father? Explain how you worked it out.

_____

_____

_____

_____ [3 marks]

B2b revision checklist

I know:

what enzymes are and what their functions are

☐ enzymes are proteins that act as biological catalysts, speeding up chemical reactions

☐ each enzyme works at an optimum temperature and pH

☐ they are involved in respiration, photosynthesis, protein synthesis and digestion

☐ enzymes are used in washing powders and in industry

how our bodies keep internal conditions constant

☐ blood sugar levels are controlled by the pancreas, which makes insulin to bring down blood sugar levels

☐ sweating cools the body down and helps to maintain a steady body temperature

☐ waste products e.g. carbon dioxide and urea must be removed from the body

some human characteristics show a simple pattern of inheritance

☐ some inherited characteristics are controlled by a single gene

☐ different forms of a gene are called alleles; in homozygous individuals the alleles are the same, in heterozygous individuals they are different

☐ in mitosis each new cell has the same number of identical chromosomes as the original

☐ sex chromosomes determine the sex of the offspring (male XY, female XX)

☐ stem cells can specialise into many types of cells

Atomic structure 2

1 a The diagram shows the structure of an atom.
Complete the labels on the diagram by choosing words from the box.

| electron | nucleus | neutron |
| proton | ion | |

[4 marks]

the centre of the atom is called the ....

b The table shows some information about the particles in an atom.

Particle	Charge	Mass
A	-1	Very small
B	+1	1
C	0	1

Name particles **A**, **B** and **C** by choosing words from the box.

| electron | nucleus | neutron | proton | ion |

A _____

B _____

C _____ [3 marks]

Electronic structure

2 The table shows some information about how electrons are arranged in a sodium atom.

a Complete the table to show the same information for magenesium and fluorine.

Periodic table element	²³₁₁ **Na** sodium	²⁴₁₂ **Mg** magnesium	¹⁹₉ **F** fluorine
Number of electrons	11	12	
Electron arrangement	Na	Mg	F
Notation	2, 8, 1		

[5 marks]

b Sodium loses an electron to form a stable ion. What is the electronic structure of a sodium ion?
Draw a ring round the **correct** answer.

| 2, 8 | 2, 8, 8 | 2, 8, 2 | 1, 8, 1 | [1 marks] |

Mass number and isotopes

1 Fluorine and chlorine are in Group 7 of the periodic table. The diagram shows the atomic structures of fluorine and chlorine.

Use ideas about electrons to explain why fluorine and chlorine are both in the same group.

_____

[1 mark]

Ionic bonding

2 Lithium is in Group 1.

a Lithium has one electron in its outer shell. Name **one** other atom that has one electron in its outer shell. Draw a ring around the **correct** answer.

      sodium   **magnesium**   **fluorine**   [1 mark]

b Complete the sentences by crossing out the **incorrect** word in each box.

Lithium is a
| metal |
| non-metal | .

A lithium ion has a
| negative |
| positive | charge.

When a lithium atom forms a lithium ion it
| loses |
| gains | an electron.   [2 marks]

3 The diagram shows what happens when a calcium atom reacts with a chlorine atom.

a What is the symbol for the calcium ion that forms? Draw a ring around the **correct** answer.

    Ca   **Ca⁺** Ca^{2+} **Ca⁻** Ca^{2-} [1 mark]

b What is the formula for the calcium chloride that forms?

_____

[1 mark]

Ionic compounds

1 The table shows some information about some substances.

Substance	Appearance	Melting point	Does it conduct electricity when solid?	Does it conduct electricity when molten?
A	White solid	Very high	No	Yes
B	White solid	Melts at 90 °C	No	No
C	Shiny grey solid	Very high	Yes	No

a Which of the substances, A, B or C is most likely to be a **metal**? Give reasons for your answer.

Substance _____

Reasons _____ [3 marks]

b Which of the substances, A, B or C is most likely to be an ionic compound? Give reasons for your answer.

Substance _____

Reasons _____ [3 marks]

c Substance B has the weakest forces between its molecules. How does the information in the table show that this is true?

_____ [1 mark]

Covalent bonding

2 The diagram shows the covalent bonds in a molecule of water. Only the outer shell of oxygen is shown.

a Draw a ring around **one** covalent bond in the molecule. [1 mark]

b How many electrons does an oxygen atom have in its outer shell **before** it forms bonds?

_____ [1 mark]

3 Look at the diagrams to show the bonding in chlorine and oxygen. Only the outer shell electrons are shown.

chlorine   oxygen

a The bond between the chlorine atoms is a **single** bond. What type of bond joins the oxygen atoms together?

_____ [1 mark]

b Explain how you can tell.

_____ [1 mark]

c Explain why the oxygen atoms are not joined together by a single bond.

_____

_____ [2 marks]

Simple molecules

1 The diagram shows some simple molecules.

Complete the sentences by crossing out the **incorrect** word in each box.

Simple molecules contain | a few / many | atoms.

The bonds between the atoms are | ionic / covalent |.

These bonds are very | strong / weak |.

Simple covalent compounds are often | gases / solids | at room temperature. [3 marks]

Giant covalent structures

2 The diagram shows the structure of diamond. Diamond is a very big molecule known as a macromolecule. What type of structure does diamond have? Draw a ring around the **two correct** words.

diamond

giant　**simple**　**covalent**　**ionic**　**metallic**

[2 marks]

3 Diamond and graphite are both macromolecules that contain carbon atoms. The diagram shows the structure of graphite.

bonds between atoms in layers are strong

bonds between the layers are weak

graphite

Compare the diagram of graphite to the diagram of diamond in question **2**. Explain why graphite is soft and slippery but diamond is not.

_____

_____

_____ [3 marks]

Metals

1 Read the article about mercury.

> **MERCURY**
> Mercury used to be called 'quicksilver' because it is a very shiny, silver metal. It has such a low melting point that it is a liquid at room temperature. Liquid mercury forms small drops that are very slippery and so flow quickly over surfaces. Drops of mercury sink if they drop into water, because it is very dense.

a Use information in the article to help you to answer these questions.

   i Give reasons why mercury is an unusual metal.

_____

_____ [2 marks]

   ii Give **one** property of mercury that is typical of all metals.

_____

_____ [2 marks]

G–E

---

2 The diagram shows the structure of a metal.

a Label the diagram by choosing words from the box.

anion	negative ion	proton	positive ion	electron

[2 marks]

b Use the diagram to help you to explain what happens when a metal conducts electricity.

_____ [1 mark]

c The layers of metal atoms can slip over each other. What property of metals does this cause?

  A metals are very strong

  B metals expand when they are heated

  C metals can be bent and shaped

  D metals conduct heat [1 mark]

D–C

Alkali metals

3 Sodium reacts with chlorine to make sodium chloride. This is the equation for the reaction:

      sodium + chlorine → sodium chloride

a Name the substance in the equation that is an **alkali metal**.

_____ [1 mark]

Name the substance in the equation that is an **ionic compound**.

_____ [1 mark]

b The formula of a sodium ion is Na^+. What happens to a sodium atom when it forms a sodium ion?

_____ [1 mark]

G–E

Halogens

1 The table shows some information about the halogens. Complete the table by filling in the missing information.

Name of halogen	Formula	State at room temperature	Colour
Fluorine		Gas	Colourless
Chlorine	Cl_2	Gas	
Bromine	Br_2		Dark red
Iodine	I_2	Solid	Black

[3 marks]

2 The diagram shows the electron arrangement in a fluorine atom.

fluorine atom   **fluoride ion**

a Complete the second diagram to show the electron arrangement in a fluoride ion. [1 mark]

b What is the charge on a fluoride ion?

_____ [1 mark]

c Which of the following elements will react with fluorine gas to produce fluoride ions? Draw a ring around the **correct** answer.

chlorine   **oxygen**   **sodium**   **helium**   [1 mark]

Nanoparticles

3 A new type of fabric contains nanoparticles. It is much harder to wear than normal fabrics and does not get dirty as easily. A manufacturer uses the fabric to make a new waterproof coat. The coat is more expensive than coats made using normal fabric.

a Complete the sentences about nanoparticles by crossing out the **incorrect** word in each box.

Nanoparticles contain | hundreds / millions | of atoms, which means they are very | big / tiny | .

They have a very high | surface area / volume | for their size. [2 marks]

b Explain the **advantages** to the customer of paying more for a coat made with the new fabric.

_____

_____ [2 marks]

c Companies spent millions of pounds on the research that was needed to develop nanoparticles. Explain how those companies benefit now that nanoparticles have been developed.

_____

_____

_____ [3 marks]

Smart materials

1 Smart materials have a wide range of uses. **A**, **B**, **C** and **D**, are different types of smart materials.

A shape changing alloy   **B** photochromic polymer
C thermochromic material   **D** electroluminescent material

What type of smart material is used for each of the following uses?

Spectacle frame made from smart material type _____

Dials in a car light up when a current passes
through them. They contain smart material type _____

The plastic visor goes dark when it is very sunny.
It is made from smart material type _____

The pictures on this mug change when it is filled with a hot drink.
It is coated with smart material type

_____

[4 marks]

Compounds

2 The diagrams show the formulae and
arrangement of molecules in oxygen
and methane gas.

oxygen gas   methane gas

a **i** How many atoms are in each
molecule of oxygen? _____

[1 mark]

ii How can you tell that oxygen is an **element**?

_____

[1 mark]

b **i** How many atoms are in each molecule of methane?

_____

[1 mark]

ii Give the names of the **two** types of atom in **methane**.

_____ and _____

[2 marks]

3 The table shows some information about hydrogen, oxygen and water.

Name	Hydrogen	Oxygen	Water
Formula	H_2	O_2	H_2O
Properties	Very flammable gas	Very reactive gas	Unreactive liquid

a Explain why hydrogen is an **element** but water is a **compound**.

_____

[2 marks]

b A mixture of hydrogen and oxygen react together explosively when they are lit.
The product is water. Explain the differences between a **mixture** and a
compound using this information.

_____

_____

_____

[3 marks]

Percentage composition

G–E

1 Oxygen has a relative atomic mass of 16. The table shows some information about two different molecules that contain oxygen. Complete the table by filling in the missing relative formula masses.

	Oxygen molecule	Ozone molecule
Formula	O_2	O_3
Relative formula mass M_r		

[2 marks]

D–C

2 Harvey works out the percentage of oxygen in two compounds. Here are his results.

> Percentage of oxygen in compound A = 57%
> Percentage of oxygen in compound B = 72%

a What is the formula for compound A and compound B? Choose from this box.

O_2	CO	CO_2	O_3	CH_4

Compound A _____   Compound B _____   [2 marks]

b Explain your reasoning.

_____

[1 mark]

Moles

G–E

3 Sulfur dioxide is a pollutant gas that comes from power stations. It is made when sulfur impurities in fuel burn. Some people are protesting against the amount of sulfur dioxide coming from a power station.

> Sulfur in fuel produces twice the amount of sulfur dioxide when it burns.

a Laura does some calculations to check if this protest sign is true. Complete her calculations by filling in the missing number.

> Atomic mass of sulfur, S = 32
> Mass of one mole of sulfur = 32g
>
> Atomic mass of oxygen, O = 16
> Relative molecular mass of sulfur dioxide SO_2 = _____
>
> Mass of one mole of sulfur dioxide = _____

[2 marks]

b The protest sign says that sulfur produces twice as much sulfur dioxide when it burns. Use your calculation to explain whether or not you think this is true.

_____

_____

[2 marks]

D–C

c Laura calculates the percentage mass of sulfur in sulfur dioxide. Complete her calculation.

> Atomic mass of sulfur, S = 32
>
> Atomic mass of oxygen, O = 16
>
> Percentage mass of sulfur in SO_2 =
>
> _____ %

[2 marks]

Yield of product

1 Rose makes magnesium oxide by heating up some magnesium in air.

  a Rose calculates the maximum amount of magnesium oxide she can make from the magnesium she starts with.
What is the name for this calculated amount? Draw a ring around the **correct** answer.

           percentage yield   **actual yield**   **theoretical yield**   [1 mark]

  b This is how Rose carries out her experiment.

Rose is disappointed when she weighs her product. The amount of magnesium oxide she has made at the end of the experiment is much less than expected.

Rose heats some magnesium   She weighs it on a balance

Give **two** reasons why this might have happened.

1 _____

2 _____   [2 marks]

2 Harry works in a factory that makes a chemical compound that is added to painkiller tablets.

  a Which of the following statements is **not** true for a process with a high yield?

    A The process has a high percentage yield.

    B The actual yield is as high as possible.

    C The actual yield is higher than the theoretical yield.

    D There is a high atom economy in the process.   [1 mark]

> Our method for making this compound has a very high yield. This is important because it makes the process more profitable for the company.

  b Give **two** reasons why having a high yield makes the process more profitable.

_____

_____   [2 marks]

Reversible reactions

3 Hydrated cobalt chloride is a different colour to anhydrous cobalt chloride.

hydrated cobalt chloride ⇌ anhydrous cobalt chloride

pink   blue

  a What does the symbol ⇌ mean?

_____   [1 mark]

  b Ethan carries out an experiment. The flow chart shows what he does. Fill in the boxes to show the colours of the cobalt chloride at each step.

	Start		**Step 1**		**Step 2**
hydrated cobalt chloride →	colour _____	dry in an oven →	colour _____	add drops of water →	colour _____

                                                                            [2 marks]

Equilibrium 1

1 Hydrogen and iodine react together in a reversible reaction to make hydrogen iodide.

  a Complete the equation for this reaction. Use the sign for a reversible reaction in your equation.

    hydrogen +   [2 marks]

  b Explain why reversible reactions do not give 100% yield.

    _____

    _____ [2 marks]

2 Ethanol is a solvent used to make perfume. One way of making ethanol is by reacting ethene gas with water vapour in a reactor.

    ethene + water vapour ⇌ ethanol

> **Conditions**
> **High temperature:** 300 °C
> **High pressure:** 70 atm
> **Catalyst:** phosphoric acid

  a Which of the reaction conditions will change the rate of reaction?

    A all of them

    B only temperature

    C temperature and pressure

    D only the catalyst   [1 mark]

  b Which of the reaction's conditions will **not** affect the yield?

    _____ [1 mark]

  c When the gases come out of the reactor, the ethanol is separated out. The rest of the gases are recycled back into the reactor again. Explain why this is necessary.

    _____ [1 mark]

Haber process

3 The Haber process is used to make ammonia.

  a Complete the equation for the Haber process by filling in the empty box.

☐	+ hydrogen	⇌	ammonia
N_2	+ $3H_2$	⇌	$2NH_3$

                                           [1 mark]

  b What is the main raw material that is used to make hydrogen for the Haber process? Draw a ring around the **correct** answer.

    coal   **limestone**   **methane**   **air**   [1 mark]

  c The process uses a catalyst and is carried out at high pressure.
    i Give **one** reason why a catalyst is used.

    _____ [1 mark]

    ii Give **two** reasons why a high pressure is used.

    _____ [2 marks]

C2a revision checklist

I know:

how sub-atomic particles help us to understand the structure of substances

☐ atoms consist of sub-atomic particles called protons, neutrons and electrons

☐ an element's mass number is the number of protons plus the number of neutrons in an atom

☐ an element's atomic number is the number of protons in an atom

☐ the arrangement of electrons can be used to explain what happens when elements react and how atoms join together to form different types of substances

how structures influence the properties and uses of substances

☐ ionic bonding is the attraction between oppositely charged ions

☐ ionic compounds are giant lattice structures with high melting points that conduct electricity when molten or dissolved

☐ non-metal atoms can share pairs of electrons to form covalent bonds

☐ giant covalent structures are macromolecules that are hard, have high melting points but do not conduct electricity

☐ nanoparticles are very small structures with special properties because of their unique atom arrangement

how much can we make and how much we need to use

☐ the relative masses of atoms can be used to calculate how much to react and how much we can produce, because no atoms are gained or lost in chemical reactions

☐ the percentage of an element in a compound can be calculated from the relative masses of the element in the formula and the relative formula mass of the compound

☐ high atom economy (atom utilisation) is important for sustainable development and economic reasons

☐ in some chemical reactions, the products of the reaction can react to produce the original reactants; they are called reversible reactions

Rates of reactions

1 Salma adds a few drops of pH Indicator to some dilute
hydrochloric acid. She then adds a piece of magnesium ribbon.
She waits for a few minutes until the reaction has finished.

Look at Salma's results.

hydrochloric acid +
pH indicator

magnesium

at the start   at the end

Time taken for reaction to finish	2 minutes
Observations	1 _____
	2 _____
	3 _____

a Complete the table to show **three** observations that Salma sees during the reaction. [2 marks]

b i How can you tell that one of the products of the reaction is a gas?

_____ [1 mark]

ii How can you tell that the acid is used up in the reaction?

_____ [1 mark]

c Salma does the experiment again. This time she heats the acid to a higher
temperature before she starts. Give **two** differences that Salma will see when
she adds the magnesium to hotter acid.

_____

_____ [2 marks]

Following the rate of reaction

2 Callum is investigating the rate of reaction of marble chips with acid. He knows that the reaction
produces carbon dioxide gas. He set up this experiment to find out the rate of reaction.

gas syringe

acid
marble chips

a What **two** measurements will Callum need to make to follow the rate of reaction?

1 _____

2 _____ [2 marks]

b How will Callum be able to tell when the reaction has stopped?

_____ [1 mark]

c What **other** change could Callum measure to follow the rate of reaction?
Draw a ring around the **correct** answer.

colour   **mass**   **volume of acid**   **size of chips** [1 mark]

Collision theory

1 The diagram in box A shows acid particles reacting with a large lump of zinc.

acid
particle

zinc

Box A

zinc

Box B

a Draw a similar diagram in Box B to show a **more concentrated** acid reacting with the zinc.

[1 mark]

b Give **two other** ways that the rate of reaction between zinc and acid can be made faster.

1 _____

2 _____ [2 marks]

c Complete the sentences by crossing out the **incorrect** word in each box.

For a reaction to happen, acid particles must | collide / move | with zinc.

The reaction will only occur if the particles have enough | volume / energy | . [1 mark]

Heating things up

2 Carbon dioxide is made when copper carbonate reacts with hydrochloric acid. David investigates the reaction. The diagram shows how he set up his experiment.

cotton wool bung

conical flask

hydrochloric acid and lumps of copper carbonate

151.95g

a What would you expect to happen to the mass of the flask during the experiment?
Explain your reasoning.

_____

_____

_____ [3 marks]

b David carries out his experiment at different temperatures. Use ideas about collisions and energy to explain why the reaction is faster at a higher temperature.

_____

_____

_____ [3 marks]

Grind it up, speed it up

1 Isabel carried out some experiments. She reacted zinc with acid in a flask. She used different conditions. The table shows the conditions she used.

Experiment	A	B	C	D
Size of zinc lumps	Large lumps	Small lumps	Large lumps	Small lumps
Concentration of acid	Low	High	High	Low
Temperature	20 °C	50 °C	50 °C	20 °C

a Which experiment, **A**, **B**, **C** or **D**, has the **fastest** rate of reaction?

_____ [1 mark]

b Which experiment, **A**, **B**, **C** or **D**, has the **slowest** rate of reaction?

_____ [1 mark]

c What reaction condition changes when small lumps are used instead of large lumps? Draw a ring round the **correct** answer.

concentration   surface area   pressure   catalyst [1 mark]

d During the experiment, hydrogen gas is made. What will happen to the mass of the flask during the experiment?

_____

_____ [2 marks]

Concentrate now

2 Leo does an experiment (experiment 1). He reacts some acid with some marble chips. He measures the change in mass as the reaction happens. He stops measuring when no more carbon dioxide is given off. Leo repeats the experiment. He uses the same amount of acid and marble chips, but this time he adds some water to the flask (experiment 2).

at the start   at the end

a What has been used up when the reaction stops?

_____ [2 marks]

b Give **two** ways that Leo can tell when the reaction has finished.

1 _____

2 _____ [2 marks]

c The reactions for both experiments were fast at the beginning and then slowed down. Explain why.

_____

_____

_____ [3 marks]

Catalysts

1 Hydrogen peroxide is used to bleach hair. It slowly decomposes to form water and oxygen gas.

  a Write a word equation for this reaction.

  _____ [1 mark]

  b Sophie does some experiments using hydrogen peroxide. The diagram shows how she set up her first experiment and her result.

Result		
Time taken to produce 20 cm³ of gas		
hours	minutes	seconds
45	30	45

  gas syringe

  hydrogen peroxide

  00.0 cm³

  Give **one** reason that it is easier to use a datalogger in this experiment rather than to keep taking readings.

  _____ [1 mark]

  c Sophie repeated the experiment. This time she added some manganese(IV) oxide powder to the hydrogen peroxide.
  i How would the results for this experiment be different from the first one?

  _____ [1 mark]

  ii Explain why the results would be different.

  _____ [1 mark]

2 The Haber process is used to make ammonia for making fertilisers. Iron is used as catalyst in the reaction.

  a The iron catalyst is packed into the reaction tower. It is only replaced about every 15 years.
  i Why does the iron catalyst have such a long lifetime?

  _____ [1 mark]

  ii Many metal catalysts are very expensive. Give **two** reasons why they make industrial processes cheaper in the long run.

  1 _____

  2 _____ [2 marks]

Energy changes

3 a The table shows some information about temperature and energy changes during three reactions. Complete the table by filling in the boxes.

Reaction	Temperature change	Exothermic or endothermic?
Dissolving ammonium nitrate in water	Decreases	
Adding zinc powder to copper		Exothermic
Adding magnesium ribbon to an acid	Increases	

[2 marks]

  b Which of these reactions is likely to be exothermic? Draw a ring around **two correct** answers.

  oxidation   **thermal decomposition**   **neutralisation**   **evaporation**   [2 marks]

Equilibrium 2

1 Joseph does an experiment on ice and water. He keeps a mixture of ice and water at exactly 0 °C.

    a Joseph knows that this is a reversible change. Draw the **correct** symbol in the box to show a reversible change.

        ice [　　　　] water

    b What would happen to the mixture if the temperature rose above 0 °C?

    _____

          — thermometer
          — ice
          — water

[2 marks]

Industrial processes

2 The flow chart shows how ammonia is made in the Haber process.

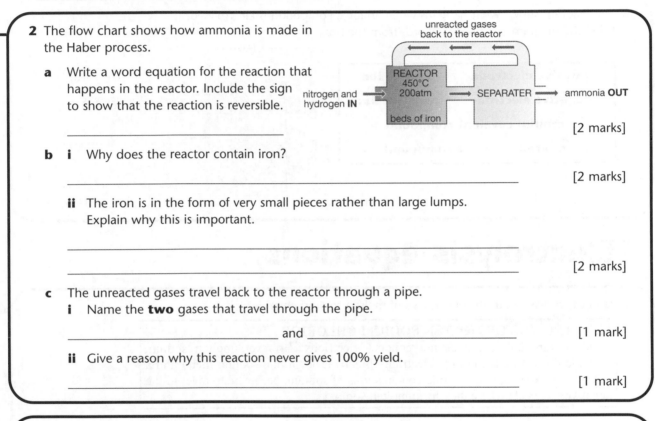

    a Write a word equation for the reaction that happens in the reactor. Include the sign to show that the reaction is reversible.

    _____

[2 marks]

    b **i** Why does the reactor contain iron?

        _____

[2 marks]

        ii The iron is in the form of very small pieces rather than large lumps. Explain why this is important.

        _____

        _____

[2 marks]

    c The unreacted gases travel back to the reactor through a pipe.
        i Name the **two** gases that travel through the pipe.

        _____ and _____

[1 mark]

        ii Give a reason why this reaction never gives 100% yield.

        _____

[1 mark]

3 The table shows some data about the yield of ammonia from the Haber process under different conditions.

Temperature (°C)	Pressure (atm)	Yield
400	200	40%
400	100	25%
500	100	15%

    a Predict the yield of ammonia at 400 °C and 50 atm pressure. Draw a ring around the **correct** answer.

    70%   **50%**   **30%**   **10%**   [1 mark]

    b Complete the sentences about the Haber process by choosing the **correct** words from the box.

    | **increases**   **decreases**   **slower**   **faster** |

    When the temperature is higher, the yield _____ .

    When the pressure is higher, the yield _____ .

    At low temperatures the rate of reaction is _____ .

    At higher pressures the rate of reaction is _____ .   [3 marks]

Free ions

1 The diagram shows the ions in a solid.

a Give the names of the two ions in this solid

_____ ion and _____ ion   [2 marks]

b Which of the formulae show this compound in its solid form?
Draw a ring around the **correct** answer.

LiCl(aq)   **LiCl(g)**   **LiCl(l)**   **LiCl(s)**   [1 mark]

---

2 The diagram shows what happens when an ionic compound is electrolysed.
Label the diagram by choosing words from the box.

positive electrode	positive ion
negative electrode	negative ion
molten covalent compound	
molten ionic compound	

....

....

....

[3 marks]

Electrolysis equations

3 Read the article about the electrolysis of molten sodium chloride.

ELECTROLYIS OF MOLTEN SODIUM CHLORIDE
Solid sodium chloride does not conduct electricity. However, molten sodium chloride does conduct, but the electricity breaks it down into sodium metal and a gas. The molten sodium chloride contains ions of sodium, Na^+, which travel to the negative electrode where they form molten sodium.

a Why do sodium ions, Na^+ ions travel to the negative electrode?

_____

_____   [2 marks]

b What is the name of the gas that forms during the electrolysis?

_____   [1 mark]

c The sentences below are about what happens to sodium ions, Na^+, during the electrolyis of
sodium chloride. Complete the sentences by crossing out the **incorrect** answer in each box.

At the negative electrode, sodium ions | gain / lose | one electron.

The charge on the sodium is now | −1 / 0 | .

The sodium ion has been | oxidised / reduced | .   [2 marks]

Uses for electrolysis

1 Three products form when sodium chloride solution is electrolysed.

a Complete the table to show where each product forms by choosing words from the box.

hydrogen	chlorine	sodium hydroxide

At the positive electrode	At the negative electrode	Left behind in solution

[2 marks]

b Why does sodium metal **not** form when sodium chloride solution is electrolysed?

A only gases can be made by electrolysis   **B** sodium is too reactive

C there are no sodium ions in the solution   **D** the ions cannot move in the solution

[1 mark]

2 Copper is purified by electrolysis.

a Complete the sentences to explain what happens during electrolysis by choosing the **correct** words from the box.

copper	carbon	positive	negative	gaining	losing

When copper is purified, both electrodes are made from _____ .

A block of impure copper is used as the _____ electrode. Copper ions

from the solution form pure copper by _____ two electrons.   [3 marks]

b The solution used during the electrolysis is usually copper sulfate solution. The copper sulfate solution is blue because it contains Cu^{2+} ions. Explain why the blue colour stays the same during the electrolysis.

_____

_____

[2 marks]

Acids and metals

3 The table shows the reactivity of some metals.

magnesium	more reactive
zinc	
lead	
(hydrogen)	
copper	
silver	less reactive

a Name **two** metals that will react with an acid to make hydrogen gas.

_____ and _____   [1 mark]

b Where would gold appear in the list? Explain your reasoning.

_____

_____

[2 marks]

c Holly reacts some iron with an acid. One of the products is iron chloride.
Which acid did she use? Draw a ring around the **correct** answer.

sulfuric acid   **nitric acid**   **carbonic acid**   **hydrochloric acid**   [1 mark]

Making salts from bases

1 Ella makes some zinc sulfate by reacting together an acid and a base.

a What is the name of the acid Ella needs to use?

_____ [1 mark]

b What **base** could Ella use to make zinc sulfate? Draw a ring around the **correct** answer.

zinc chloride   **zinc oxide**   **zinc sulfate**   **zinc nitrate**   [1 mark]

c Ella adds the base to the acid until no more base reacts. She wants to remove the left over solid base from the solution she has made. How can she do this?

_____ [1 mark]

2 Karim makes a salt by reacting copper oxide with hydrochloric acid.

a Complete the equation for the reaction by filling in the boxes.

copper oxide   +   hydrochloric acid   →   [   ]   +   [   ]

CuO + $2HCl$ → $CuCl_2$ +   sulfur dioxide   [2 marks]

b Karim keeps adding copper oxide until no more dissolves. Explain why he does this.

_____

_____ [2 marks]

Acids and alkalis

3 a Max makes sodium chloride. He puts some sodium hydroxide solution (an alkali) in a flask and adds hydrochloric acid from a burette.

He adds litmus solution to the sodium hydroxide to act as an indicator.

He adds hydrochloric acid to the sodium hydroxide until he reaches the end point when all the sodium hydroxide has been neutralised. The table shows the litmus at different pHs.

pH below 7	pH above 7
red	blue

burette

2 mol/dm³ hydrochloric acid

sodium hydroxide solution and litmus indicator

a What colour change will the litmus indictor show at the end point of the experiment?

From _____ to _____ [1 mark]

b Why could litmus indicator **not** be used to measure the exact pH of the sodium hydroxide?

_____ [1 mark]

c Max wants to make a pure solution of sodium chloride without an indicator.

   i Describe how he can do this.

_____

_____

_____ [3 marks]

   ii How can Max make sodium chloride crystals from his solution?

_____ [1 mark]

Neutralisation

1 The table shows the pH of some household substances.

Substance	pH
Sour milk	6
Fresh milk	8
Toothpaste	8.5
Lemon juice	3
Pure water	7
Oven cleaner	10

a Which of the substances is the **most** acidic?

_____ [1 mark]

b Which of the substances is the **weakest** alkali?

_____

_____ [2 marks]

c Tammy has some Universal Indicator at home. She wants to check if some milk is sour before she uses it on her breakfast cereal. Write three instructions to tell her how to carry out the test.

1_____

2_____

3_____ [3 marks]

Precipitation

2 Lead chromate is a bright yellow, solid salt that is used to make yellow paint for road markings. The diagram shows how lead chromate can be made.

a Use the substances named on the diagram to answer the following questions.

potassium chromate

lead nitrate

solid **lead chromate**

i Give the name of a **soluble** substance.

_____ [1 mark]

ii Give the name of a **precipitate**.

_____ [1 mark]

b When the reaction happens, lead chromate and another compound form. What is the name of the other compound? Draw a ring around the **correct** answer.

lead nitrate   **potassium chromate**   **potassium nitrate**   [1 mark]

3 Read the article about removing pollutants from water.

> **REMOVING POLLUTANTS FROM WATER**
> Most pollutants can be removed from water by precipitation reactions. Positively charged metal ions are usually removed by reacting them with sodium carbonate or sodium hydroxide. Negative ions containing non-metals can be removed by adding calcium compounds. Once the precipitates are formed, they are easy to remove from the water.

a Answer the questions by choosing words from the box.

lead nitrate	lead carbonate	calcium phosphate	phosphoric acid

i What compound is formed when lead ions are removed from water?

_____ [1 mark]

ii Which compound is formed when phosphate ions are removed from water?

_____ [1 mark]

b How can precipitates be 'easily removed' from water?

_____ [1 mark]

I know:

how we can control the rates of chemical reactions

☐ the rate of a chemical reaction can be found by measuring the amount of a reactant used or the amount of product formed over time

☐ chemical reactions can be speeded up by increasing the: temperature; concentration of a solution; pressure of a gas, surface area of a solid; and by using a catalyst

☐ particles must collide with sufficient energy in order to react; the minimum energy required is the activation energy

whether chemical reactions always release energy

☐ chemical reactions involve energy transfers

☐ exothermic reactions give OUT energy; endothermic reactions take IN energy

☐ if a reversible reaction is exothermic in one direction, it is endothermic in the opposite direction. The same amount of energy is transferred in each case

how can we use ions in solutions

☐ when molten or dissolved in water, ions in ionic compounds are free to move around

☐ passing an electric current through an ionic compound breaks it down into its elements: this is called electrolysis

☐ at the negative electrode, positively charged ions gain electrons (reduction) and at the positive electrode, negatively charged ions lose electrons (oxidation)

☐ electrolysis of sodium chloride solution makes hydrogen, chlorine and sodium hydroxide; copper can be purified by electrolysis

☐ metal oxides and hydroxides are bases and react with acids to form salts

☐ soluble salts can be made from acids and insoluble salts can be made by mixing solutions of ions

☐ in neutralisation reactions, hydrogen ions from acids react with hydroxide ions to produce water

See how it moves!

1 The distance-time graph shows the motion of a bus.

  a Describe the motion of the bus at

    i section A _____

    _____

    [2 marks]

    ii section B _____

    [1 mark]

  b In which section did the bus travel fastest? _____ [1 mark]

2 a A coach driver travels 90 km in 2 hours. Calculate his **average speed**.

_____

_____ km/h   [3 marks]

  b Explain why the coach does not always travel at this average speed.

_____

_____ [2 marks]

Speed isn't everything

3 a What is **velocity**?

_____

_____ [2 marks]

  b A runner runs once around a circular track. Describe how his velocity changes
  i as he starts to run.

_____ [1 mark]

  ii as he runs around the corner.

_____ [1 mark]

  iii as he reaches the end.

_____ [1 mark]

4 The diagram shows Simon firing an arrow.

  a What is the **speed** of the arrow just before it leaves the bow?

_____ m/s

[1 mark]

  b After 0.5 s, the arrow reaches its top speed of 150 m/s. Calculate the **acceleration** of the arrow.

_____

_____

_____ m/s^2 [3 marks]

acceleration

Velocity-time graphs

1 The velocity-time graph shows the motion of Alice on her bicycle.

a Describe Alice's motion at

i section A _____

[1 mark]

velocity

A   B
          C
              D

time

ii section B _____

[1 mark]

b In which section is Alice travelling slowest? _____ [1 mark]

c How could you tell if Alice stopped moving?

_____

[1 mark]

2 The table gives information about the speed of a runner.

a Plot the velocity-time graph for this runner.

[4 marks]

b Explain how you could use your graph

i to calculate the distance that the runner ran.

_____

[1 mark]

ii to calculate the acceleration of the runner at a certain time.

_____ [1 mark]

iii to calculate the speed of the runner at a certain time.

_____

[1 mark]

Time (s)	Speed (m/s)
0	0
1	1.5
2	3
3	4.5
4	4.5
5	4.5

Let's force it!

3 a The diagram shows the forces acting on a car moving forwards.

i What is the size of the **resultant force**?

_____

[1 mark]

backward force = 400N   forward force = 800N

ii Describe the **motion** of the car including its direction.

_____

_____

[2 marks]

b The driver reduces the force from the engine until is the same size as air resistance. Complete the sentence by choosing the **correct** word from the box.

balanced	unbalanced	constant	changing

i The forces are _____ .

[1 mark]

ii Describe the **motion** of the car now.

_____

_____

[2 marks]

Force and acceleration

G–E

1 a What can you say about the resultant force on a car if it is
  i accelerating?

_____

_____ [2 marks]

  ii travelling at a steady speed?

_____ [1 mark]

b Max is pushing a shopping trolley. He puts more shopping in.
  i What happens to the **force** needed to make the trolley move faster?

_____ [1 mark]

D–C

  ii The mass of the trolley is 20 kg. What force is needed to make it accelerate
  by 4 m/s^2? Include the unit in your answer.

_____

_____

_____ [3 marks]

G–E

2 Kamal is investigating the acceleration caused by different forces.   Here is a table of his results.

  a What equipment should he use to measure the force?

_____

_____ [1 mark]

Force	Acceleration
1 N	2 m/s^2
2 N	4 m/s^2
3 N	6 m/s^2
4 N	7 m/s^2
5 N	10 m/s^2

D–C

  b Explain whether his results proved that acceleration
  is proportional to force used.

_____

_____

_____ [2 marks]

  c Explain which one of his results should be repeated.

_____

_____ [2 marks]

Balanced forces

G–E

3 a What are **balanced forces**?

_____

_____ [2 marks]

b A van is trying to pull another car out of a ditch. They are not moving.
  i Are the forces balanced or unbalanced?

_____ [1 mark]

  ii What can you say about the size of the force from the van and the size of
  the force from the car?

_____ [1 mark]

c A tractor is used instead of the van. It pulls the car out. How does the force
from the tractor compare with the force from the car?

_____ [1 mark]

Terminal velocity

1 The cyclist is travelling at a steady speed. Friction is one force that slows him down.

a What is the name of another force that slows him down?

_____

[1 mark]

b Describe how the size of this force changes as he travels faster.

_____
[1 mark]

c What is **terminal velocity**?

_____
[1 mark]

d The cyclist starts to pedal harder and accelerates until he reaches a faster terminal velocity. Describe how the forces on him change as his speed changes.

_____

_____

_____

[3 marks]

2 Skydivers open parachutes to help them land safely.

a What **force** increases when the parachute is open?

_____
[1 mark]

b Explain why this helps the parachutist to slow down.

_____

_____
[2 marks]

Stop!

3 a Explain what the **stopping distance** of a car is.

_____
[1 mark]

b Stopping distance can be split into two parts. What are these called?

_____

_____
[2 marks]

c Nicole is taking her first driving lesson. Why will her thinking distance be longer than her driving instructor's?

_____
[1 mark]

d On her second lesson, Nicole was very tired. How will this affect her stopping distance?

_____

_____
[2 marks]

e Nicole's car has been fitted with new tyres. Explain how these will help to reduce her stopping distance.

_____

_____
[2 marks]

G–E

D–C

D–C

G–E

D–C

Moving through fluids

1 Charlotte is comparing the speed that a marble falls through different liquids.

a What is the name of the force that slows the marble down as it falls through the liquid?

_____

[1 mark]

b First she drops the marble in a cylinder of water. Explain why it reaches its terminal velocity.

_____

_____

[2 marks]

c Next, she drops the marble through thick honey in the same sized cylinder. Explain whether the terminal velocity will be larger or smaller.

_____

_____

[2 marks]

d She repeats the experiment using a piece of plasticine the same mass as the marble. The plasticine is **streamlined**. How does the plasticine's streamlined shape affect its terminal velocity?

_____

[1 mark]

Energy to move

2 a Complete the sentences.

A hairdryer is designed to transfer electrical energy as _____

energy to _____ energy. [2 marks]

b The kinetic energy in a motorcycle comes from _____ energy in the fuel. [1 mark]

c What is the main energy change in a light bulb?

_____ to _____ [2 marks]

3 James is racing his electric car.

a Where did the kinetic energy in the car come from?

_____

[1 mark]

b The batteries ran out and the car stopped. Explain where the energy from the batteries have gone.

_____

_____

[2 marks]

c Why did the batteries run out more quickly when he drove the car on a rough surface?

_____

_____

[2 marks]

Working hard

1 Muhammed lifted an apple. The work he did was 12 joules.

a How much **energy** was transferred to the apple?

_____ J   [1 mark]

b What force was he working against when he lifted the apple up?

_____   [1 mark]

c Muhammed dragged a box of apples 2 metres along the ground. He measured the force needed as 25 N. How much **work** did he do?

_____

_____

_____ J   [3 marks]

d When he stopped dragging the box, it did not move. Explain what happened to its kinetic energy.

_____

_____   [2 marks]

G–E

D–C

How much energy?

2 Ryan is playing rugby. He runs fast holding the ball, giving the ball kinetic energy.

a Explain what is meant by **kinetic energy**.

_____   [1 mark]

b What should Ryan do to reduce the amount of kinetic energy the ball has?

_____   [1 mark]

c Ryan kicks the ball 10 m in the air. Calculate the amount of **potential energy** it gains if its weight is 3 N.

_____

_____

_____ J   [3 marks]

G–E

D–C

3 Complete the spaces in the sentences.

Andrea has a large elastic harness attached to her when she does a bungee jump

from the top of a bridge. As she jumps, the _____ energy she

has on the bridge changes into _____ energy as she falls.

At the bottom of the jump, the rope is fully extended and it gains

_____ energy. Eventually she stops moving because all the

energy has spread as _____ energy to the surroundings.

[4 marks]

D–C

Momentum

1 a Complete the sentences by choosing the **correct** words from the box.

velocity	speed	mass	force

    i One example of a scalar quantity is _____ [1 mark]

    ii One example of a vector quantity is _____ . [1 mark]

b What is the **momentum** of a ball? Its mass is 0.5 kg and it travels at 8 m/s.

_____

_____

_____ kg m/s [3 marks]

2 The diagram shows two trolleys colliding.

Inelastic collision

before   after

a Is the collision elastic or inelastic?

_____ [1 mark]

b Calculate the total momentum before the collision.

_____

_____

_____ kg m/s [3 marks]

c Write down the total momentum after the collision.

_____ kg m/s [1 mark]

d Calculate the speed that the joined-up trolley moves off with.

_____

_____

_____ [3 marks]

Off with a bang!

3 Jake blows up a balloon and lets go of it.

a The balloon moves forwards. In what direction does the air inside move?

_____ [1 mark]

b Explain how momentum is conserved as it starts moving.

_____

_____ [2 marks]

c Use this idea to explain why a swimmer moves forward in the water.

_____

_____ [3 marks]

Keep it safe

1 a What is needed to make a moving object change direction or speed?

_____ [1 mark]

G–E

b Two vehicles are driving at the same speed. The van stops quickly to avoid hitting a child on the road and the car stops more slowly.

i Which vehicle feels the biggest braking force? _____ [1 mark]

ii Which vehicle feels the braking force for longest? _____ [1 mark]

c A car seat belt is slightly elastic. If the car is involved in a crash the belt stretches slightly and the passenger is less likely to be hurt.

D–C

i The elastic in the seat belt lets the momentum of the car change over a longer time. How does this help passengers avoid getting hurt?

_____

_____ [2 marks]

ii What is the momentum of a car that has stopped? _____ [1 mark]

2 Adam drops his mobile phone onto a concrete path and it breaks. Sarah drops her mobile phone onto a carpet and it does not break.

G–E

a What is the momentum of each mobile phone after it lands?

_____ [1 mark]

b Each mobile phone has the same change in momentum. Explain why Sarah's phone did not break.

D–C

_____

_____

_____ [3 marks]

Static electricity

3 Owen rubs a balloon on some cloth. The balloon becomes charged.

G–E

a What type of particles are negatively charged?

_____ [1 mark]

b Explain how rubbing the balloon makes it negatively charged.

_____

_____

_____ [3 marks]

c The negatively charged balloon is held near another negatively charged balloon. Describe what happens.

D–C

_____

_____ [2 marks]

Charge

1 a What is meant by **electrical current**?

_____

_____ [2 marks]

b Damien is rubbing a balloon with a piece of woollen cloth.
i Is the balloon a conductor or an insulator?

_____ [1 mark]

ii Use your answer to explain why a **static charge** builds up on the balloon.

_____

_____

_____ [3 marks]

2 a What can you say about the number of positive and negative charges on a **neutral object**?

_____ [1 mark]

b Write down **one** piece of equipment that can be used to detect an electric charge.

_____ [1 mark]

Van de Graaff generator

3 a Are the belt and rollers of a Van de Graaff generator made from conductors or from insulators?

_____ [1 mark]

b Explain how the belt and rollers create a charge.

_____

_____

_____ [2 marks]

c Explain why the equipment must stay on an insulated mat.

_____ [1 mark]

4 During a thunderstorm, clouds lose their electric charge as lightning. Many buildings have lightning conductors. Explain how a lightning conductor protects the building during a thunderstorm.

_____

_____

_____

_____

_____ [3 marks]

Sparks will fly!

1 a Do static charges build up on conductors or on insulators?

_____ [1 mark]

b How can you tell if a charged object has lost its charge suddenly?

_____ [1 mark]

c Why can it be dangerous to let static charges build up in an oil refinery?

_____

_____

_____ [3 marks]

G–E

2 A photocopier uses static electricity to transfer negatively charged toner to a sheet of paper. First, the toner passes over a positively charged drum.

a Why does the toner stick to the drum?

_____ [1 mark]

b Paper passes over the drum and the toner sticks to the paper. Should the paper have a positive or negative charge?

_____ [1 mark]

D–C

3 Monique's shoes are made from plastic which is an insulator. When she walks across a woollen carpet she becomes charged.

a Explain why a negative static charge builds up.

_____

_____

_____ [3 marks]

b Monique reaches for a door metal handle and a spark jumps from her hand to the handle.

i What is this process called? _____ [1 mark]

ii Explain why this process occurs.

_____

_____ [2 marks]

G–E

4 When paint is sprayed from a can it becomes electrically charged.

a Explain why the paint becomes positively charged.

_____

_____

[2 marks]

b The particles spread thinly and evenly on the car's surface. Explain why the particles spread out.

_____

_____ [2 marks]

D–C

P2a revision checklist

I know:

how we can describe the way things move

☐ distance-time graphs show how an object moves

☐ velocity-time graphs show how the velocity changes; the slope represents acceleration, and the area under the graph represents distance travelled

how we make things speed up or slow down

☐ to change the speed of an object, an unbalanced force must act on it

☐ forces can add up or cancel out to give a resultant force; when the resultant force is not zero, an object accelerates

resultant force (N) = mass (kg) × acceleration (m/s²): $F = ma$

☐ an object falling through a fluid accelerates until it reaches a terminal velocity, when the resultant force is zero

☐ the stopping distance of a car is the thinking distance plus the braking distance. This increases as the speed increases

what happens to the movement energy when things speed up or slow down

☐ when a force causes an object to move, energy is transferred and work is done

☐ every moving object has kinetic energy that can be transformed into other forms

☐ the kinetic energy of a body depends on its mass and its speed

what momentum is

☐ every moving object has momentum that depends on its mass and its velocity:

momentum (kg m/s) = mass (kg) × velocity (m/s)

☐ momentum has size and direction and is conserved in collisions and explosions

what static electricity is, how it can be used and the connection between static electricity and electric currents

☐ rubbing electrical insulators together builds up static electricity because electrons are transferred

☐ if an object gains electrons it has a negative charge; if it loses electrons it has a positive charge

☐ electrostatic charges can be used in photocopiers, smoke precipitators and paint sprayers

☐ when electrical charges move we get an electrical current

Circuit diagrams

1

a How could you change the resistance of component 1?

_____

[1 mark]

b A diode only allows current to flow in one direction. Which symbol shows a diode?

_____

[1 mark]

c What is the name of component 4?

_____

[1 mark]

2 The circuit shows the wiring of a circuit in a reading lamp.

a What is the name of component X?

_____

[1 mark]

b Explain how the **brightness** of the light can be increased using the slider on component X.

_____

_____

_____

[3 marks]

c A second bulb is added to the circuit in series with the first bulb. Write down **two disadvantages** of including the second bulb in series with the first.

_____

_____

[2 marks]

Resistance 1

3 a What is meant by **electrical current**?

_____

_____

[2 marks]

b How is the resistance in a conductor different from the resistance of an insulator?

_____

_____

[2 marks]

c Give **one** reason why metals are better conductors of electricity than plastics.

_____

_____

[1 mark]

Resistance 2

1 a Leo measures the resistance of several wires. How does the resistance change when

    i the length of the wire increases?

_____ [1 mark]

    ii the width of the wire increases?

_____ [1 mark]

b i The resistance of a wire depends on collisions inside the wire. Write down the **three** main points of the **collision theory** of resistance.

_____

_____

_____ [3 marks]

    ii Use this idea to explain why different materials have a different resistance.

_____ [1 mark]

Ohm's Law

2 The graph shows the results of an experiment to find out how current changes when the voltage changes in a wire.

    a What pattern does the graph show?

_____

_____

_____

_____ [2 marks]

    b Use the graph to write down the current when the voltage is 6 V.

_____ [1 mark]

    c Use your answer to part **b** to calculate the resistance of the wire at 6 V.

_____

_____

_____ [3 marks]

    d The length and thickness of the wire stay the same. Write down **two** other factors to keep constant for the resistance of the wire to stay the same.

_____

_____ [2 marks]

    e The graph shows the experiment repeated using a filament bulb. Write down **one** reason why this graph is a different shape from the graph in part **a**.

_____

_____

_____

_____

[2 marks]

More components

1 a How could you change the resistance of the thermistor?

_____

_____

thermistor

[1 mark]

b What happens inside the thermistor when it is heated up?

_____

_____

[2 marks]

c Write down **two** uses for a circuit containing a thermistor.

_____

_____

[2 marks]

d How does the resistance of the light dependent resistor change if a light shines onto it?

_____

_____

[1 mark]

e A circuit contains a light dependent resistor. How does the current in the circuit change as the light intensity gets lower?

_____

[1 mark]

f What type of material is a light dependent resistor made from?

_____

[1 mark]

Components in series

2 The circuit shows three bulbs wired in series.

a Fred measures the current in four places in the circuit. What can you say about the readings?

_____

[1 mark]

b Each cell provides 1.5 V. What is the size of the voltage supplied to the circuit?

_____

[1 mark]

c Write down the voltage across each of the three bulbs.

1_____ 2_____ 3_____

[1 mark]

d Fred turns one of the cells around. What is the size of the voltage supplied to the circuit?

_____

[1 mark]

Grades

G–E

D–C

Components in parallel

1 The circuit shows three bulbs wired in parallel. The circuit is switched on.

a Write down the voltage across each of the three bulbs.

_____ [1 mark]

b Each cell provides 1.5 V. What is the size of the voltage supplied to the circuit?

_____ [1 mark]

c One of the bulbs breaks. What happens to the other two bulbs?

_____ [1 mark]

d Write down **one** reason to wire the bulbs in parallel rather than in series.

_____ [1 mark]

The three-pin plug

2 The diagram shows a plug connecting the lead from an iron.

a Which colour wire is attached to the neutral terminal?

_____

_____ [1 mark]

b Why are the cable and casing made of plastic?

_____

_____ [1 mark]

c Explain why the fuse must be connected in series to the live wire.

_____

_____ [2 marks]

d Describe the structure of the cable that connects the plug to the equipment.

_____

_____

_____ [3 marks]

e Explain why the cable grip is an important part of the plug.

_____

_____ [2 marks]

Domestic electricity

1   X   Y

G–E

a   What type of current do you get from a **battery**?

_____   [1 mark]

b   Which diagram shows the trace you would get from a battery?

_____   [1 mark]

c   Write down **one advantage** of using alternating current for electricity supplied to homes instead of batteries.

_____   [1 mark]

D–C

d   What equipment is used to show the picture of changing voltage in a circuit?

_____   [1 mark]

Safety at home

2   Brad's radio stopped working until his dad changed the fuse for him.

G–E

a   How does the fuse protect the radio from damage?

_____

_____

_____   [2 marks]

b   Why is a different fuse needed for Brad's electric heater?

_____   [1 mark]

D–C

c   Explain why Brad's electric heater needs to be earthed but his radio does not.

_____   [1 mark]

d   Explain how earthing the heater protects the user.

_____

_____

_____   [3 marks]

e   Explain why a circuit breaker protects someone more than a fuse.

_____

_____   [2 marks]

Which fuse?

1 a Nathan has to put a fuse into the plug for the radio. He has a choice of fuses rated at 3 A, 5 A or 13 A. The current flowing through the radio is 0.4 A.

i Which fuse should he use?

_____ [1 mark]

ii One fuse has "13 A" printed on it. What does this tell you about the fuse?

_____

_____

_____ [3 marks]

iii The radio has a power rating of 200 W. How much energy does it use per second?

_____ J [1 mark]

b i Nathan also has a heater. Its power rating is 2000 W. Calculate the size of the current flowing through the heater, if mains electricity is supplied at 230 V.

_____

_____ A [3 marks]

ii What would happen if Nathan put a fuse in with too low a rating?

_____

_____ [2 marks]

Radioactivity

2 a In an atom

i which particle has no charge?

_____ [1 mark]

ii which particles are found in the nucleus?

_____ [2 marks]

b What is an ion?

_____ [1 mark]

c i Write down the mass number for beryllium.

_____ [1 mark]

ii How many protons has boron got in its nucleus?

_____ [1 mark]

iii How many neutrons has helium got in its nucleus?

_____ [1 mark]

Element	Atomic No.	Notation
Hydrogen	1	$^{1}_{1}H$
Helium	2	$^{4}_{2}He$
Lithium	3	$^{7}_{3}Li$
Beryllium	4	$^{9}_{4}Be$
Boron	5	$^{11}_{5}B$
Carbon	6	$^{12}_{6}C$

d Carbon has several isotopes.

i What is different for different isotopes of carbon?

_____ [1 mark]

ii What is the same for all isotopes of carbon?

_____ [1 mark]

Alpha, beta and gamma rays 2

1 a Why is it not safe to use a cardboard box to store a sample which emits **gamma radiation**?

_____

[1 mark]

b What **equipment** can you use to detect radiation?

_____

[1 mark]

2 a Complete this table by showing the mass and the charge of different types of radiation.

Type	Mass	Charge
Alpha		+2
Beta	Negligible	
Gamma		0

[3 marks]

b Alpha radiation is the most **ionising** type of radiation. Explain what ionising means.

_____

_____

[2 marks]

Background radiation 2

3 The diagram shows the main sources of background radiation.

a What is meant by **background radiation**?

_____

_____

_____

_____

[2 marks]

Fallout from nuclear weapons testing 0.5%
Work-related 0.4%
Total discharge from all nuclear power industry 0.1%
Medical, mainly from X-rays 11.5%
Cosmic rays from outer space 14%
Other – mainly air travel 0.5%
All living things give off radiation from their bodies and from the food they eat 17%
Gamma rays from rocks and soil 19%
Radon and thoron gas is released from soil, rocks and building materials 37%

b Write down **one** source of background radiation from outer space.

_____

[1 mark]

c Write down **one** source of man-made radiation.

_____

[1 mark]

d Is most background radiation from natural sources or from man-made sources?

_____

[1 mark]

4 Radon gas is radioactive and comes from rocks like granite.

a Why do some parts of the United Kingdom have a higher background radioactivity than others?

_____

[1 mark]

b Explain whether we should allow people to live in places where there is granite rock.

_____

_____

_____

[3 marks]

Inside the atom

1 The diagram shows one of the first models of the atom.

   a What is this model called?

   _____

   _____

                                   [1 mark]

positive 'jelly'   embedded electrons

   b Another model of the atom is the nuclear model.
       Write down the **three** main points about this model.

   _____

   _____

   _____ **[3 marks]**

   c What evidence is there that inside an atom
      i there is a positively charged nucleus?

   _____ **[1 mark]**

      ii the nucleus is very small?

   _____ **[1 mark]**

     iii electrons orbit outside the nucleus?

   _____ **[1 mark]**

Nuclear fission

2 a What is nuclear fission?

   _____

   _____ **[2 marks]**

   b Write down **one** use for nuclear fission.

   _____ **[1 mark]**

   c The picture shows a chain reaction.
      i How many neutrons are produced from each reaction?

   _____

   _____

                        [1 mark]

      ii Why does a chain reaction make the size of the reaction
         increase?

   _____

   _____ **[2 marks]**

   d Explain why only certain elements can undergo fission.

   _____

   _____ **[2 marks]**

Nuclear power station

1 The diagram shows a nuclear power station.

a What type of reaction takes place in the uranium rod?

_____ [1 mark]

b Why are the uranium rods kept inside a graphite moderator?

_____ [1 mark]

c What part of the reactor controls the speed of the fission reaction?

_____ [1 mark]

d What energy change takes place in the generator?

_____ [1 mark]

e How does the nuclear reactor
i create steam?

_____

_____ [2 marks]

ii generate electricity?

_____

_____ [2 marks]

Nuclear fusion

2 a What is **nuclear fusion**?

_____

_____ [2 marks]

b Write down **one** place where nuclear fusion takes place.

_____ [1 mark]

c One nuclear fusion reaction involves hydrogen nuclei.
i What is created when hydrogen nuclei fuse?

_____

_____ [2 marks]

ii Why are large amounts of heat and pressure needed for nuclear fusion to take place?

_____

_____ [2 marks]

iii Explain why only certain elements can undergo fusion.

_____

_____ [2 marks]

P2b revision checklist

I know:

what the current through an electrical current depends on

☐ the symbols for components shown in circuit diagrams

☐ resistance is increased in long, thin, heated wires

☐ the current through a component depends on its resistance. The greater the resistance the smaller the current for a given p.d. across the component

☐ **potential difference = current × resistance**
       (volt, V)   **(ampere, A)**   **(ohm, Ω)**

☐ in a series circuit: the total resistance is the sum of the resistance of each component; the current is the same through each component and the total p.d. of the supply is shared between the components

☐ in a parallel circuit: the total current through the whole circuit is the sum of the currents through the separate components; the p.d. across each component is the same

what mains electricity is and how it can be used safely

☐ mains electricity is an a.c. supply of 230 V and has a frequency of 50 Hz; it is very dangerous

☐ fuses and earth wires protect appliances from damage and people from harm or death

☐ three-pin plugs must be wired correctly and hold the correct fuse

why we need to know the power of electrical appliances

☐ the power of an electrical appliance is the rate at which it transforms energy:

$$\text{power (W)} = \frac{\text{energy transformed (J)}}{\text{time taken (s)}}$$

☐ power, potential difference and current are related by the equation:

      power = current × potential difference
    (watt, W)   **(ampere, A)**   **(volt, V)**

☐ most appliances have their power and the potential difference of the supply they need printed on them so we can calculate the current and fuse required

what happens to radioactive substances when they decay

☐ isotopes (elements with the same number of protons but a different number of neutrons) with unstable nuclei emit energy as particles or rays called radiation

☐ the Rutherford and Marsden scattering experiment revealed the structure of the atom

☐ background radiation comes from rocks, soil, cosmic rays, living things and medical X-rays

what nuclear fission and nuclear fusion are

☐ nuclear fission is the splitting of an atomic nucleus; it is used in nuclear reactors

☐ nuclear fusion is the joining of two smaller nuclei to form a larger one; stars release energy by nuclear fusion

Periodic table

Key

relative atomic mass
atomic symbol
name
atomic (proton) number

Example:
1
H
hydrogen
1

Group 1	Group 2												Group 3	Group 4	Group 5	Group 6	Group 7	Group 8
																		4 He helium 2
7 Li lithium 3	9 Be beryllium 4												11 B boron 5	12 C carbon 6	14 N nitrogen 7	16 O oxygen 8	19 F fluorine 9	20 Ne neon 10
23 Na sodium 11	24 Mg magnesium 12												27 Al aluminium 13	28 Si silicon 14	31 P phosphorus 15	32 S sulfur 16	35.5 Cl chlorine 17	40 Ar argon 18
39 K potassium 19	40 Ca calcium 20	45 Sc scandium 21	48 Ti titanium 22	51 V vanadium 23	52 Cr chromium 24	55 Mn manganese 25	56 Fe iron 26	59 Co cobalt 27	59 Ni nickel 28	63.5 Cu copper 29	65 Zn zinc 30		70 Ga gallium 31	73 Ge germanium 32	75 As arsenic 33	79 Se selenium 34	80 Br bromine 35	84 Kr krypton 36
85 Rb rubidium 37	88 Sr strontium 38	89 Y yttrium 39	91 Zr zirconium 40	93 Nb niobium 41	96 Mo molybdenum 42	[98] Tc technetium 43	101 Ru ruthenium 44	103 Rh rhodium 45	106 Pd palladium 46	108 Ag silver 47	112 Cd cadmium 48		115 In indium 49	119 Sn tin 50	122 Sb antimony 51	128 Te tellurium 52	127 I iodine 53	131 Xe xenon 54
133 Cs caesium 55	137 Ba barium 56	139 La* lanthanum 57	178 Hf hafnium 72	181 Ta tantalum 73	184 W tungsten 74	186 Re rhenium 75	190 Os osmium 76	192 Ir iridium 77	195 Pt platinum 78	197 Au gold 79	201 Hg mercury 80		204 Tl thallium 81	207 Pb lead 82	209 Bi bismuth 83	[209] Po polonium 84	[210] At astatine 85	[222] Rn radon 86
[223] Fr francium 87	[226] Ra radium 88	[227] Ac* actinium 89	[261] Rf rutherfordium 104	[262] Db dubnium 105	[266] Sg seaborgium 106	[264] Bh bohrium 107	[277] Hs hassium 108	[268] Mt meitnerium 109	[271] Ds darmstadtium 110	[272] Rg roentgenium 111								

Elements with atomic numbers 112–116 have been reported but not fully authenticated.

* The Lanthanides (atomic numbers 58–71) and the Actinides (atomic numbers 90–103) have been omitted.
Cu and Cl have not been rounded to the nearest whole number.

Notes

Notes

Notes

Notes

Answers

B1a Human biology
Page 148 Coordination
1 a Brain *(1)*; spinal cord *(1)*

b i Motor *(1)*
ii

(2)

iii Neurones carry information in the form of **electrical** impulses. Sensory neurones carry impulses from **receptors** to the **central** nervous system. *(3)*

Page 148 Receptors
2

1	2	3	4
D	B	A	C

(4)

3 a

1	2	3	4
D	C	A	B

(4)

b i The heart *(1)*
ii Adrenaline provides more oxygen to muscles (so you can run faster) *(1)*

Page 149 Reflex actions
1 a C *(1)*

b i Transmitter substance / Acetylcholine *(1)*
ii Transmitter substance (contents of X) is released into the cleft and diffuses across the cleft and sets up impulse in the next neurone *(2)*

Page 149 In control
2 a A *(1)*

b 1 from the skin in sweat *(1)*; 2 in urine *(1)*; 3 from the lungs in breath *(1)*

c To replace (sodium and chloride ions) lost in sweat *(1)*; to maintain their concentration in the blood/so that cells can work effectively *(1)*

d Makes the liver take glucose from the blood storing it as glycogen *(2)*

Page 150 Reproductive hormones
1 a B *(1)*
b B *(1)*

Page 150 Controlling fertility
2 a A *(1)*

b i The older the woman, the lower the success rate *(1)*
ii The success rate is too low *(1)*; uses up money better spent on other health treatments *(1)*; possible problems for children of (very) elderly parents *(1)*

3 a A *(1)*

b *(Any 2:)* Could encourage people to have more sexual partners; could encourage people to have sex more often; could mean people stop using condoms

Page 151 Diet and energy
1 a We get our energy from the food that we eat, especially **fats**, carbohydrates and proteins. A **balanced** diet contains the right amount of energy. People who eat a poor diet are said to be **malnourished**. *(3)*

b i The rate of chemical reactions in cells / the body *(1)*
ii Katie's metabolic rate is faster *(1)*; so she uses up energy faster *(1)*
iii Olivia could do more exercise *(1)*

Page 151 Obesity
2 a France *(1)*
b 28 % *(1)*
c In Russia, there are more obese women than obese men, in Canada, the percentages obese men and obese women are the same *(1)*
d Obese people put more, stress / weight, on their joints *(1)*; cause damage to the joint surface / wear away cartilage *(1)*

Page 152 Not enough food
1 a Lack of food reduces resistance to (infectious) diseases
b i Energy taken in depends on the total amount eaten / what matters is the balance between energy taken in and energy used / weight is lost only if energy taken in is less than energy used
ii *(Any 2:)* They will not be able to maintain this weight loss; weight will easily go back on again they may feel ill; weak (after losing so much weight so quickly) they may have lost a lot of water

Page 152 Cholesterol and salt
2 a B *(1)*
b The blood carries **cholesterol** in two forms, HDL and LDL. It is better to have **more** HDL than LDL. Eating **monounsaturated** fats can help to reduce blood cholesterol. *(3)*
c *(Any 3:)* Plaques form; in walls of arteries; from cholesterol; clots can form; heart muscle does not get a good supply of blood; so does not receive enough oxygen; muscle can, stop working / die

Page 153 Drugs
1 a

1	2	3	4
A	D	C	B

(4)

b You cannot manage without it *(1)*; if you stop taking the drug you experience withdrawal symptoms *(1)*

Page 153 Trialling drugs
2 a Tested on animals / human cells (in laboratory) *(1)*
b i A (mean age in years was the same) *(1)*
ii B, C, D and E *(1)*
iii To control a variable *(1)*; every subject took a pill *(1)*; to provide a comparison point with those who took the drug *(1)*
iv Ensures there is only one independent variable *(1)*; removes possibility of observer bias *(1)*

Page 154 Illegal drugs
1 a C *(1)*
b i Syringe may be contaminated *(1)*; with viruses *(1)*
ii May become addicted *(1)*; suffer withdrawal symptoms if they stop taking it *(1)*

Page 154 Alcohol
2 a Alcohol *(1)*
b Alcohol *(1)*
c No, we only know how many were admitted to hospital *(1)*; perhaps heroin is more likely to cause hospital admission than cocaine *(1)*
d No, we only know how many were admitted to hospital *(1)*; perhaps women take less of the drug than men / are less likely to be made ill by taking the drug *(1)*
e Liver cells are damaged as they try to break down alcohol *(1)*

Page 155 Tobacco
1 a

1	2	3	4
D	A	B	C

(4)

b i 1945 / 1946 *(1)*
ii There is a time lag between smoking and getting cancer *(1)*; early increase in deaths from lung cancer may not have been noticed took a while for people to make a link between the two *(1)*
iii Some other factor could be causing both increases

Page 155 Pathogens
2 a A *(1)*
b *(Any 2:)* He realised that the ward with the most deaths was visited by doctors *(1)*; he suggested they were passing disease-causing substances between patients *(1)*; he asked doctors to wash their hands (before and after touching patients) *(1)*

Page 156 Body defences

1 a White **blood** cells help to defend us against **pathogens**. Some of them **ingest** bacteria and viruses. Others produce chemicals called **antibodies** that destroy particular bacteria and viruses. *(4)*
b i Phagocytosis *(1)*
ii It is destroyed / killed *(1)*; digested by enzymes *(1)*

Page 156 Drugs against disease

2 a Antibiotics kill (only) bacteria *(1)*; antibiotics do not kill viruses *(1)*
b C *(1)*

Page 157 Arms race

1 Bacteria can become **resistant** to antibiotics by the process of **natural** selection. One example is **MRSA**. We can help to prevent this happening by using **fewer** antibiotics. *(4)*
2 D *(1)*

Page 157 Vaccination

3 a A *(1)*
4 a 9000 *(1)*
b No, because we cannot be sure the drop in cases was caused by the vaccination *(1)* some other factor might have caused them to fall *(1)*; but there is a correlation between the use of the vaccine and the fall in the number of cases *(1)*
c Not everyone was vaccinated straight away *(1)*; only 10–14 year olds were vaccinated *(1)*; as time went on a bigger proportion of people had immunity (accept other well-argued suggestions) *(1)*
d *(Any 4:)* Reference to antigens; white blood cells respond; lymphocytes; secrete antibodies; specific to, antigen / TB / bacterium; multiply; form memory cells; on next infection bacteria killed immediately

B1b Evolution and environment
Page 159 Hot and cold

1 a

1	2	3	4
B	D	A	C

(4)

2 a 38 °C *(1)*
b It loses heat to the cold, air / ground *(1)*
c The cold blood from the feet is warmed by the warm blood in the artery as it flows upwards *(1)*

Page 159 Adapt or die

3 Plants and animals are **adapted** for survival in their habitat. Some plants have **poisons** in their leaves, which harm insects that eat the leaves. Some plants have warning colours to deter **predators**. *(3)*
4 A *(1)*

Page 160 Two ways to reproduce

1 a In asexual reproduction, the offspring are all **genetically** identical. They are **clones**. In sexual reproduction, the offspring get a mixture of **genes** from both their parents. *(3)*
b

1	2	3	4
D	A	C	B

(4)

Page 160 Genes and what they do

2 B *(1)*
3 a Nose shape and hair colour *(1)*
b Kim could be heavier because she eats more or exercises less / Kim could be better at maths because she works harder at it *(1)*
c They could have inherited different (combinations of) genes from their parents *(2)*

Page 161 Cuttings

1 D *(1)*
2 a *(Any 2:)* They might have different genes; they might be growing in different kinds of soil; one might have more water than the other; one might have more light than the other
b Acorns are produced by sexual reproduction, so they are genetically different *(1)*; cuttings only have genes from one parent, so they are genetically identical *(1)*

Page 161 Clones

3 a Fertilisation happens when a **sperm** and an egg fuse together. They form a new cell called a **zygote**. This divides to form a ball of cells. The cells in the ball can be split apart. Each cell grows into an embryo, which can be put into a different **mother**. *(3)*
b They are all formed from one zygote so they all have the same genes *(1)*; they are genetically identical *(1)*

Page 162 Genetic engineering

1 A *(1)*
2 The bacteria contain the human gene for making insulin *(1)*; the gene carries the code/instructions for making human insulin *(1)*

Page 162 Theories of evolution

3 a D *(1)*
b C *(1)*
c Only the ones with long necks reproduce *(1)*; they pass on their genes to their offspring; all offspring contain the genes for long necks *(1)*

Page 163 Natural selection

1 Organisms **compete** for survival. Only the best **adapted** ones live long enough to **reproduce**. They pass on their **genes** to their offspring. Over many **generations**, most of the organisms will contain these advantageous genes. *(3)*
2 a D *(1)*
b A *(1)*

Page 163 Fossils and evolution

3 a D *(1)*
4 a From fossils *(1)*
b *(Any 2:)* Idea that they were adapted to their environment; more toes better for running on wet ground; climate became dryer; horses with fewer toes more likely to survive; and pass on their genes for fewer toes to their offspring; over time/many generations, all horses had only one toe

Page 164 Extinction

1 B *(1)*
2 *(Any 2:)* Competition with the introduced crayfish / introduction of new competitor; compete for limited resources / not enough food for all; new disease may kill them; not enough new crayfish born to replace the ones that die

Page 164 More people, more problems

3 As the human **population** increases, we use more and more non-renewable fuels such as **oil**. We also use more **metals**, which we extract from ores. We cause more **pollution**. *(4)*
4 a The right hand part beyond the current year *(1)*
b About 2 billion *(1)*
c By about three times *(1)*

Page 165 Land use

1

1	2	3	4
D	A	C	B

(4)

Page 165 Pollution

2 Could catch diseases by taking in viruses / bacteria / pathogens *(1)*; examples – cholera, polio, gasroenteris, Weil's disease, allergic alveiolitis *(1)*
3 a *(Any 2:)* Extra nitrate made plants grow more; they shaded out plants beneath them; many plants died; bacteria fed on them (so the bacterial population increased)
b The bacteria respired *(1)*; aerobically *(1)*
4 a Concentration of DDT increases up the food chain *(1)*; greatest concentration in the animals at the end of the chain *(1)*
b *(Any 2:)* DDT persists for a long time; in the environment / in the bodies of animals; perhaps DDT is still being used in other countries

Page 166 Air pollution

1 a B *(1)*
b Cause bronchitis / trigger asthma attacks (but not actually cause asthma) *(1)*

Page 166 What causes acid rain?
2 a Carbon dioxide *(1)*
b It prevents the blood from carrying oxygen *(1)*
c They damage the lungs *(1)*; cause breathing problems / coughing / asthma attacks *(1)*
The gases dissolve in water droplets react to form acids *(1)*; rain droplets containing acid fall to the ground when it rains *(1)*
e Kills trees by damaging their leaves / roots *(1)*; washing away nutrients kills fish by making water acidic / toxic chemicals washed in from soil *(1)*
f The limestone neutralises the acid by reacting with it (to form salts and water) *(1)*; pH is raised *(1)*

Page 167 Pollution indicators
1 a Chironomid larva *(1)*
b The pH is less than 5.4 / about 5.1 *(1)*
2 a *(Any 2:)* Use the animals as indicators; different animals can live in different oxygen concentrations; find out which animals are living in the water; match them to the oxygen concentration in which they are known to live
b *(Any 2:)* Rat-tailed maggots; they can survive in the lowest oxygen concentration; raw sewage reduces the oxygen concentration in the water.

Page 167 Deforestation
3 C *(1)*
4 A *(1)*

Page 168 The greenhouse effect – good or bad?
1 A *(1)*
2 Carbon dioxide traps energy that is **radiated** from the Earth. This stops the energy from escaping into **space**. It keeps the Earth **warmer** than it would otherwise be. *(3)*
3 A *(1)*

Page 168 Sustainability – the way forward?
4 a Less electricity / less fuel needed to heat the house *(1)*
b i *(Any 2:)* Making sure we do not use resources faster than they are being replaced; using the environment in such a way that we do not spoil it for the future; improving the quality of people's lives without damaging the chances of our survival in the future
ii *(Any 2:)* Less petrol used; so less oil needs to be taken from the ground; less carbon dioxide released to the atmosphere (by the combustion of fuels); so less contribution to global warming

C1a Products from rocks
Page 170 Elements and the periodic table
1 H; Na; Be   *All three correct = (2) 2 correct = (1)*

Page 170 Atomic structure 1
2 C *(1)*
3 a i 16; 16; 16   *All three correct = (2) 2 correct = (1)*
ii Electrons have no / negligible mass *(1)*
b i Cl *(1)* accept Cl2
ii O *(1)* accept O2

Page 171 Bonding
1

1	2	3	4
D	B	C	A
(4)

Page 171 Extraction of limestone
2 a Calcium carbonate *(1)*; marble *(1)*
3 a *(Any 2:)* Only allow traffic during the day; only operate quarry during the day; use rail to transport limestone; building banks around the quarry to act as 'noise shadows'
b Employment / jobs / income for local shops *(1)*

Page 172 Thermal decomposition of limestone
1

Name of atom	Symbol	Number of atoms in formula
Calcium	**Ca**	1
Carbon	C	1
Oxygen	**0**	3
(3)

2 a Calcium carbonate → **calcium oxide** *(1)* + **carbon dioxide** *(1)*
$CaCO_3$ → CaO + CO_2 *(1)*
b B *(1)*

Page 172 Uses of limestone
3 Mortar *(1)*; glass *(1)*   **4** Stronger *(1)*

Page 173 The blast furnace
1

1	2	3	4
C	D	B	A
(4)

2 Carbon monoxide; oxygen; reduction
All three correct = (2) 2 correct = (1)

Page 173 Using iron
3 a 4% *(1)*   **b** It is softer *(1)*; easier to shape *(1)*

Page 174 Using steel
1 C *(1)*
2 a *(Any 3:)* Wrought iron contains the same type of atoms; atoms in wrought iron are in regular rows; atoms in steel are of different elements; sizes of atoms are different; rows are not regular in steel
b *(Any 2:)* Steel is harder; does not rust; steel is stronger; steel has a shinier finish

Page 174 Transition metals
3 Copper *(1)*; manganese *(1)*
4 a Positive ions *(1)*; electrons *(1)*
b Electrons are involved in conduction *(1)*; they move *(1)*

Page 175 Aluminium
1

1	2	3	4
B	D	C	A
(4)

2 Reactive; negative; gain   *All three correct = (2) 2 correct = (1)*

Page 175 Aluminium recycling
3 a *(Any 2:)* Spoiling of habitats / tree felling / ecosystem damage; burning trees leads to climate change; loss of living area for local people; health problems due to mining e.g. dust; use of old mines as dumps
b *(Any 2:)* Thrown away with other waste; difficulty of sorting / collecting; attitude of people to recycling

Page 176 Titanium
1 a Lower density; this means that the bikes will be lighter; corrosion resistant; this means that the frame will not rust or corrode *(3)*
b i Aluminium oxide is an ionic compound and titanium oxide is a covalent compound *(1)*; ionic compounds conduct electricity when they are molten, covalent compounds do not *(1)*
ii Titanium has a melting point above 900 °C *(1)*

Page 176 Copper
2 a Copper *(1)*; sulfur *(1)*   **b** Oxidation *(1)*
3

1	2	3	4
D	C	A	B
(4)

Page 177 Smart alloys
1 a

1	2	3	4
B	D	C	A
(4)

2 a (Similar because:) Contain mainly metal atoms *(1)*; contain mixture of elements *(1)*; can be bent and stretched into different shapes *(1)*; (different because:) have a shape memory *(1)*
b Heating *(1)*; then cooling *(1)*

Page 177 Fuels of the future
3 a Petrol is made from crude oil. Crude oil will run out because it is **non-renewable**. When petrol burns it makes **carbon dioxide** and **water**. Carbon monoxide comes from car exhaust due to **incomplete** combustion. *(4)*
b More cars in cities, less chance for gases to spread out and away *(1)*

Page 178 Crude oil

1

1	2	3	4
C	B	D	A

(4)

Page 178 Alkanes

2 **a i** A *(1)*

   ii Does not have all single bonds / has a double bond *(1)*; contains oxygen / does not contain just hydrogen and carbon *(1)*

b i B *(1)*

   ii

(1)

Page 179 Pollution problems

1

1	2	3	4
D	B	C	A

(4)

Page 179 Reducing sulfur problems

2 **a** 5 *(1)*

b Damage to trees / vegetation *(1)*; test the pH of the lake *(1)*; look for dead fish *(1)*

3 C only *(2)*

4 **a** Stops problems with acid rain *(1)*

b Limestone must be quarried which can cause environmental damage *(1)*

C1b Oils, Earth and atmosphere

Page 181 Cracking

1 **a** Carbon *(1)*; hydrogen *(1)*

b i Used for fuels *(1)*; fuels are needed in much larger quantities than road building materials or other chemicals *(1)*

   ii Makes shorter chain molecules *(1)*; which have more uses / needed for fuels / used in larger amounts *(1)*

2 C *(1)*

Page 181 Alkenes

3 **a** C_5H_{10} *(1)*

b

True for alkanes	True for alkenes
B	A
C	
D	

All four correct = (3) 3 correct = (2) 1 or 2 correct = (1)

Page 182 Making ethanol

1 Ethene has a double bond *(1)*; ethanol has OH / contains an oxygen atom *(1)*

2 **a** Fermentation *(1)*; using yeast *(1)*

b Renewable means that the supply will not run out *(1)*; more sugar beet can be grown to produce more fuel *(1)*; it is called a biofuel because it is made from living things *(1)*

Page 182 Plastics from alkenes

3 **a**

1	2	3	4
B	A	C	D

(4)

b i Structure B *(1)*

   ii Contains a double bond *(1)*

Page 183 Polymers are useful

1 **a** B *(1)*

b D *(1)*

c *(Any 2:)* A has shorter chains; and so is more runny; molecules can move over each other more easily

2 **a** Does not soften at high temperatures *(1)*; very slippy so gives pans a non-stick coating *(1)*

b Melts at 100 °C *(1)*; which is fine for hot drinks, but would melt on a cooker *(1)*

c Waterproof *(1)*; flexible whereas polyethene would crack *(1)*

Page 183 Disposing of polymers

3

1	2	3	4
D	B	C	A

(4)

Page 184 Oil from plants

1 **a** Nuts *(1)*; seeds *(1)*

b A *(1)*

2 Dissolves; distilled; boiling points

All 3 correct = (2) 2 correct = (1)

Page 184 Green energy

3 **a** Mixed with other oils / fuels (fuels that are mixed in are usually petrol) / fuels from crude oil *(1)*

b *(Any 2:)* Saves fossil fuels; idea that fossil fuels are not renewable / will run out; biofuels are renewable; they are 'greenhouse neutral'

c Petrol is non-renewable / will run out *(1)*; when it becomes scarce it will be more expensive *(1)*

Page 185 Emulsions

1

1	2	3	4
D	C	A	B

(4)

2 **a**

Type of emulsion	Dispersed phase	Continuous phase
Shaving foam	**Air/gas**	**Liquid**
After shave cream	**Water/liquid**	**Oil/liquid**

(2)

b Do not mix together *(1)*

Page 185 Polyunsaturates

3 **a** Contains all single bonds / contains no double bonds *(1)*

b Contains lots of / many double bonds *(1)*

c

Type	Olive oil	Saturated fat	Polyunsaturated oil
Observations	Orange to colourless	**No change**	**Orange to colourless**

(2)

Page 186 Making margarine

1 **a** It is a runny liquid *(1)*

b Hydrogenation *(1)*

2 **a**

Box 2 *(2)*

b Oil is heated (to 60 °C) *(1)*; a nickel catalyst is used *(1)*

Page 186 Food additives

3 **a**

1	2	3	4
B	A	C	D

(4)

b B *(1)*

Page 187 Analysing chemicals
1 **a** Solvent should be pure / should not use a mixture of ink and water *(1)*; ink spot should not be below level of solvent *(1)*; need to stop before the solvent reaches the top of the paper *(1)*
  b i R *(1)*
    ii B *(1)*; C *(1)*
    iii Chromatography *(1)*

Page 187 The Earth
2 **a**

1	2	3	4
B	D	A	C

(4)

Page 188 Earth's surface
1 D *(1)*
2

1	2	3	4
A	C	B	D

(4)

Page 188 Earthquakes and volcanoes
3 **a**

1	2	3	4
D	C	A	B

(4)
  b The continental plate is less dense because it rises on top of the oceanic plate *(1)*

Page 189 The air
1

1	2	3	4
A	C	D	B

(4)
2 B *(1)*

Page 189 Evolution of the air
3 **a i** The volcano is at over 1000 °C / at a very high temperature *(1)*
    ii The Earth cooled *(1)*; water vapour condensed *(1)*; to form the seas *(1)*
  b i Carbon dioxide *(1)*
    ii Dissolving in sea water *(1)*; photosynthesis *(1)*
  c *(Any 2:)* Does not contain oxygen; does not contain carbon dioxide; contains large amounts of sulphur dioxide / hydrogen / carbon monoxide

Page 190 Atmospheric change
1 Increased; melt; rise; extreme
    All four correct = (3) 3 correct = (2) 2 correct = (1)
2 **a**

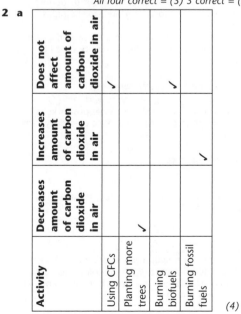

Activity	Decreases amount of carbon dioxide in air	Increases amount of carbon dioxide in air	Does not affect amount of carbon dioxide in air
Using CFCs			✓
Planting more trees	✓		
Burning biofuels			✓
Burning fossil fuels		✓	

(4)

3 **a** Crude oil *(1)*; limestone *(1)*
  b Short lived plants decay and return carbon dioxide to the air *(1)*; trees live a long time and the carbon dioxide is 'locked up' in the wood for a long time while it rots away slowly *(1)*
  c If the seas warm up, not as much carbon dioxide will dissolve in them *(1)*; therefore, more carbon dioxide would be released into the air *(1)*

P1a Energy and electricity
Page 192 Heat energy
1 **a** Heat transfer is the movement of heat energy *(1)*; from one place to a cooler place *(1)*
  b Heat energy travels from a warmer place *(1)*; to a cooler place *(1)*
2 **a** Travel at speed of light *(1)*; travel through vacuum *(1)*; travel in straight lines *(1)*
  b i Drawing the curtains stops heat radiation coming into the house *(1)*; through the glass *(1)*
    ii Paint the house white *(1)*; white reflects heat *(1)*; reduces heat transfers into the house *(1)*

Page 192 Thermal radiation
3

1	2	3	4
C	A	B	D

(4)
4

1	2	3	4
A	D	B	C

(4)

Page 193 Conduction and convection
1

1	2	3	4
C	D	A	B

(4)

2 **a**

1	2	3	4
D	A	B	C

(4)
  b i Heat conduction *(1)*
    ii As each particle becomes warmer it vibrates more *(1)*; the vibration is passed from particle to particle *(1)*; this transmits the energy *(1)*
  c The best conductor heats up quickest *(1)*; touch the end of each one to feel which gets hot first *(1)*

Page 193 Heat transfer
3 **a** The carpet has trapped air pockets *(1)*; air is a poor heat conductor / good insulators *(1)*; these stop heat losses by conduction *(1)*
  b Free electrons are not bound to any atoms *(1)*; can travel between them, moving through the material *(1)*
  c Free electrons carry heat energy from one part of the substance to another *(1)*; aluminium contains free electrons so it conducts heat well *(1)*; wool has no free electrons so cannot conduct heat well *(1)*

Page 194 Types of energy
1

1	2	3	4
A	D	C	B

(4)

Page 194 Energy changes
2 **a** Chemical energy *(1)* → kinetic energy *(1)*
  b Heat *(1)*; sound *(1)*
  c It spreads to the surroundings *(1)*; making them warm up a bit *(1)*
  d Chemical energy from fuel changes into (gravitational) potential energy *(1)*; and kinetic energy *(1)*; and then into heat (in the brakes) *(1)*; also wasted forms of energy: sound/heat *(1)*
3 **a** A *(1)*
  b C *(1)*
  c D *(1)*
  d B *(1)*

Page 195 Energy diagrams

1

1	2	3	4
C	D	A	B

(4)

2 **a** 700J (1)

   b Heat (1)

   c Sankey diagram (see diagram) with one input which splits into two output arrows (1); equal width (1) input labelled 1000J, each output labelled 500J (1); input labelled chemical energy and output labelled kinetic energy and heat (1)

Page 195 Energy and heat

3 **a** C (1)

   b A (1)

   c B (1)

   d D (1)

Page 196 Energy, work and power

1

1	2	3	4
D	A	B	C

(4)

2 **a** Work done = force x distance (1)
              500 x 60 (1)
              30 000 J (1)

   b Power = work done ÷ time taken (1)
              30 000 ÷ 75 (1)
              400 W (1)

   c (Any 1:) Energy is lost as heat / sound to surroundings; the ski lift carries more people / chairs than Molly

   d (Any 2:) Efficiency measures the proportion of energy usefully transferred (1); the new lift wastes less energy doing the same work (1); more of the input energy is usefully transformed (1)

Page 196 Efficiency

3 **a** B

   b D

   c A

   d C

Page 197 Using energy effectively

1 **a** Saves money (1); good for the environment (1)

   b **i** Reduces electricity used (1)

       ii Stops heat escaping / reduces heating bills (1)

   c Keep doors closed – only heat ed rooms used; close windows – reduces costs of heating; don't fill kettles full – saves electricity (2)

2 **a** Time taken (1) for the savings to match the costs (1)

   b (Any 2:) Reduces heating bills; stop draughts; easy to install; cheap to install

   c (Any 3:) Annual savings are smallest (1); over several years in the same home, savings using other methods will be greater (1); over 10 years, draught proofing saves 150 – 45 = £105 but loft insulation saves more (1); £600 – 240 = £360 (1)

Page 197 Why use electricity?

3

1	2	3	4
C	A	B	D

(4)

Page 198 Electricity and heat

1 **a** Electricity → heat (1)

   b Current too large (1); fuse melts (1); breaks the circuit (1)

   c 15 A fuse wire is thicker (1); can carry larger current before melting (1)

Page 198 The cost of electricity

2 **a** Energy transferred (1); per second (1)

   b 2400 W (1)

   c **i** 2 x 4 = 8 kWh (2)

       ii $1.5 \times \frac{1}{4} = 0.375$ kWh (3)

3

1	2	3	4
B	A	D	C

(4)

Page 199 The National Grid

1

1	2	3	4
B	C	D	A

(4)

2

1	2	3	4
D	A	C	B

(4)

3 **a** 230 V (1)

   b Reduces the size of the voltage (1)

   c High voltage means small current is needed (1); less energy is wasted as heat (1)

Page 199 Generating electricity

4 **a** **i** Wheels have kinetic energy which changes into electrical energy in the dynamo (1)

       ii Gets brighter (1)

Page 200 Power stations

1 **a** Fuel is burned (1) and used to heat the water (1)

   b **i** Chemical → heat → kinetic (2)

       ii kinetic → electrical (1)

   c (Any 1:) Reliable; abundant; concentrated energy source

   d (Any 2:) Changed in to heat energy; spreads to surroundings; surroundings become warmer

2

1	2	3	4
D	A	C	B

(4)

Page 200 Renewable energy

3 **a** There is no electricity produced if it is calm or stormy (1)

   b **i** A barrage / dam is built across an estuary / river mouth (1); as tidal water moves in or out through pipes in the wall (1); it forces turbines in the pipes to spin (1)

       ii Only a few suitable sites / floods local area / effect on local ecosystems / only provides electricity when the tide in or out (1)

Page 201 Electricity and the environment

1 **a** Release of greenhouse gases (1); causes acid rain (1)

   b (Any 2:) Wind turbines can be stand-alone; power lines from the National Grid are not needed; produce electricity directly from the wind

Page 201 Making comparisons

2 **a** **i** Coal fired power station (1)

       ii Coal / nuclear fuels (1)

   b (Any 3:) Availability of fuel; transport links; local population; land costs and availability

3 **a**

1	2	3	4
B	A	D	C

(4)

4 **a** Cost of fuel (1); cost of building power stations (1)

   b No fuel costs for Hydroelectricity (1); ongoing fuel costs for oil (1) (relatively cheap at the moment, but varies and will rise as oil starts to run out)

P1b Radiation and the Universe

Page 203 Uses of electromagnetic radiation
1 a Continuous range of values *(1)*
  b All travel at same speed *(1)*; can travel through vacuum *(1)*; travel in straight lines *(1)*

Page 203 Electromagnetic spectrum 1
2 a 1 Microwave *(1)*; 2 visible *(1)*; 3 X-rays *(1)*
  b As frequency increases, energy increases *(1)*

3

1	2	3	4
D	B	A	C

(4)

4 a

(1)

  b

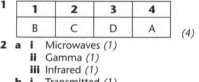

Wave smaller, constant wavelength = *(1)*
same amplitude = *(1)*

Page 204 Electromagnetic spectrum 2
1

1	2	3	4
B	C	D	A

(4)

2 a i Microwaves *(1)*
    ii Gamma *(1)*
    iii Infrared *(1)*
  b i Transmitted *(1)*
    ii Absorbed *(1)*
    iii Reflected *(1)*

Page 204 Waves and matter
3 a X-rays *(1)*
  b X-rays pass through soft tissue *(1)*; they are absorbed by bones *(1)*; and detected / absorbed by photographic plate under the person *(1)*
  c X-rays can ionise / damage cells *(1)*

Page 205 Dangers of radiation
1

1	2	3	4
A	D	C	B

(4)

2 a Ultraviolet *(1)*
  b Sun burn *(1)*; premature skin aging / wrinkling *(1)*; skin cancer *(1)*
  c Dark skin cells on the surface absorb the UV *(1)*; so it does not penetrate *(1)*; to more vulnerable cells under the surface *(1)*

Page 205 Telecommunications
3 a Analogue *(1)*
  b Infrared / visible pulses *(1)*; sent down optic fibres *(1)*
  c Disruption digging up roads better alternatives e.g. satellite links *(1)*
  d *(Any 2:)* Links possible to countries further away; information travels very rapidly; can reach remote areas; very clear signals possible

Page 206 Fibre optics: digital signals
1 a Pulses of light *(1)*; repeatedly reflect from the inside surfaces of the optic fibre *(1)*; emerge from the other end *(1)*
  b

Diagram should show reflections off inside of fibre = (1) emerges from far end = (1) reflected angles approximately equal to incident angles = (1)
    ii Total internal reflection
    iii *(Any 3:)* Cheaper; thinner cable so more can be packed into the same space; quality of transmission is better; thousands of simultaneous signals can be sent; no fire hazard

Page 206 Radioactivity
2

1	2	3	4
A	D	B	C

(4)

3 a A; B *(1)*
  b 6 *(1)*
  c 14 *(1)*
  d This is when particles / bursts of radiation *(1)*; are emitted from the nucleus *(1)*

Page 207 Alpha, beta and gamma rays 1
1 a i Gamma *(1)*
    ii Alpha *(1)*
    iii Alpha / beta *(1)*
2 a Helium nucleus / 2 protons and 2 neutrons *(1)*; emitted from nuclei *(1)*
  b They cannot penetrate through skin *(1)*
  c Smoke detector *(1)*; localised destruction of cancer cells *(1)*
  d Gamma rays have no charge *(1)*; no mass / or do not interact with atoms easily *(1)*

Page 207 Background radiation 1
3 a C *(1)*
  b B *(1)*
  c A *(1)*
  d D *(1)*

Page 208 Half-life
1 a D *(1)*
Note: the multiple choice options to C and D should be 'three quarters' instead of 'half'. This means the correct answer is D: about three quarters of the atoms will decay.
  b A *(1)*
  c B *(1)*
  d C *(1)*

Page 208 Uses of nuclear radiation
2 a Using radioactive materials to treat illnesses *(1)*
  b *(Any 2:)* High doses of gamma needed to kill cells *(1)*; cancer cells are targeted to receive high doses *(1)*; healthy cells are targeted to receive lower doses *(1)*
  c *(Any 2:)* Radioactive sample injected into blood stream *(1)*; detector used outside the body *(1)*; to measure level of radioactivity in different places *(1)*
  d Sterilising equipment / producing images of some organs *(1)*
3 a Maintains quality of product / stops wastage *(1)*
  b Can't be detected through paper *(1)*
  c Less radiation means the thickness is too big *(1)*; so the rollers should come closer *(1)*

Page 209 Safety first
1 a D *(1)*
  b B *(1)*
  c A *(1)*

Page 209 Searching space
2 a C *(1)*
  b A lens *(1)*
  c Large diameter *(1)*; to collect more light *(1)*
  d Telescope in orbit / in space *(1)*; transmits images back to earth *(1)*
  e *(Any 2:)* Less distortion due to pollution; better quality images; very large telescope so can collect very faint images

Page 210 Gravity
1

1	2	3	4
B	A	D	C

(4)

2 a Effect of gravity *(1)*
  b Effect increased *(1)*

Page 210 Birth of a star

3

1	2	3	4
B	D	A	C

(4)

4

1	2	3	4
C	A	D	B

(4)

Page 211 Formation of the Solar System

1

1	2	3	4
B	A	C	D

(4)

2 a The dust was flung outwards *(1)*
b Rocks are heavier than dust particles *(1)*; are attracted more strongly to the Sun *(1)*

Page 211 Life and death of a star

3 a The mass is lost as energy given out by the Sun *(1)*
b Red giant *(1)* → white dwarf *(1)* → black dwarf *(1)*
c Gravity *(1)*; pressure due to fusion reactions *(1)*

4

1	2	3	4
C	D	B	A

(4)

Page 212 In the beginning

1 a Billions of years ago *(1)*
b Universe began from a very small initial point *(1)*; that exploded apart in an instant *(1)*
c Quarks join to form atomic particles *(1)*; these join to form hydrogen atoms *(1)*; hydrogen fuses to form heavier atoms *(1)*

Page 212 The expanding Universe

2 a Light from a moving source *(1)*; is redder than expected / shifted to the red end of the spectrum *(1)*
b They are moving away from us *(1)*
3 a B *(1)*
b C *(1)*
c A *(1)*
d D *(1)*

B2a Discover Martian living!
Page 214 Cells

1 a A cell membrane
C vacuole
E nucleus
F cell wall   *All four correct = (3) 3 correct = (2) 2 correct = (1)*
b A, D, E *(1)*
c It has chloroplasts *(1)*
d Ribosomes – where proteins are made *(1)* Mitochondria – where (aerobic) respiration takes place *(1)*

Page 214 Specialised cells

2 a When it fuses with a sperm cell, the new cell (the zygote) will end up with two sets *(1)*
b To provide energy for cell division *(2)*
c i Sweep mucus (containing trapped bacteria and dirt) up from the lungs *(1)*
ii They have cilia which can move in unison *(1)*; they have many mitochondria to provide energy for the movement of the cilia *(1)*

Page 215 Diffusion 1

1 a Diffusion is the spreading of the **particles** of a **gas**, or of any substance in solution. This results in a **net** movement from a region where they are in a **higher** concentration. *(4)*
b Particles are moving faster / have more kinetic energy *(1)*; so they spread out faster / bump into each other more often *(1)*

2 a Animal, because it does not have a cell wall (absence of chloroplasts does not mean it is an animal cell, because many plant cells do not have chloroplasts) *(1)*
b The cell is respiring *(1)*; which uses oxygen *(1)*
c Oxygen is able to pass through it *(1)*
d Mitochondria *(1)*
e Respiration happens faster / oxygen used up more rapidly resulting in a lower concentration of oxygen in the cell so there is a greater concentration gradient (between the cell and outside) *(2)*

Page 215 Diffusion 2

3 a Carbon dioxide *(1)*
b Carbon dioxide is used in photosynthesis so there is a lower concentration inside the leaf than in the air *(1)*; there is a concentration gradient from the air to the leaf *(1)*

Page 216 Osmosis 1

1 Osmosis is the **diffusion** of **water** from a **dilute** to a more concentrated solution, through a **partially** permeable membrane. *(4)*
2 a Beaker / dilute sugar solution, tubing / concentrated sugar solution *(1)*
b It rises *(1)*
c *(Any 3:)* Water molecules can go through the membrane sugar; molecules are too large to go through; water molecules diffuse / move by osmosis from the more dilute solution to the concentrated one; idea of random movement

Page 216 Osmosis 2

3 a Cell membrane *(1)*
b It will burst *(1)*
c The strong cell wall stops it bursting *(1)*
d *(Any 3:)* It will lose water; by osmosis; because the cytoplasm is a less concentrated solution than the sugar solution; contents will shrink; may pull away from the cell wall

Page 217 Photosynthesis

1 a (The remaining words should be:) Carbon dioxide; water; glucose; oxygen   *All three correct = (2) 1/2 correct = (1)*
b *(Any 2:)* Absorbs energy from light; the energy is used to make glucose; from carbon dioxide and water

Page 217 Leaves

2 a *(Any 4:)* Tiny pores (stomata) on the underside; allow carbon dioxide to diffuse into the leaf; broad flat surface large surface; area to absorb, sunlight / carbon dioxide; veins; bring water from the roots; thin, so carbon dioxide can diffuse in quickly / light can reach all cells
b i Near the upper surface so light can easily reach them *(1)*; so they get a lot of light *(1)*
ii Cell in epidermis (either upper or lower) *(1)*

Page 218 Limiting factors

1 a 32.0 *(1)*
b The number of bubbles given off per minute *(1)*
c The greater the light intensity, the greater the rate of photosynthesis *(1)*
d *(Any 2:)* The number of bubbles increases with each reading; suggesting the plant was still responding to the increase in light intensity; she should have waited until the plant was photosynthesising steadily
e *(Any 3:)* Yes, the more light the plant is given, the faster it photosynthesises; suggesting that lack of light was stopping it from photosynthesising faster; however, at the highest light intensities the rate of photosynthesis seems to be levelling off; suggesting some other factor is beginning to be the limiting factor.

Page 218 Healthy plants

2 a Plants need nitrate ions for producing **amino** acids, which are then used to form **proteins**. They need **magnesium** ions for making chlorophyll. *(3)*
b Through their roots *(1)*; into root hairs by active transport *(1)*
c i It cannot make proteins which are needed for growth *(1)*; forming new cells *(1)*
ii It cannot make chlorophyll, which is green *(1)*

Page 219 Food chains
1 a phytoplankton → krill → penguins or fish → seals → killer whales *(1)*
b The arrows show energy flowing along the food chain *(1)*
c Phytoplankton *(1)*
d All except primary consumer *(1)*
2 a As chemical potential energy (in food) *(1)*
b *(Any 2:)* Some light does not hit their leaves; some light passes straight through the leaves; some light is reflected from the leaves; chlorophyll does not absorb all wavelengths (colours) of light

Page 219 Biomass
3 a The mass of living organisms *(1)*
b *(Any 2:)* Lions do not eat all of the antelopes; they do not eat all parts of the antelopes; the lions do not digest all of the antelopes so some remains are lost in their faeces
4 *(Any 2:)* Glucose broken down by respiration; to provide heat energy (to keep warm); this is lost to the environment so there is less energy for the next organism in the chain

Page 220 Food production
1 a Apple tree → Karen *(1)* Wheat (or other plant) → chicken → chicken → Jo *(1)*
b A shorter food chain *(1)*; so less energy lost / energy is lost at each transfer *(1)*
c *(Any 2:)* The chickens do not need to produce heat in their bodies; which involves respiration; and uses up food; so more of their food becomes chicken meat
2 Transport vehicles emit pollutants *(1)*; especially carbon dioxide, which may cause global warming *(1)*; (there are other pollutants that could be described, e.g. sulfur dioxide causing acid rain, nitrogen oxides causing acid rain, carbon monoxide reducing oxygen carriage in the blood, particulates causing damage to the lungs)

Page 220 The cost of good food
3 a *(Any 2:)* There are greater energy losses; because the chickens use energy moving around; so fewer eggs produced per, chicken / square metre / quantity of food provided
b *(Any 2:)* They may think they taste better; they may think they are better for health; they may dislike the idea of hens being reared in battery cages

Page 221 Death and decay
1 a Bacteria *(1)*; fungi *(1)*
b Many microorganisms need oxygen *(1)*; for (aerobic) respiration *(1)*
2 a Temperature in a fridge at about –4 °C *(1)*; temperature at which spores are killed at 120 °C *(1)*
b *(Any 2:)* Microorganisms are less active at lower temperatures (but they are not killed); because their metabolic reactions take place more slowly; reference to optimum temperature of enzymes

Page 221 Cycles
3 a Organisms such as earthworms, which eat dead leaves and other plant remains, are called **detritus** feeders. They help to recycle the materials in the plant remains, so that they become available to other members of the **community** of organisms in the ecosystem. For example, they release some of the carbon in the leaves back into the air, in the form of carbon dioxide, by the process of **respiration**. *(3)*
b All of them *(1)*
c They eat food that has been produced by other organisms *(1)*; they need organic nutrients / they need carbohydrates, fats and proteins; they do not photosynthesise *(1)*

Page 222 The carbon cycle 1
1 a Photosynthesis *(1)*
b Respiration *(1)*
c Carbohydrates; fats; proteins
All three correct = (2) 1/2 correct = (1)

Page 222 The carbon cycle 2
2 a *(Any 2:)* Microorganisms; break down carbon compounds in dead bodies / wastes; use them in respiration (which produces carbon dioxide)
b 70 billion tons *(1)*
c Plants take in more carbon than they give out; so they help to remove carbon dioxide from the atmosphere; reducing the concentration of carbon dioxide; carbon dioxide traps heat and is contributing to a global rise in temperature
d Carbon came from animals / plants / waste materials *(1)*; decomposer break them down parts of the decomposers / dead bodies / faeces remain in the soil *(1)*
e Tiny plants / phytoplankton remove carbon dioxide for photosynthesis *(1)*; they and sea animals release carbon dioxide by respiration *(1)*

B2b Discover DNA!
Page 224 Enzymes – biological catalysts
1 Enzymes are **biological** catalysts. They are **protein** molecules. Each kind of **enzyme** only works on a particular kind of substrate, which fits perfectly into a fold in the enzyme called the **active** site *(4)*
2 a 40 °C *(1)*
b Particles moving faster / have more kinetic energy (as temperature increases) *(1)*; more frequent collisions *(1)*; more energetic collisions between enzyme and substrate *(1)*
c *(Any 3:)* Enzymes are not alive so cannot be killed; the enzyme is denatured (at high temperatures); the enzyme molecule loses its shape; so its substrate cannot fit into its active site

Page 224 Enzymes and digestion
3 a Mouth *(1)*; pancreas *(1)*
b It digests starch to maltose *(1)*
c Any value below 5 (it is actually 2) *(1)*
d They are neutralised / the pH is raised *(1)*

Page 225 Enzymes at home
1 a *(Any 2:)* Haemoglobin is a protein; proteases break down proteins; turns them into amino acids; that can be washed away
b This is the optimum temperature for the enzymes in them *(1)*; enzymes are denatured if they get too hot *(1)*
c Hands / cells, contain proteins *(1)*; the enzymes could start to digest these proteins *(1)*

Page 225 Enzymes and industry
2 a Proteases predigest protein in some baby foods *(1)*; carbohydrases convert starch into sugar syrup *(1)*; isomerase changes glucose to fructose *(1)*
b *(Any 2:)* Babies may not be able to make all the enzymes they need; so some of their food would be undigested / not be absorbed; would remain as, large / insoluble, molecules; pre-digested food contains small molecules that can be absorbed
c Fructose is sweeter than glucose *(1)*; so less needs to be used *(1)*

Page 226 Respiration and energy
1 a glucose + oxygen → carbon dioxide + water *(2)*
b Respiration releases energy from glucose, so that the cell can use it *(1)*
c Lungs *(1)*
d *(Any 2:)* Aerobic respiration takes place in mitochondria; so many mitochondria supply a lot of energy; sperm need a lot of energy for swimming

Page 226 Removing waste: lungs

2 a

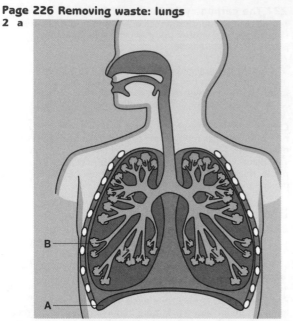

(2)

b In the blood (mostly) *(1)*; dissolved in the blood plasma *(1)*

3 a Oxygen in exhaled air around **16 %** *(1)*; Nitrogen in inhaled air around **79 %** *(1)*

b Body cells use oxygen in respiration; so blood arriving at the lungs is low in oxygen

c *(Any 2:)* Oxygen moves between the lungs and the blood by diffusion; down a concentration gradient; diffusion only occurs until concentrations are equal (another reason is that not all of the air that is breathed into and out of the lungs ever reaches the alveoli)

Page 227 Removing waste: liver and kidneys

1 Urea is a waste product that is made in the **liver** from excess **amino** acids. The urea dissolves in the blood and is carried to the **kidneys**, where it is excreted from the body dissolved in a liquid called **urine**. *(4)*

2 a There is more urea in the urine than in the blood because the kidneys remove urea from the blood *(1)*

b Glucose is useful to the body *(1)*; it provides energy *(1)*

c If there are too few sodium or chloride ions, the kidneys can keep most of them in the blood *(1)*; if there are too many, they can excrete more of them in the urine *(1)*

Page 227 Homeostasis

3 a Water would enter the cells by osmosis *(1)*; could make them burst *(1)*

b *(Any 2:)* In sweat; in breath; in faeces

c *(Any 2:)* More sweat is lost on hot days; so there is less water in the body; so the kidneys conserve water.

Page 228 Keeping warm, staying cool

1 a Sweating; water in sweat evaporates taking heat from the skin *(1)*

b *(Any 2:)* Enzymes; are denatured by high temperatures; so cannot catalyse metabolic reactions

Page 228 Treating diabetes

2 a Just over 90 mg per 100 cm3 *(1)*

b It fell rapidly then rose again and levelled off *(2)*

c Take care about when he eats foods containing carbohydrates, and how much carbohydrate he eats *(1)*

3 a So that the only source of 'pancreas extract' was the injections they gave *(1)*

b It was given an injection of glucose *(2)*

c *(Any 1:)* The blood glucose concentration was very high; because the dog had no pancreas / there was no insulin; that would normally lower the blood glucose concentration (in normal circumstances there is no glucose in the urine)

d *(Any 2:)* Yes, the hypothesis is supported; in a dog with no pancreas the blood glucose concentration became too high; injecting pancreas extract lowered the blood glucose level

Page 229 Cell division – mitosis

1 a 23 *(1)*

b 2 *(1)*

c They must duplicate (make copies of themselves) *(1)*

2 a

Each cell should contain two chromosomes inside its nucleus = (2)

b Genetically *(1)*

Page 229 Gametes and fertilisation

3 a So that after fertilisation the new cell has the correct number of 46 *(2)*

b A variety of a gene *(1)*

c *(Any 3:)* Some characteristics are determined by genes; the parents may have different mixtures of alleles; so the gametes also have different mixtures of alleles; so the offspring have different mixtures of alleles (and therefore characteristics)

Page 230 Stem cells

1 a

1	2	3	4
D	B	C	A

(4)

2 a Embryo stem cells can form every different kind of cell *(1)*; bone marrow stem cells can only form blood cells *(1)*

b Embryo stem cells could be made to produce new nerve cells *(1)*; which could help to mend the spinal cord *(1)*

Page 230 Chromosomes, genes and DNA

3 a DNA *(1)*

b Amino acids *(1)*

c Ribosomes / cytoplasm *(1)*

d The sequence of bases in the DNA of the gene determines the sequence of amino acids that will be linked together to form the protein *(2)*

Page 231 Inheritance

1 Some characteristics, such as flower colour in pea plants, are controlled by a single **gene**. Genes often have different forms, called **alleles**. If an allele produces a characteristic only when a different allele is not present, it is said to a **recessive** allele. *(3)*

2 a T *(1)*; its characteristic appears even when the other allele is present *(1)*

b t and t *(1)*

Page 231 How is sex inherited?

3 a The X chromosome and the Y chromosome *(1)*

b A man has XY *(1)*; a woman has XX *(1)*

c *(Any 2:)* The father's sperm; which could carry either an X chromosome or a Y chromosome; an egg always carries an X chromosome

4 a Two *(1)*

b Female, they have two X chromosomes *(1)*

Page 232 Inherited disorders

1 a It is caused by genes *(1)*; DNA passed on from parent to child *(1)*

b Dominant *(1)*

2 a The cystic fibrosis allele is recessive *(1)*; Angela and Sam both have this allele but they both also have the normal allele *(1)*; they are carriers *(1)*; their child inherited a cystic fibrosis allele from each parent *(1)*

b She could have a child with cystic fibrosis if she is a carrier and her partner is also a carrier but she may not be a carrier, in which case she cannot have a child with the disorder *(3)*

Page 232 DNA fingerprinting

3 a Because everyone's DNA is unique (unless they are an identical twin) (2)
  b i (See diagram) (1)
    ii (See diagram) (1)

possible mother child possible
father A   father B

  iii (Any 3:) Possible father B is the actual father; the child has some bands that have not come from its mother; these bands must have come from its father; not all of these bands are present in possible father A; they are all present in possible father B

C2a Discover Buckminsterfullerene!
Page 234 Atomic structure 2

1 a

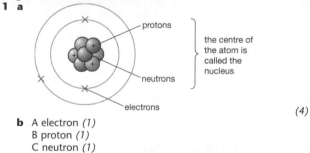

protons

the centre of the atom is called the nucleus

neutrons

electrons

(4)

  b A electron (1)
    B proton (1)
    C neutron (1)

Page 234 Electronic structure

2 a

Periodic table element	$^{23}_{11}$Na sodium	$^{24}_{12}$Mg magnesium	$^{19}_{9}$F fluorine
Number of electrons	11	12	9
Electron arrangement	Na	Mg	F
Notation	2, 8, 1	**2, 8, 2**	**2, 7**

For Mg: draw correctly on diagram = (1) written correctly as 2, 8, 2 = (1)
For F: correct number of electrons = (1)
draw correctly on diagram = (1)
written correctly as 2, 7 = (1)

  b 2, 8 (1)

Page 235 Mass number and isotopes

1 Both same number / 7 electrons in the outer shell (1)

Page 235 Ionic bonding

2 a Sodium (1)
  b Metal; positive; loses   All three correct = (2) 2 correct = (1)
3 a Ca^{2+} (1)
  b $CaCl_2$ (1)

Page 236 Ionic compounds

1 a C (1); because it is grey and shiny (1); and it conducts electricity when it is a solid (1) (a high melting point does not show that it is a metal, because other types of solids also have high melting points)
  b A (1); because it conducts electricity when it is melted (1); but does not conduct when it is a solid (1) (The fact that it is a white solid also gives a clue but it isn't the 'best' answer.)
  c Because it has a low melting point

Page 236 Covalent bonding

2 a

      Ring around a pair of electrons shared between oxygen and hydrogen = (1)

  b 6 (1)
3 a Double (1)
  b 4 shared electrons (1)
  c Oxygen atoms have 6 electrons in their outer shell (1); so need to gain two electrons to form a stable arrangement (1)

Page 237 Simple molecules

1 A few; covalent; strong; gases
    All four correct = (3) 3 correct = (2) 2 correct = (1)

Page 237 Giant covalent structures

2 Giant (1); covalent (1)
3 Graphite has weak bonds between layers (1); layers can break off (1); all bonds in diamond are very strong (1)

Page 238 Metals

1 a i (Any 2:) Low melting point; liquid at room temperature; flows over surfaces
    ii Very dense (2)
2 a

positive ion

electron

(2)

  b Electrons can move (1)
  c C (1)

Page 238 Alkali metals

3 a Sodium (1); sodium chloride (1)
  b Loses one electron (1)

Page 239 Halogens

1 a

Name of halogen	Formula	State at room temperature	Colour
Fluorine	F_2	Gas	Colourless
Chlorine	Cl_2	Gas	**Pale yellow/green**
Bromine	Br_2	**Liquid**	Dark red
Iodine	I_2	Solid	Black

(3)

2 a

8 electrons drawn in outer shell = (1)
b –1 *(1)*
c Sodium *(1)*

Page 139 Nanoparticles
3 a Hundreds; tiny; surface area
All three correct = (2) 2 correct = (1)
b *(Any 2:)* The coat is harder wearing and does not get dirty; therefore it will last longer than a normal coat; and will not need to be washed / cleaned as often
c Companies invest in new research that might make profits in the future; nanoparticles have many uses that can be sold to make money; (examples of uses of nanoparticles:) biosensors; harder wearing / stain resistant materials; information processors; catalysts

Page 240 Smart materials
1 A spectacles; D car dials; B helmet; C mugs
All correct = (4) 2/3 correct = (2) 1 correct = (1)

Page 240 Compounds
2 a i 2 *(1)*
ii Only contains one type of atom / all the atoms are the same *(1)*
b i 5 *(1)*
ii Carbon *(1)*; hydrogen *(1)*
3 a Elements contain only one type of atom *(1)*; compounds contain more than one type of atom chemically joined together *(1)*
b *(Any 3:)* In a mixture the elements are not chemically joined together; in a compound the elements are chemically joined together; mixtures have different properties to the compounds of the same elements; for example, hydrogen and oxygen are gases, water is a liquid

Page 241 Percentage composition
1 a 32 *(1)*; 48 *(1)*

2 a Compound A is CO *(1)*; compound B is CO_2 / CO / CO_2 *(1)*
b Compound B contains a higher percentage than A *(1)*

Page 241 Moles
3 a 64 *(1)*; 64 g *(1)*
b Yes it is true *(1)*; because there is twice as much SO_2 as S in the calculation: 64 ÷ 32 = 2 *(1)*
c Mass of oxygen in SO_2 = 32
Percentage mass of sulfur in SO_2 = (64 ÷ 32) x 100 *(1)*
= 50 % *(1)*

Page 242 Yield of product
1 a Theoretical yield *(1)*
b *(Any 2:)* Some of the magnesium oxide has been lost; not all the magnesium has reacted; the original magnesium may not have been pure; Rose may have made some errors in her measurements; detail of where the magnesium oxide is lost e.g. during heating / when Rose carried the crucible to the balance
2 a C *(1)*
b *(Any 2:)* Makes more product; less reactants wasted; saves energy / fuel

Page 242 Reversible reactions
3 a Reversible *(1)*
b Pink; blue; pink
All three correct = (2) 2 correct = (1)
allow (1) mark for blue pink blue

Page 243 Equilibrium 1
1 a hydrogen + **iodine** ⇌ **hydrogen iodide**
Iodine and hydrogen iodide in correct places = (1)
correct sign for reversible reaction = (1)
b Reaction goes in both directions *(1)*; so there is always a mixture of products and reactants / always some reactants left *(1)*

2 a A *(1)*
b Catalyst *(1)*
c *(Any 2:)* Reaction is reversible; do not get 100% yield; recycling gases means more will react / better yield

Page 243 Haber process
3 a Nitrogen *(1)*
b Methane *(1)*
c i Speeds up the reaction *(1)*
ii Speeds up the reaction *(1)*; gives a higher yield *(1)*

C2b Discover electrolysis!
Page 245 Rates of reactions
1 a Solution goes blue; bubbles form; magnesium is used up
All three = (2) 1 or 3 correct = (1)
b i Bubbles / fizzing *(1)*
ii pH indicator goes blue / is no longer red *(1)*
c Idea that reaction is faster; (example of what will be seen:) faster colour change / faster fizzing / magnesium used up faster *(1)*

Page 245 Following the rate of reaction
2 a Time *(1)*; volume of gas *(1)*
b No more gas will be made *(1)*
c Mass *(1)*

Page 246 Collision theory
1 a

More acid particles shown than in box A = (1)
b *(Any 2:)* increase temperature; smaller zinc pieces / increased surface area of zinc. *Allow use a catalyst,* but this is not the best answer here
c Collide; energy *Both correct = (1)*

Page 246 Heating things up
2 a Flask loses mass / gets lighter *(1)*; because reaction produces carbon dioxide *(1)*; which is a gas / leaves the flask / is lost *(1)*
b *(Any 3 at a higher temperature:)* particles have more energy / move about faster; collisions are more frequent; more particles have enough energy to react; there are more successful collisions

Page 247 Grind it up, speed it up
1 a B *(1)*
b A *(1)*
c Surface area *(1)*
d Mass falls *(1)*; flask gets lighter *(1)*

Page 247 Concentrate now
2 a The acid *(1)*
b No more fizzing *(1)*; no more mass change *(1)*
c Acid is being used up; concentration of acid falls; reactions are slower at lower concentrations *(3)*

Page 248 Catalysts
1 a Hydrogen peroxide → water + oxygen *(1)*
b Experiment takes a very long time / would have to sit and watch for a very long time *(1)*
c i Reaction would be faster *(1)*
ii Manganese (IV) oxide acts as a catalysts *(1)*

2 a i It is not used up *(1)*
ii *(Any 2:)* They have a very long lifetime / are not used up; they make reactions faster; (link to cost:) more products made quickly / do not need to buy more catalyst very often / works out cheaper over time

Page 248 Energy changes

3 a

Reaction	Temperature change	Exothermic or endothermic?
Dissolving ammonium nitrate in water	Decreases	**Endothermic**
Adding zinc powder to copper	**Increases**	Exothermic
Adding magnesium ribbon to an acid	Increases	**Exothermic**

All three correct = (2) 1 or 2 correct = (1)

b Oxidation *(1)*; neutralisation *(1)*

Page 249 Equilibrium 2

1 a ⇌ *(1)*
b Ice melts *(1)*

Page 249 Industrial processes

2 a Nitrogen + hydrogen ⇌ ammonia
Correct names = (1) reversible sign correct = (1)
b i Acts as a catalyst *(2)*
ii Increased surface area *(1)*; makes the reaction faster *(1)*
c i Nitrogen; hydrogen *Both correct = (1)*
ii It is reversible / it reaches equilibrium *(1)*
3 a 10% *(1)*
b Decreases; increases; slower; faster
All four correct = (3) 2 or 3 correct = (2) 1 correct = (1)

Page 250 Free ions

1 a Lithium *(1)*; chloride *(not chlorine)* *(1)*
b LiCl(s) *(1)*
2 a

negative electrode

molten ionic compound

positive ion

(3)

Page 250 Electrolysis equations

3 a Idea of attraction *(1)*; between opposite charges *(1)*
b Chlorine *(1)*
c Gain; 0; reduced *All three correct = (2) 2 correct = (1)*

Page 251 Uses for electrolysis

1 a

At the positive electrode	At the negative electrode	Left behind in solution
Chlorine	Hydrogen	Sodium hydroxide

All three in correct places = (2) 1/2 in correct places = (1)

b B *(1)*
2 a When copper is purified, both electrodes are made from **copper**. A block of impure copper is use as the **positive** electrode. Copper ions from the solution form pure copper by **gaining** two electrons. *(3)*
b The concentration of Cu^{2+} ions stays the same *(1)*; because they go into solution from the impure copper electrode and leave the solution at the pure electrode *(1)*

Page 251 Acids and metals

3 a *(Any 2:)* Magnesium; zinc; lead
b At the bottom *(1)*; it is very unreactive / less reactive than silver *(1)*
c Hydrochloric acid *(1)*

Page 252 Making salts from bases

1 a Sulphuric acid *(1)*
b Zinc oxide *(1)*
c Filtering *(1)*
2 a Copper chloride *(1)*; water H2O *(1)*
b Idea that this adds excess / too much *(1)*; to make sure all the acid is used up / has reacted *(1)*

Page 252 Acids and alkalis

3 a From **blue** to **red** *(1)*
b It only has two colours / needs to be different colours at different pHs idea *(1)*
c i Use the indicator to find out how much acid he needs to add *(1)*; repeat without indicator *(1)*; using same amounts of acid and alkali *(1)*
ii By evaporating the solution *(1)*

Page 253 Neutralisation

1 a Lemon juice *(1)*
b Fresh milk *(1)*
c Take a sample of milk *(1)*; add Universal Indicator solution / dip in some UI paper *(1)*; compare the colours to the chart / if it is orange-yellow it is sour / if it is blue-green it is not sour *(1)*

Page 253 Precipitation

2 a i Lead nitrate / potassium chromate *(1)*
ii Lead chromate *(1)*
b Potassium nitrate *(1)*
3 a i Lead carbonate *(1)*
ii Calcium phosphate *(1)*
b The water can be filtered to remove the precipitate *(1)*

P2a Discover forces!

Page 255 See how it moves!

1 a i Travelling forward *(1)*; at a steady speed *(1)*
ii Stopped *(1)*
b Section C *(1)*
2 a $\text{Average speed} = \dfrac{\text{total distance}}{\text{time}}$ *(1)*

$= \dfrac{90}{2}$ *(1)*

$= 45$ km/h *(1)*
b He will slow down at junctions / traffic lights *(1)*; and speed up on clear roads / his speed is constantly changing *(1)*

Page 255 Speed isn't everything

3 a Speed *(1)*; in a certain direction *(1)*
b i Velocity increases as he starts to run faster *(1)*
ii Changes because he changes direction *(1)*
iii Decreases as he slows down at the end *(1)*
4 a 0 m/s *(1)*
b $\text{Acceleration} = \dfrac{\text{change in speed}}{\text{time taken}}$ *(1)*

$= \dfrac{150}{0.5}$ *(1)*

$= 300$ m/s^2 *(1)*

Page 256 Velocity-time graphs

1 a i Steady speed *(1)*
ii Accelerating *(1)*
b Section D *(1)*
c The line would reach the x-axis *(1)*
2 a

Both axes labelled correctly = (2)
points correctly plotted = (1) line drawn = (1)

b i Area under graph (1)
  ii Gradient of graph at that point (1)
  iii Read value off the y-axis (1)

Page 256 Let's force it!
3 a i 400 N (1)
  ii Accelerating (1); forwards (1)
 b i Balanced (1)
  ii Steady speed (1); forwards (1)

Page 257 Force and acceleration
1 a i Positive (1); acting in forwards direction (1)
  ii It is zero (1)
 b i It increases (1)
  ii Force = mass x acceleration (1)
        = 20 x 4 (1)
        = 80 N (1)
2 a Newtonmeter / force meter (1)
 b Yes, it does prove this (1); as force doubles, acceleration doubles (1)
 c 4N reading (1); doesn't follow the pattern (1)

Page 257 Balanced forces
3 a Points of forces that are equal in size (1); opposite in direction (1)
 b i Balanced (1)
  ii They are the same (1)
 c Bigger (1)

Page 258 Terminal velocity
1 a Air resistance (1)
 b It increases (1)
 c Top speed reached when forces balance (1)
 d (Any 3:) Forward force gets greater causing acceleration (1); air resistance increases(1); because his speed increases (1); at his top speed, forces are balanced (1)
2 a Air resistance (1)
 b Air resistance acts in the opposite direction to forces causing motion (1); and makes him decelerate (1)

Page 258 Stop!
3 a Distance travelled while the car is stopping / between the driver seeing a hazard and the car stopping (1)
 b Braking distance (1); thinking distance (1)
 c Slower reaction time / less experience / concentrating on car control not road conditions (1)
 d Longer distance (1); slower reaction time (1)
 e More friction (1); so shorter braking distance (1)

Page 259 Moving through fluids
1 a Drag / friction / allow upthrust (1)
 b As it falls it accelerates and the drag increases with speed (1); until drag equals weight (1)
 c Thicker fluid causes more drag (at lower speeds) (1); so the terminal velocity is slower (1)
 d Increase the terminal velocity (1)

Page 259 Energy to move
2 a Heat (1); kinetic (1)
 b Chemical (1)
 c Electrical energy (1); light energy (1)
3 a Batteries (1)
 b The energy spreads to surroundings and moving parts of car (1); which become hotter (1)
 c More work done (1); against friction (1)

Page 260 Working hard
1 a 12 J (1)
 b Weight (allow gravity) (1)
 c Work done = force x distance (1)
        = 25 x 2 (1)
        = 50 J (1)
 d Changed into heat and sound energy (1); as work was done against friction (1)

Page 260 How much energy?
2 a Energy something has because it is moving (1)
 b Slow down / stop (1)
 c PE = weight x height (1)
    = 3 x 10 (1)
    = 30 J (1)
3 Andrea has a large elastic harness attached to her when she does a bungee jump from the top of a bridge. As she jumps, the **gravitational potential** energy she has on the bridge changes into **kinetic** energy as she falls. At the bottom of the jump, the rope is fully extended and it gains **elastic potential** energy. Eventually she stops moving because all the energy has spread as **thermal/heat** energy to the surroundings. (4)

Page 261 Momentum
1 a i Speed / mass (1)
  ii Velocity / force (1)
 b Momentum = mass x velocity (1)
        = 0.5 kg x 8 m/s (1)
        = 4 kg m/s (1)
2 a Inelastic (1)
 b Momentum = mass x velocity (1)
        = 2 x 3 (1)
        = 6 kgm/s (1)
 c 6 kgm/s (1)
 d Velocity = $\frac{momentum}{mass}$ (1)

      = $\frac{6}{4}$ (1)

      = 1.5 m/s (1)

Page 261 Off with a bang!
3 a Backwards (1)
 b Momentum of balloon equals momentum of air (1); the air and balloon move in opposite directions so total momentum remains zero (1)
 c Swimmer pushes water backwards (1); so she must move forwards (1); to conserve momentum (1)

Page 262 Keep it safe
1 a A force or similar answer (1)
 b i The van (1)
  ii The car (1)
 c i Same change in momentum (1); but less force is felt (1); because the momentum takes longer
  ii Zero
2 a Zero (1)
 b Smaller force (1); due to longer time of impact (1); carpet absorbs some KE (1)

Page 262 Static electricity
3 a Electrons
 b Electrons (1); are rubbed off the cloth (1); and move onto the balloon (1)
 c The two balloons move apart (1); because like charges repel (1)

Page 263 Charge
1 a A flow (1); of electrons (1)
 b i Insulator (1)
  ii Electrons rubbed off materials (1); cannot flow away / through an insulator (1); so they build up (1)
2 a They are the same (1)
 b Gold leaf electroscope (1)

Page 263 Van de Graaff generator
3 a Insulators (1)
 b As the belt moves, it rubs against the roller (1); transferring electrons (1)
 c To stop the charge flowing away through the bench / prevent discharge (1)
4 (Any 3:) The sharp point of lightning conductor is higher than surrounding buildings (1); lightning is attracted to conductor instead of to the building (1); the lightning conductor is made from metal (1); so it is easy for lightning charge to flow down it (1)

Page 264 Sparks will fly!

1 a Insulators *(1)*
  b You see a spark *(1)*
  c Flammable gases *(1)*; can explode *(1)*; if there is a spark from a discharge *(1)*
2 a Opposite charges attract / the toner and drum are oppositely charged *(1)*
  b Positive *(1)*
3 a Shoes rub *(1)*; electrons off *(1)*; the carpet *(1)*
  b i Discharge *(1)*
    ii *(Any 2:)* Metal is a conductor *(1)*; connected to Earth *(1)*; so electrons can flow off Monique *(1)*
4 a Electrons are rubbed off the paint particles *(1)*; as they rub on the sides of the spray gun *(1)*
  b Particles have the same charge *(1)*; so repel each other *(1)*

P2b Discover nuclear fusion!

Page 266 Circuit diagrams

1 a Shine a light on it / cover it up *(1)*
  b 2 *(1)*
  c Fuse *(1)*
2 a Variable resistor *(1)*
  b Adjusting the variable resistor's setting reduces its resistance *(1)*; which increases the current *(1)*; which increases the brightness *(1)*
  c Can't control bulbs separately *(1)*; bulbs are dimmer *(1)*

Page 266 Resistance 1

3 a Flow *(1)*; of electrons *(1)*
  b The resistance of a conductor is low *(1)*; the resistance of an insulator is high *(1)*
  c Electrons free to move around *(1)*

Page 267 Resistance 2

1 a i It increased *(1)*
    ii It decreased *(1)*
  b i Resistance increases if electrons *(1)*; have more collisions *(1)*; with nuclei *(1)*
    ii Different materials have an atomic structure / arrangement so there is a different chance of collisions *(1)*

Page 267 Ohm's Law

2 a As voltage increases *(1)*; current increases / they are proportional to each other *(1)*
  b 1.5 A *(1)*
  c $\text{Resistance} = \dfrac{\text{voltage}}{\text{current}}$ *(1)*
    $= \dfrac{6}{1.5}$ *(1)*
    $= 4$ ohms *(1)*
  d Type of material *(1)*; temperature *(1)*
  e Temperature of wire increases / is not constant *(1)*; or the resistance of the wire is not constant *(1)*

Page 268 More components

1 a Heat it up / cool it down *(1)*
  b More electrons are released *(1)*; so more electrons can flow in the circuit *(1)*
  c Electronic thermometer *(1)*; smoke alarm *(1)*
  d It decreases *(1)*
  e The current falls *(1)*
  f Semiconductor *(1)*

Page 268 Components in series

2 a They are the same *(1)*
  b 3V *(1)*
  c 1 IV 2 IV 3 IV *(1)*
  d 0V *(1)*

Page 269 Components in parallel

1 a 3V *(1)*
  b 3V *(1)*
  c The other bulbs stay on *(1)*
  d Can control bulbs separately / brighter than in a series circuit *(1)*

Page 269 The three-pin plug

2 a Blue *(1)*
  b Plastic is an insulator (allow for safety) *(1)*
  c *(Any 2:)* If the fuse blows; the current will stop; it isolates the live wire
  d Three metal wires *(1)*; each enclosed in plastic *(1)*; all covered in one plastic / insulating layer *(1)*
  e It prevents wires being pulled out by accident *(2)*

Page 270 Domestic electricity

1 a Direct current *(1)*
  b Y *(1)*
  c A higher voltage can be supplied / the cost is less per unit supplied / a continuous supply of energy *(1)*
  d (Cathode ray) Oscilloscope (or CRO) *(1)*

Page 270 Safety at home

2 a When the current was too big *(1)*; the fuse blew / melted / broke *(1)*; and broke the circuit *(1)*
  b Different equipment uses different sizes of current *(1)*
  c Radio double insulated / covered in plastic / no metal parts accessible (or vice versa for heater) *(1)*
  d If the metal part becomes live *(1)*; earth wire provides safe / low resistance route for charge to flow away *(1)*; which blows the fuse *(1)*
  e Responds more quickly if there is a fault *(1)*; responds to much lower current differences *(1)*

Page 271 Which fuse?

1 a i 3 A fuse *(1)*
    ii It will blow / the fuse wire will melt *(1)*; if the current is higher than *(1)*; 13 A *(1)*
    iii 200 J *(1)*
  b i $\text{Current} = \dfrac{\text{power}}{230\text{V}}$ *(1)*
    $= \dfrac{2000}{230}$ *(1)*
    $= 8.7$ A *(1)*
    ii Fuse would keep blowing *(1)*; so the equipment would stop working *(1)*

Page 271 Radioactivity

2 a i Neutron
    ii Neutron *(1)*; proton *(1)*
  b Charged particle / atom which has lost / gained an electron *(1)*
  c i 9 *(1)*
    ii 5 *(1)*
    iii 2 *(1)*
  d i The number of neutrons in the nucleus *(1)*
    ii The number of protons in the nucleus *(1)*

Page 272 Alpha, beta and gamma rays 2

1 a Gamma penetrates cardboard / is only stopped by thick lead *(1)*
  b Geiger Muller tube / Geiger counter *(1)*
2 a

Type	Mass	Charge
Alpha	+4	+2
Beta	Negligible	−1
Gamma	0	0

(3)

  b Ionising radiation knocks electrons off *(1)*; nearby atoms leaving them as charged ions *(1)*

Page 272 Background radiation 2

3 a Radiation that surrounds us *(1)*; all the time *(1)*
  b Cosmic rays *(1)*
  c Medical e.g. X-rays / fall-out from nuclear weapon testing / discharge from nuclear power / air travel *(1)*
  d Natural sources *(1)*
4 a Granite is more radioactive than other rocks *(1)*; granite is only found in certain places *(1)*
  b Yes *(1)*; levels of radioactivity are low / have been naturally present for many years *(1)*; and can be reduced with vents in homes / do not pose a great health risk *(1)*

Page 273 Inside the atom

1 a Plum pudding model *(1)*
  b Positively charged *(1)*; central nucleus *(1)*; surrounded by negatively charged electrons *(1)*
  c i Some positive charged alpha particles are deflected during Rutherford's experiment *(1)*
    ii Many alpha particles passed straight through *(1)*
    iii Ions can be produced from neutral atoms *(1)*

Page 273 Nuclear fission

2 a When one large nucleus splits *(1)*; into two or more smaller parts releasing energy *(1)*
  b Nuclear power stations / nuclear bombs *(1)*
  c i 3 *(1)*
    ii At each stage a bigger number of nuclei get involved *(1)*; each neutron produces more than one more neutron from the reaction *(1)*
  d The nucleus must be large *(1)*; and unstable for fission to take place *(1)*

Page 274 Nuclear power station

1 a Nuclear fission *(1)*
  b The moderator absorbs neutrons *(1)*
  c Boron rods *(1)*
  d Kinetic energy → electrical energy *(1)*
  e i Heat is given out by nuclear fission *(1)*; which changes water into steam *(1)*
    ii Steam spins the turbine *(1)*; which spins the generator *(1)*

Page 274 Nuclear fusion

2 a Nuclear fission is when two small nuclei *(1)*; join up releasing heat *(1)*
  b In a star *(1)*
  c i Helium *(1)*; energy *(1)*
    ii The heat and pressure is needed to overcome strong forces inside the atom *(1)*; which repel the positively charged nuclei *(1)*
    iii Only small nuclei undergo fusion *(1)*; because the temperature / pressures needed are too high for larger nuclei *(1)*